Solvents in Common Use

Health risks to workers

Health risks to workers from the following commonly used solvents:

acetone
carbon disulphide
diethyl ether
1,4-dioxane
ethyl acetate
methanol
nitrobenzene
pyridine
toluene
xylene

Prepared by an expert sub-committee of the Health, Safety and Environment Committee of the Royal Society of Chemistry for the Commission of the European Communities, Directorate-General Employment, Social Affairs and Education, Health and Safety Directorate, Division for Industrial Medicine and Hygiene.

Commission of the European Communities

Solvents in Common Use

Health risks to workers

ROYAL
SOCIETY OF
CHEMISTRY

Publication No. EUR 11553 of the
Commission of the European Communities,
Scientific and Technical Communications Service,
Directorate-General Telecommunications, Information
Industries and Innovation, Luxembourg

British Library Cataloguing in Publication Data
Solvents in common use: health risks to workers
    1. Industrial chemicals: Solvents.
    Safety measures
    I. Royal Society of Chemistry. *Health,*
    Safety and Environment Committee
    II. Commission of the European Communities
    661'.807

    ISBN 0-85186-088-5

Typeset by Method Ltd, Epping, Essex.
Printed in Great Britain by
Whitstable Litho Printers Ltd, Whitstable, Kent.

Contents

INTRODUCTION

Preamble

This study was carried out by the Royal Society of Chemistry (RSC) under contract from the Commission of the European Communities (CEC). The work was overseen by the Society's Health, Safety and Environment Committee (HSEC) and performed largely by the secretariat of the committee. The HSEC itself consists of members of the RSC whose affiliations include industry, the trade unions, government and independent consultants.

The solvents studied were selected by the CEC partly on the basis of their widespread commercial usage.

Sources of Information

The basis of these reports is largely authoritative reviews and textbooks. The Committee was of the opinion that, in general, sufficient reliable information was available concerning the solvents considered to obviate the necessity of consulting original work (primary sources). However, these have been consulted where thought appropriate or to remove anomalies. The information has been compiled selectively, with material of doubtful validity or relevance being omitted.

Although reputable reviews can generally be taken to underwrite the sources of the information they review, their use has had two principal disadvantages. First, very recent studies may have been omitted because they were not recorded in the tertiary sources that were examined. (However such studies may not have been validated.) Second, the precise interpretation of primary data as reported in tertiary sources has not always been clear. In certain instances the Committee has drawn attention to such problems of interpretation. In these cases the Committee's comments have been distinguished from the text by the use of square brackets. Such comments are not meant to imply any criticism of either the primary source or the tertiary source in which it was reported.

In addition to the aforementioned sources of information a number of others have been used to a lesser extent. Such sources, which include government publications and trade statistics, have been identified in the usual way.

Finally, certain parts of the reports consist of original material prepared by the Committee itself. Such material has also been identified by the use of square brackets and is included because of the lack of suitable published data in some areas and the need to make judgements on certain matters.

Medical / Health Surveillance[1]

Exposure to organic solvents may produce acute and chronic effects. Consideration should therefore be given to the establishment of appropriate medical/health surveillance. This book does not set out to describe in detail occupational medical practice but to indicate briefly some points of relevance to each solvent. Correlation of information on environmental exposures with medical results is essential. Adverse effects of organic solvents indicate that both pre-employment and periodic medical examination should include consideration of previous exposure to organic solvents and an assessment of current potential exposure. In addition to the hazards to respiratory, digestive, cardiovascular and renal systems and the blood, long-term neurotoxicity effects may arise from some solvent exposures. On the basis of our investigations into neurotoxicity we consider that the evidence currently available is insufficient to justify specific control measures and specialized medical screening. However because of the possibility of long-term adverse neurotoxic effects further research is urgently required.

The Committee considers it essential that information on medical/health surveillance should reflect current professional occupational medical practice. It is grateful to those doctors and other professionals who have provided relevant information.

The Committee draws attention to the disparity between normal professional practice and the recommendations for extensive medical and other tests in some of the published reviews on individual substances. This discrepancy is partly explained by the fact that many of the tests recommended might be justified if continued or sporadic *excessive* exposure due to faulty preventive measures could occur in the processes to which the reviews referred (published documents often refer to the use of such additional tests "where potentially hazardous exposure may occur"). The medical/health surveillance sections in this book have been drafted so as to present information on standard practice in situations where such excessive exposures would *not* normally be expected to occur. Where doubt arises the Committee recommends that expert advice should be obtained (*e.g.*, from an occupational physician) in order to decide the scope of medical/health surveillance required in any particular work situation.

The Committee has not addressed the question of maintenance of records of medical surveillance. Similarly neither the preparation of reports on excessive exposures, nor legal requirements for medical examination or notification of industrial diseases or accidents are considered because these may vary in the light of the statutory requirements of individual member states.

Preamble to First Aid Section

The following general first aid recommendations have been abstracted from authoritative sources including published advice from health and safety authorities, manufacturers and poisons reference centres. The recommendations are repeated in the sections dealing with each individual solvent together with certain additional information. The application of first aid must take into account the intrinsic toxicity of the substance(s) concerned and the degree of exposure.

First Aid

The following general recommendations are made for first aid in case of hazardous exposure to any of the solvents listed.

General

The casualty should be removed from the danger of further exposure into the fresh air. Rescuers should ensure their own safety from inhalation and skin contamination where necessary. Conscious casualties should be asked for information about what has happened. It should be borne in mind that the casualty may lose consciousness at any time. Where necessary continued observation and care should be ensured. First aiders should take care not to become contaminated.

Inhalation

The conscious casualty should be kept at rest. The unconscious casualty should be placed in the recovery position and an open airway maintained. If breathing or heart beat stops, resuscitation should be commenced immediately. Medical aid should be obtained or the casualty removed to hospital immediately.

Skin Contamination

Remove contaminated clothing. Wash with copious amounts of water, or soap and water, for at least 15 minutes. If necessary, seek medical aid. In the case of persistent skin irritation refer for medical advice.

Eye Contamination

Irrigate with copious amounts of water for at least ten minutes but avoid further damage to the eye from the use of excessive force. Seek medical aid. Remember the possible presence of contact lenses which may be affected by some solvents and may impede decontamination of the eye.

Ingestion

If the lips or mouth are contaminated rinse thoroughly with water. Do not induce vomiting. Give further supportive treatment as for inhalation. In the case of ingestion of significant amounts wash out the mouth with water. Obtain immediate medical attention. Contact a hospital or poisons centre at once for advice. Remember that symptoms may develop many hours after exposure so continued care and observation may be necessary.

In all cases note information on the nature of exposure and give this to ambulance or medical personnel. Information for doctors is provided in some technical literature issued by manufacturers. Antidotes may be available for retention at the place of work for use by persons competent to treat casualties. The Committee did not consider it appropriate to deal in detail with the specialized treatment measures that might be applied by medical or toxicological experts. Access to specialist advice on such measures is available according to arrangements in different states from professional scientific bodies, Government authorities, poison reference centres, or the manufacturers.

Odour Thresholds[2]

Minimum concentrations for detection of the solvents considered are listed below. However it should be noted that chemical contaminants and individual olfactory perceptions may cause large variation in thresholds of perception.

Nasal detection must never be relied upon as a means of hazard warning or estimating exposure. Some substances induce olfactory fatigue and may lead to false perception that the concentration has reduced.

Table of Odour Thresholds and RSC Recommended OELs

	Odour thresholds	RSC recommended OELs		
Solvent	*Minimal detection* (ppm)	RSC TWA (ppm)	RSC STEL (ppm)	
Acetone	20	250	500	
Carbon disulphide	0.008	10	20	Sk
Diethyl ether	0.3	400	500	
1,4-Dioxane	0.003	25	50	Sk
Ethyl acetate	0.006	400	400	
Methanol	8	200	250	Sk
Nitrobenzene	0.004	1	3	Sk
Pyridine	0.003	5	15	Sk
Toluene	3	100	150	
Xylene	0.1	100	150	

Sk — this notation indicates a substance which may produce serious toxic effects by absorption through the skin. (Minimal detection levels are based on ref. 2.)

RSC Recommended Occupational Exposure Limits[3]

The following note outlines the philosophy of Occupational Exposure Limits (OELs) as applied to the derivation of recommended limits for the solvents considered here. Although the main occupational exposure to such materials is likely to be *via* inhalation the Committee recognized that certain materials may pose a significant extra risk by absorption through the intact skin. The limits proposed relate to exposures by inhalation only. If skin absorption occurs a suitable safety margin should be allowed. The Committee felt unable to recommend meaningful 'limits' for skin absorption analogous to OELs for atmospheric concentration. Therefore it has simply drawn attention to the risk of skin absorption where it exists. Another phenomenon which should be noted is that some of the solvents considered can de-fat the skin. This in turn can lead to more ready adsorption of other toxic materials through the skin.

The dose–response relationship is the cornerstone of modern toxicology. It has been used over a period of some thirty years by the American Conference of Governmental Industrial Hygienists (ACGIH) in its pioneering work on Threshold Limit Values (TLVs). While there have recently been alternative approaches the Committee believes that it still represents the best method for use with the particular substances examined here.

Some biological responses are trivial and well within the body's normal capacity to adapt. Some responses require compensation mechanisms within the body which are compatible with normal biological function. However if environmental exposures are increased beyond this compensatory range impairment and disability will result. The assessment of health risk revolves about our ability to detect and understand these changes. Some are easy to detect, others are not.

The Occupational Exposure Limits recommended by the RSC represent exposure concentrations averaged over a ten minute (STEL) or an eight hour period. It is believed that few workers will suffer adverse health effects as a result of prolonged exposure to such averaged concentrations. RSC OELs are not designed nor intended to protect all workers as a few workers may be more susceptible than others to the effects of exposure to certain chemicals (this matter is considered further in the sections dealing with medical/health surveillance).

The reasons for unusual susceptibility may include inherited genetic factors, nutritional deficiencies, parasitic diseases, pre-existing diseases. Alcohol or drug consumption, and smoking may also increase susceptibility. It is important as far as practicable to determine whether unusual susceptibility exists in order to protect vulnerable workers.

Carcinogens differ from other chemicals in causing irreversible changes which may lead to the manifestation of overt malignancy only after a long latent period. Although some chemical carcinogens appear to show a dose dependent response in experimental animals, the existence of a 'no-effect' or 'threshold level' for some or all carcinogens is at present a matter of dissension among experts.

Much of the assessment of human risk from exposure to chemicals derives from experimental data in animals. The available data should be evaluated critically. Particular attention should be given to reliable epidemiological or case data when assessing human risk from exposure to chemicals. Although it is possible to extrapolate toxicity data from animals to man it is necessary to take account of basic species differences which may be relevant. In general, it is important to try to understand the various factors which operate before making such extrapolations. Such factors may include for example metabolic pathways, protective mechanisms, activation, co-carcinogenicity, sensitivity and perhaps inhibition.

Finally, in deriving its recommended OELs the Committee first examined existing exposure limits. These were then reviewed in the light of the toxicological data contained in these reports. The Committee recognized that the OELs (by whatever names they are known) set by different countries reflect not only toxicological data but also economic practicability and public opinion as expressed in political pressure (perhaps most clearly illustrated in the case of asbestos). Thus there are often clear differences between values in Eastern and Western Europe, and between those in the developed and underdeveloped worlds.

Thus the Committee took account not only of toxicological information but also of these other factors reflected in published OELs. The end result is inevitably a value judgement reflecting as far as possible the philosophy described above.

There is no scientific basis for Action Levels. However, the Committee considered that it would be prudent for those responsible for the working environment to consider carefully any exposures likely to exceed 30–50% of the OEL and take appropriate action depending on the circumstances.

In practice the concentrations to which individuals are exposed will constantly vary. Therefore, some upper limit for short periods of exposure (STELs) may need to be defined. The Committee has attempted to provide guidance on such limits. The STEL values given should be regarded as peak values in the breathing zone although in practice they will often have to be evaluated over a ten minute period for analytical reasons. Where direct reading instruments or techniques sampling over very short periods of time are available, the figures so obtained should be used. In any case, instantaneous values in the breathing zone should not be permitted substantially to exceed the short-term limit whatever sampling time is chosen. N.B. Where the TWA is thus exceeded for short periods of time, subsequent exposure should be below the TWA for such time as to ensure that the average exposure over eight hours does not exceed the TWA.

The Committee recognized that adequate monitoring for contaminants in workplace air is essential if OELs are to be of any value. Establishing a suitable

Industrial Usage

Solvent	EC production/consumption (year to which figures relate)/kilotonnes	EUR12 Estimate of numbers potentially exposed in industrial processes
Xylenes	540(83)/615(83)	20,000
Acetone	710(83)/700(83)	15,000
Methanol	2530(83)/3870(83)	15,000
Toluene	780(83)/990(83)	15,000
Carbon disulphide	150(83)/150(83)	2,000
Diethyl ether	50(84)/45(84)	2,000
1,4-Dioxane	70(84)/70(84)	2,000
Ethyl acetate	140(80)/ –	2,000
Nitrobenzene	500(83)/505(83)	2,000
Pyridine	9(82)/22(82)	2,000

Note: Many persons including members of the public may experience undetermined secondary exposure to certain of these solvents. This arises from their widespread use.

monitoring system may involve complex considerations on which it may be necessary to seek expert guidance.

Commercial grades of some of the solvents considered may contain additives or unintentional impurities. Because of the variability of such mixtures they have in general not been considered in this report. However, it should be noted that these impurities may be both volatile and toxic and as such should be monitored individually. The relative vapour concentrations of the solvent and its impurities may not however reflect their proportions in the liquid.

Waste Disposal

Where practicable, solvents should be recovered and recycled. Where this is not practicable they should be disposed of in a safe, approved manner. This may be, for example, by disposal to landfill or by incineration.

References

1. National Institute for Occupational Safety & Health. NIOSH Current Intelligence Bulletin 48: 'Organic Solvent Neurotoxicity'. DHHS (NIOSH) Publication No. 87-104. US Department of Health & Human Services, 1987.
2. Ruth, J.H. *American Industrial Hygiene Association Journal*, 47(1986):A-142. 'Odor thresholds and irritation levels of several chemical substances: a review'.
3. Lewis, T.R. *Applied Industrial Hygiene*, **1**, No.2(1986):66. 'Identification of sensitive subjects not adequately protected by TLVs'.

GLOSSARY

Primary sources	–	Original papers.
Secondary sources	–	Abstracts, indexes and databanks of primary source information.
Tertiary sources	–	Reviews and reports of studies relating to a specific subject or substance. Reference works such as encyclopaedias, textbooks, *etc.*

Abbreviations

ACGIH	–	American Conference of Governmental Industrial Hygienists
ACS	–	American Chemical Society
ADH	–	alcohol dehydrogenase
AL	–	acid labile
APHA	–	American Public Health Administration
API	–	American Petroleum Institute
ASTM	–	American Society for Testing and Materials
BS	–	British Standard
B.P.	–	British Pharmacopoeia
BTLV	–	biological TLV
BTX	–	benzene–toluene–xylene
BUN	–	blood urea nitrogen
ca.	–	*circa*, approximately
CEC	–	Commission of the European Communities
CEFIC	–	Conseil Européen des Fédérations de l'Industrie Chimique (European Council of Chemical Manufacturers Federation)
CHD	–	coronary heart disease
CIIT	–	Chemical Industry Institute of Toxicology (US)
CLV	–	ceiling value
The Committee	–	an Expert Sub-Committee of the Health, Safety and Environment Committee of the RSC
CNS	–	central nervous system
DMBA	–	7,12-dimethylbenz(α)anthracene
DNA	–	deoxyribonucleic acid
EC	–	European Communities
ECDIN	–	Environmental Chemicals Data and Information Network
ECETOC	–	European Chemical Industry Ecology and Toxicology Centre
EEG	–	electroencephalograph
EPA	–	Environmental Protection Agency (US)
ETA	–	equivocal tumorigenic agent
EUR12	–	the twelve countries of the EC
gc	–	gas chromatography, gas chromatograph
gc-ms	–	gas chromatography–mass spectrometry

GHS	–	glutathione
GIT	–	gastrointestinal tract
GOT	–	glutamate oxaloacetate transaminase
gpg	–	guinea pig
GPT	–	glutamic-pyruvic transaminase
HEAA	–	hydroxyethoxyacetic acid
hplc	–	high performance liquid chromatography
HSE	–	Health and Safety Executive (UK)
HSEC	–	Health, Safety and Environment Committee of the RSC
IDLH	–	immediately dangerous to life or health
IHD	–	ischaemic heart disease
ihl	–	inhalation
ILO	–	International Labour Office
i.p., ipr	–	intraperitoneal
ir	–	infrared radiation
IUPAC	–	International Union of Pure and Applied Chemistry
LC_{50}	–	calculated lethal concentration, exposure to which for a specified length of time is expected to cause 50% mortality (inhalational or aquatic exposure)
LCLo	–	lowest published lethal concentration (inhalational or aquatic exposure)
LD	–	lethal dose
LD_{50}	–	calculated lethal dose expected to cause 50% mortality (all routes of exposure except inhalation)
LDLo	–	lowest published lethal dose (all routes of exposure except inhalation)
MAC	–	maximum allowable concentration
ms	–	mass spectrometry
MTBE	–	methyl t-butyl ether
MTD	–	maximum tolerated dose
mus	–	mouse
NAD	–	nicotinamide adenine dinucleotide
NADH	–	reduced form of NAD
NADP	–	nicotinamide adenine dinucleotide phosphate
NADPH	–	reduced form of NADP
NCI	–	National Cancer Institute (US)
NFPA	–	National Fire Protection Association (US)
NIOSH	–	National Institute for Occupational Safety and Health (US)
NMR	–	nuclear magnetic resonance
NRC	–	National Research Council (US)
NTP	–	National Toxicology Program (US)
OEL	–	occupational exposure limit
orl	–	oral
OSHA	–	Occupational Safety and Health Administration (US)
PAN	–	peroxyacetyl nitrate
PBzN	–	peroxybenzoyl nitrate
PEL	–	permissible exposure limit

PTFE	–	polytetrafluoroethene
PVA	–	polyvinyl alcohol
PVC	–	polyvinyl chloride
RBC	–	red blood cells
rbt	–	rabbit
RNA	–	ribonucleic acid
RSC	–	Royal Society of Chemistry (UK)
RTECS	–	Registry of Toxic Effects of Chemical Substances (a NIOSH publication)
SAP	–	serum alkaline phosphatase
SC	–	standard conditions
SCE	–	sister-chromatid exchange
SCP	–	single cell protein
scu	–	subcutaneous
S.G.	–	specific gravity
SGOT	–	serum glutamic oxaloacetic transaminase
SGPT	–	serum glutamic pyruvic transaminase
std	–	standard
STEL	–	short-term exposure limit
TCEP	–	1,2,3-tris-(2-cyanoethoxy)propane
TCLo	–	lowest published toxic concentration (inhalational or aquatic exposure)
TDLo	–	lowest published toxic dose (all routes of exposure except inhalation)
TETD	–	tetraethylthiuram disulphide
5-THF	–	5-formyltetrahydrofolate
TLV	–	Threshold Limit Value (of the ACGIH)
TTCA	–	2-thiothiazolidine-4-carboxylic acid
TWA	–	time-weighted average
UDS	–	unscheduled DNA synthesis
UK	–	United Kingdom
US, USA	–	United States of America
USP	–	United States Pharmacopeia
UV	–	ultraviolet radiation
WBC	–	white blood cells
WHO	–	World Health Organization

Units
Prefixes

p	–	pico (10^{-12})
n	–	nano (10^{-9})
μ	–	micro (10^{-6})
m	–	milli (10^{-3})
c	–	centi (10^{-2})

d	–	deci (10^{-1})
k	–	kilo (10^3)
M	–	mega (10^6)

Units

ångstrom	–	10^{-10} m, or 0.1 nm
atm	–	atmosphere(s) (pressure of 1 atm = 760 mmHg, or 101,325 Pa)
bar	–	bar (1 bar = 10^5 Pa)
cal	–	calorie
Ci	–	curie (3.70×10^{10} disintegrations s^{-1})
d	–	day
eV	–	electron volt (1 eV $\simeq 1.602 \times 10^{-19}$ J)
ft	–	foot
g	–	gram
gal	–	gallon (British Imperial gallon unless otherwise stated)
hr	–	hour
in	–	inch(es) (1 in = 2.54 cm)
J	–	joule
kgbw	–	kilogram of body weight
kWh	–	kilowatt-hour (1 kWh = 3.6×10^6 J)
l	–	litre (1 l $\equiv$ 1 dm^3)
lb	–	pound
m	–	metre
M	–	molar
min	–	minute
mmHg	–	millimetre of mercury (1 mmHg = 1 Torr at 0 °C)
mol	–	mole ($\simeq 6.02 \times 10^{23}$ atoms, molecules, *etc.*)
N	–	newton (1 N = 1 m kg s^{-2})
p	–	poise (1 p = 1 g s^{-1} cm^{-1})
Pa	–	pascal (1 Pa = 1 N m^{-2})
ppb	–	parts per (US) billion (*i.e.*, 10^9)
ppm	–	parts per million (*Note*: Where equivalent ppm values have been calculated from mg/m^3 or similar units, a temperature of 25 °C has been assumed.)
ppt	–	parts per (US) trillion (*i.e.*, 10^{12})
psi	–	pounds per square inch (1 psi $\simeq$ 0.068 atm)
psig	–	pounds per square inch gauge
s	–	second
St	–	stokes (1 St = 10^{-4} m^2 s^{-1})
V	–	volt
vol%, % v/v	–	percentage by volume
W	–	watt (1 W = 1 J s^{-1})
wt%, % w/w	–	percentage by weight

CONTENTS

1. CHEMICAL ABSTRACTS NAME

2-propanone
(10th Collective Index)

2. SYNONYMS AND TRADE NAMES[1-6]

aceton (Dutch, German, Polish)
acétone (French)
β-ketopropane
cetona
chevron acetone
diméthylcétone (French)
dimethyl formaldehyde
dimethyl ketal
dimethyl ketone
DMK
ketone dimethyl
ketone propane
methyl ketone
OHS00140
propanone
propanone-2
pyroacetic acid
pyroacetic ether
1B-ketopropane
U002
UN 1090

3. CHEMICAL ABSTRACTS SERVICES REGISTRY NUMBER

67-64-1

4. NIOSH NUMBER

AL3150000

5. CHEMICAL FORMULA

C_3H_6O
(Molecular weight 58.09)

6. STRUCTURAL FORMULA

$$CH_3—CO—CH_3$$

7. OCCURRENCE[5,7-10]

Acetone occurs naturally as a normal microcomponent in blood and urine. It is a minor constituent in pyroligneous acid and is formed by the oxidation of alcohols and humic substances.[5] It has been detected in cigarette smoke (at 1,100 ppm), in gasoline [petrol] exhaust at 2.3–14.0 ppm (partly propionaldehyde) and in year-old leachate of an artificial sanitary landfill at 0.6 g/l. Graedel[7] gives a number of additional sources of acetone emitted to the atmosphere, including animal waste, refuse combustion, wood pulping, plastics combustion, chemical manufacturing and solvent use, fish oil manufacture, turbines, forest fires and volcanoes. Acetone is present in urban air at levels of several ppb. A study[8] of emissions from a large automotive assembly plant in Wisconsin, USA estimated that acetone was emitted from it at a rate of 74 gallons [US] per hour. In Great Britain, acetone has been identified as a micropollutant in treated water taken at waterworks treating lowland river water which contained relatively high levels of wastewater.[9] Acetone from industrial and laboratory usage may be found to have been discharged to sewers.

Graedel[7] observes that the primary fate of acetone in the troposphere is photodissociation. Both radicals formed are expected to form peroxyradicals by addition of molecular oxygen.

$$CH_3COCH_3 \quad \rightarrow \quad CH_3CO^{\cdot} + CH_3^{\cdot}$$

$$CH_3CO^{\cdot} + O_2 \quad \rightarrow \quad CH_3COO_2^{\cdot}$$

$$CH_3^{\cdot} + O_2 \quad \rightarrow \quad CH_3O_2^{\cdot}$$

A secondary reaction chain for the aliphatic ketones involves initial hydrogen abstraction by either $HO^{\cdot}$ or O, followed by peroxy formation and scission to produce lower aldehydes and oxides of carbon. Alkylperoxy radicals in the atmosphere can react to oxidize nitric oxide to nitrogen dioxide and sulphur dioxide to sulphur trioxide.

Sax[10] notes a laboratory experiment in which a mixture of acetone vapour and oxygen was irradiated with UV light of range 2,967–3,132 aångstroms, the temperature range being 100–250 °C. The main degradation products were methanol, formaldehyde, carbon dioxide and carbon monoxide. Minor products were acetic acid and methane.

8. COMMERCIAL AND INDUSTRIAL ACETONE[11]

Nelson and Webb in ref. 11 state that the IUPAC specification for acetone is representative of the purity of the commercial product. This specification states:

Purity	Not less than 99.5% v/v
S.G. (d_{20}^{20})	0.790 - 0.793 (ASTM D-891)
Odour	Passes ASTM D-1363
Distillation range	55.5–57.0 °C (ASTM D-1078)
Residue at 100 °C	Not to exceed 0.003% (ASTM D-1353)
Free acidity	Not to exceed 0.002% (calculated as acetic acid (ASTM D-1613)
Water	Not to exceed 0.3% (ASTM D-1364)
Water miscibility	Passes ASTM D-1722
Aldehydes	Passes Federal Specification QA-51F para. 4.24.2.

In addition, colour, although not a specification, is usually determined as a quality control measure by ASTM D-1209 using the APHA scale. A value of five or less is generally required.

9. SPECTROSCOPIC DATA[12]

Infrared, Raman, ultraviolet, both ^{1}H and ^{13}C NMR and mass spectral data for acetone have been tabulated.[12]

10. MEASUREMENT TECHNIQUES[13–33]

The method recommended for the determination of acetone in workshop atmospheres is the NIOSH[13] activated charcoal trapping method for collection and concentration, followed by solvent extraction of the charcoal and a gas chromatographic (gc) analysis of the extract.

The collection tube is 7 cm long with a 4 mm internal diameter. It contains a total of 150 mg of activated charcoal, (20–40 mesh) divided into a front section of 100 mg and a rear section of 50 mg separated by a small plug of urethane foam. Air is sampled at a flow-rate of 0.01–0.2 l per minute by means of a small pump for a total sample of 0.5–3 l. The entire apparatus is portable so that it may be carried in a pocket with the sampling tube in the breathing zone (normally a coat lapel). The apparatus can also be static.

Carbon disulphide is used to extract the acetone, which is separated on a 12 ft × 0.25 in glass column packed with 10% SP-2100 plus 0.1% Carbowax 500 on Supelcoport (100–120 mesh) and using flame ionization detection. For full details of the method see ref. 13.

The sampling method uses a small portable apparatus with no liquids and is therefore easy to handle and maintain. The analytical method is relatively specific with the additional advantage of flexibility of operating conditions.

Air samples are normally taken by means of suction using an electric pump capable of operating at low flow-rates (*i.e.*, 10–300 ml per minute). If higher flow-rates are used, breakthrough can occur (*i.e.*, the acetone is not completely adsorbed and some of it passes through the adsorbent). It is also convenient either to establish the trapping device in a static position, or to have it attached to the user in such a way that the breathing zone is sampled (this usually means attachment to the worker's lapel).

It should be noted that suitable personal samplers may be commercially available.

The most popular accurate measurement technique, following solvent extraction or thermal desorption, is gc. Air, containing acetone and also toluene and xylene, was passed at a flow-rate of 0.1 l per minute through a 7.5 cm × 3 mm tube containing activated charcoal and the tube was placed in a desorption chamber attached to a gc.[14] The tube was heated to 190 °C and the gases were swept by nitrogen on to a 5 m × 4 mm column packed with 10% SP-1000 on Chromosorb W AW-DMCS (100–120 mesh). The column was temperature programmed from 100 to 160 °C and detection was by flame ionization. The limit of detection was 2 ppm.

In an improvement to the method; having trapped acetone and toluene from air on activated charcoal contained in a tube, this was placed in a ceramic block attached to a gc,[15] the block being heat-regulated from room temperature to 250 °C. The gc conditions were similar to those in ref. 14.

Acetone was separated from ethyl acetate in air samples by gc at 70 °C on a 3 m × 3 mm column packed with 5% SE-30 on Chromaton N-AW.[16] Nitrogen was the carrier gas and detection was by flame ionization and the limit of detection was 4 ng of acetone.

Ten litres of air were drawn through a tube containing 150 mg charcoal at a flow-rate of 11 ml per minute and carbon disulphide was used to extract acetone, ethyl acetate, toluene and methanol.[17] The extract was examined by infrared spectrometry using CsBr windows in order to quantify the organic compounds.

Very humid air tends to deactivate the charcoal and therefore acetone would only be partially adsorbed in these conditions.[18]

If Porapak is used as a trapping medium, the usual method of obtaining the trapped compound is by thermal desorption. Porapak N was used to trap acetone from 3–8 l of air at a flow-rate of 10–15 ml per minute.[19] After thermal desorption, the acetone was determined by gc on a 10 ft × 0.125 in column packed with Porapak Q (80–100 mesh). The chromatography was carried out at 200 °C and with flame ionization detection, the suitable range being 0.1–100 ppm.

A similar method was used to determine acetone and xylene in 20–100 l of air from Paris.[20] Mass spectrometry was used to confirm identities.

Acetone from air has been adsorbed in a tube of 2.5 g of Sorbsil (Grade A) using a flow-rate of 2 l per minute.[21] The tube was heated at 60 °C for one hour and the headspace analysed by gc using flame ionization detection. The limit of detection for acetone in air was 1.5 µg per litre.

In a similar trapping method, acetone was adsorbed from air at 21 l per minute on to a column of 1 g silica gel.[22] The acetone was eluted with 5 ml water and determined either colorimetrically using salicylaldehyde at pH 13 or by gc on a 2 m × 3 mm column packed with Porapak Q (150–200 mesh) at 120 °C and using flame ionization detection.

Acetone from air was automatically estimated by passage through a moistened band of fabric which was pretreated with hydroxyammonium chloride, sodium thiosulphate and bromophenol blue. The acetone liberated hydrogen chloride which changed the colour of the indicator to yellow, which was determined spectrophotometrically.[23] In an improvement to this method, paper was used in place of fabric.[24] For the range 50–1,500 mg acetone per cubic metre of air, a 30–60 second exposure at a flow-rate of 1.5–4 l per hour was appropriate for the method.

The air over waste tip sites has been examined for acetone and other organic compounds by using a variety of trapping devices.[25] Acetone was estimated directly using a Draeger tube with an air flow-rate of 30 ml per minute.

Car exhaust fume has been examined for its ketone content by passing the fume at 600 ml per minute through two impingers in series each containing 20 ml 2M hydrochloric acid saturated with 2,4-dinitrophenylhydrazine.[26] The resulting 2,4-dinitrophenylhydrazones, including that of acetone, were extracted with chloroform and evaporated to low volume under nitrogen. Gc was carried out on the extract after the addition of anthracene as internal standard. The gc capillary glass column was 30 m long and was coated with OV-17. It was operated at 210 °C using flame ionization detection.

Exhaled air from laboratory animals, *e.g.*, rats, has been examined for acetone.[27] The feature of the method is the sampling device which incorporates a special mouthpiece connecting the animal to an elastic sampling sac as the collection device.

An electrical method has been developed for the determination of combustible gases.[28] The measuring circuit includes a small block of tin and zinc sintered oxides, maintained at approximately 250 °C by means of two internal heaters which also act as electrodes for measuring the resistance of the block. The resistance decreases when a combustible gas, *e.g.*, acetone, is present, thereby giving a sensitive detection *via* changes in electrical output. The system does not appear to have been widely used.

In an analysis of blood serum for acetone and aliphatic alcohols the serum was deproteinised, propanol added as internal standard and gc was performed on a 30 m long capillary column coated with SPB-1.[29] This column was operated at 35 °C using flame ionization detection and a precolumn was used containing 3% OV-1 on Gas Chrom Q. Acetone and aliphatic alcohols were separated in less than three minutes.

Blood serum containing acetone and other oxo-compounds was also deproteinised before reaction with 2,4-dinitrophenylhydrazine.[30] The resulting 2,4-dinitrophenylhydrazones were subjected to high performance liquid chromatography on a 30 cm × 4.6 mm column packed with μBondapak. Methanol–0.01M glycine buffer, pH 7.4 (1:3) was used as mobile phase and detection was made at 365 nm.

The purge and trap method for the collection of acetone in biological tissues prior to gc determination has been used.[31] The tissue was placed in a brass cup and purged with nitrogen for two hours while heating at 37 °C. The volatiles were trapped on Porapak Q and then thermally desorbed for gc analysis on a column containing 0.4% Carbowax 1500 on Carbopack A. The column temperature was maintained at 50 °C and detection was by flame ionization.

The acetone concentration in the blood serum of animals exposed to acetone and also that in control animals were reported.

Acetone and its metabolites, acetoacetate and 3-hydroxybutyrate were determined in blood plasma or other biological tissue after treatment of the homogenate with 4N sodium hydroxide.[32] The acetoacetate was determined as acetone after reaction with 0.6M perchloric acid at 100 °C for 90 min and the 3-hydroxybutyrate determined similarly after reaction with 0.2M potassium dichromate in 5M phosphoric acid at 100 °C for 90 min. These fractions and unmetabolized acetone were analysed by gc on a 2 m column packed with Carbowax 1500 and operated at 75 °C with flame ionization detection.

Acetone and methanol at concentrations less than 1 mM in blood serum have been determined by high-field nuclear magnetic resonance.[33]

11. CONDITIONS UNDER WHICH ACETONE IS PUT ON THE MARKET

Production[11,34-37]

Nelson and Webb in ref. 11 observe that, in the late 1950s and early 1960s, earlier production methods were overtaken by the economics of scale of both the cumene hydroperoxide route to phenol (which gives acetone as a co-product) and the dehydrogenation of isopropanol. Earlier processes had included dry distillation of calcium acetate which itself was overtaken by a process, introduced in 1920, in which cornstarch or molasses was fermented using a special bacillus.

Production of acetone by dehydrogenation of isopropanol itself had begun in the early 1920s and remained the dominant production method throughout the 1960s. By 1974 though, 65% of US acetone production was *via* the cumene hydroperoxide process. Nelson *et al.* in ref. 34 (published 1985) note that over 90% of US production was being produced *via* the cumene route.

In the isopropanol dehydrogenation process 1,218 kg of isopropanol are required to produce 1 tonne of acetone.[35] Increase of temperature favours acetone production and at 325 °C 97% conversion is theoretically possible. The reaction is endothermic, requiring 66.5 kJ/mol at 327 °C.[11]

Ref. 35 gives an example in which isopropanol vapour, preheated by heat exchange with hot effluent gases from the reactor, is passed over a brass or copper catalyst, usually at 500 °C and at 275–350 kPa. The hot reaction gases, which contain acetone, isopropanol and hydrogen, pass through a water-cooled condenser and thence to a water scrubber to remove final traces of isopropanol and acetone from the hydrogen. The aqueous solution from the scrubber is mixed with the previous condensate and the whole is then fractionated. Concentrated acetone is taken overhead and the bottoms are led to a column from the top of which water–isopropanol azeotrope (containing 91% isopropanol) is taken for recycling without further purification. Water from the bottom of this column is re-used in the scrubber.

Refs. 11 and 35 describe a lower temperature process in which the catalyst is zinc oxide (7–8%) deposited on pumice. The isopropanol feed may contain up to 12% water [approximately azeotropic proportions] without seriously affecting the reaction and it is fed with an equimolar ratio of hydrogen (recycled hydrogen can be used) in order to decrease catalyst fouling. Conversion starts at 98% per pass with fresh catalyst but then drops so that catalyst regeneration is necessary after 10 days.

This is accomplished at 500 °C with a mixture of 2% oxygen and 98% nitrogen. A batch of catalyst lasts about six months.

A large number of other catalysts have been investigated for this reaction. When brass spelter is used it is regenerated with mineral acid at intervals of 500–1,000 hours.[11]

In a variation on the above process a combined oxidation–dehydrogenation process is used. Air, saturated with isopropanol vapour, is passed over a catalytic bed of silver or copper maintained at 400–600 °C. The hot gasses are partially cooled and scrubbed with water, nitrogen being released to the atmosphere. The remainder of the process and the yield (85–90%) are similar to the dehydrogenation process.[35]

$$(CH_3)_2CHOH + \tfrac{1}{2}O_2 \text{ (air)} \rightarrow CH_3COCH_3 + H_2O$$

or

$$(CH_3)_2CHOH \rightarrow CH_3COCH_3 + H_2$$

In the cumene hydroperoxide process, benzene is alkylated to cumene. This is then oxidized to cumene hydroperoxide which in turn is cleaved under acid conditions to yield phenol and acetone. One kilogram of phenol production results in 0.6 kg of acetone.[11]

The alkylation of benzene is carried out using propylene having a low olefin content and, preferably, having a bromine index (grams of bromine per 100 kg cumene) of less than 200. Alkylation takes place at 175–225 °C and 400–600 psig over a pelletized and calcined mixture of phosphoric acid and kieselguhr. By having the benzene at a concentration several times that of propylene, polymerization of the latter is avoided and the bromine number of the cumene is kept very low. Dialkylation and other side reactions also are inhibited. Separation from propane (present in the propylene charge) and benzene enables high purity cumene (containing less than 1,000 ppm of total impurities and with a bromine index below 100) to be obtained (90% yield, based on propylene).[35]

$$C_6H_6 + CH_2CHCH_2 \rightarrow C_6H_5CH(CH_3)_2$$

Oxidation of cumene to cumene hydroperoxide (using atmospheric air or oxygen) is typically accomplished using a series of oxidizers through which the cumene [boiling point 152.5 °C, S.G. 0.86], with recycled material, passes usually at a temperature between 80 and 130 °C. In a typical reaction the temperatures decline from 115 °C in the first reactor to 90 °C in the last. At the bottom of each reactor there may be a layer of 2–3% sodium hydroxide. Compressed air is bubbled in at the bottom of each reactor and leaves from the top. The oxygen ratio, as a function of consumable oxygen, is higher in the later reactors, thus maintaining the reaction rate as high as possible while minimizing the temperature-promoted decomposition of the hydroperoxide. Total residence time in each reactor is likely to be 3–6 hours and the concentration of hydroperoxide in the last of a four-stage reactor could be 32–39%. Concentration by evaporation achieves a level of 75–85% and the yield may be 90–95%.[11]

$$C_6H_5CH(CH_3)_2 + O_2 \text{ (air)} \rightarrow C_6H_5C(CH_3)_2OOH$$

Cleavage of the cumene hydroperoxide is achieved under acid conditions in an agitated vessel at 60–100 °C. A large number of inorganic acids are noted as being useful for this purpose, including sulphur dioxide gas.[11] Mild pressure may be used.[35]

The reaction mixture obtained consists of phenol, acetone and a wide variety of other materials such as cumylphenols, acetophenone, dimethylphenylcarbinol and α-methylstyrene. Neutralization may be effected by sodium hydroxide solution or other suitable base or with ion exchange resins. The removal of inorganic salts may then be facilitated by the addition of process water. Before being passed to a distillation tower the crude product may go through a separation and wash stage.[11]

Crude acetone from the distillation tower may be further purified using one or two additional distillation columns. If two such columns are used, the first will remove such as acetaldehyde and propionaldehyde while the second will remove undesired heavies, mainly water.[11]

Bottoms from the initial distillation column are vacuum distilled, unreacted cumene and by-product α-methylstyrene appearing as overhead. The α-methylstyrene would lower the peroxidation yield if recycled with the cumene and so is either catalytically hydrogenated to cumene or is separated by careful fractionation to be available as a by-product.

The bottoms from the vacuum still are further distilled to separate acetophenone from phenol, which latter is the overhead product at 90–92% yield.[35]

Acetone is also produced by catalytic oxidation of isopropanol at 400–600 °C. The reaction is exothermic and requires careful temperature control if the yield is not to be lower than with dehydrogenation.[11]

A process for synthesizing glycerol from propylene involves a stage wherein isopropanol (from propylene and water) is oxidized to acetone plus hydrogen peroxide by reaction with oxygen at 90–140 °C. In addition, isopropanol is also reacted with acrolein at 400 °C in the presence of magnesium and zinc oxides to form allyl alcohol and acetone. The allyl alcohol and hydrogen peroxide react to form glycerol with the acetone being taken as a co-product.[11]

World production of acetone in 1983 was estimated as 2,000 kilotonnes from a manufacturing capacity of over 3,200 kilotonnes. US production in 1983 was 850 kilotonnes and its consumption was 852 kilotonnes. EC production in 1983 was 710 kilotonnes and its consumption was 700 kilotonnes.[36]

Nelson and Webb in ref. 11 note that the economics of acetone production and its consequent market position are unusual. Acetone availability is very much tied to phenol production with any shortfall being made good by (possibly under-utilised) alternative production plant.

Acetone is supplied in bottles, cans, drums and tankers.[35] Electronic and spectrophotometric grades are available.[37]

Uses[4,11,34,38,39]

Acetone is used as a solvent for vinyl or acrylic resins, fats, oils, waxes, collodion, alkyd paints, lacquers, varnishes, rubber cements and for a variety of substances in chemical syntheses. It is used in the manufacture of, *inter alia*, methyl methacrylate, diacetone alcohol, isophorone, ketene, mesityl oxide, methyl isobutyl ketone, haloforms, hexylene glycol, bisphenol A, isoprene, rayon, aeroplane dopes and photographic films. It is useful as a storage medium for acetylene gas, taking up to 24 times its own volume of acetylene. In Great Britain, acetone was recommended (1978) to be permitted for use as a solvent in food if the mesityl oxide content of the acetone did not exceed 10 ppm.

12. STORAGE, HANDLING AND USE PRECAUTIONS[11,38,40–43]

Nelson and Webb in ref. 11 observe that as modern commercial acetone is virtually a pure product, it is common practice to store it in steel tanks. Marsden and Mann[40] note (in 1963) that iron, mild steel, copper or aluminium are suitable for plant and containers but slight discolouration and perceptible corrosion may be found on copper which has been in contact with old or recovered acetone which may have developed appreciable acidity.

Acetone is highly flammable and the vapour forms explosive mixtures with air.[40] The liquid boils at 56 °C and is miscible with water in all proportions (see Section 17).

Care must be taken with storage, process and drying plant to ensure that hazard-free conditions exist regarding fire or explosion risks and exposure of personnel to acetone. Sax[41] recommends underground storage and that neither acids nor oxidizing materials should be stored nearby. Storage areas should have good ventilation and have explosive venting. An outdoor storage facility requires water-sprinkling during hot weather or protection from direct sun.

[As with other flammable solvents the organization of storage facilities has to take into account two separate requirements. These are that the stored material be protected from a fire elsewhere on the premises and that the premises be protected from a fire involving the stored materials. On an industrial scale, sufficient physical separation and water spray curtains are often used effectively. On a laboratory scale, flammable solvents need to be stored in fire resistant, thermally insulated cupboards.

It is important to appreciate that a non-combustible barrier is not synonymous with a fire barrier. A fire barrier has to protect against both conducted and radiant heat. For example, a metal cupboard is useless on both these counts.]

NIOSH[42] recommends that engineering controls such as process enclosure or local exhaust ventilation should be used to keep personnel exposures at or below permissible limits. Any ventilation system should limit accumulation or recirculation of acetone in the workplace environment and should be spark-proof. It should be regularly maintained and airflow measurements taken at least every three months to verify effectiveness. Confined spaces that have held acetone should be cleaned with water and purged with air. They should be physically isolatable to prevent accidental re-entry of the solvent.

[Personnel should not be allowed to enter areas containing potentially hazardous concentrations of acetone unless all relevant legislation and official guidelines have been strictly adhered to. Suitable respiratory protective equipment should be used.] Such exposed personnel should be able to communicate with each other and with those (*i.e.*, more than one person) monitoring the work from outside the hazardous area. Monitors should be suitably equipped to effect rescue (possibly with lifelines to exposed personnel) and should themselves ensure, so far as is possible, that one of their number remains outside the hazard area in the event of an emergency. Protective clothing should be impervious to acetone. Chemical safety goggles or face-shields (20 cm minimum[42]) should remain unaffected by the solvent. Note that rayon, which may be found in an employee's personal attire, is affected by acetone, as also may be plastic spectacles and contact lenses.[43] Prolonged or repeated topical application may cause erythema and dryness, while inhalation in large amounts can produce narcosis.[38]

N.B. the caveat in Section 13 (final paragraph).

13. FIRE HAZARDS[3,39,44−47]

Acetone is extremely flammable. It has a boiling point of 56.2 °C and a flash point of − 20 °C (closed cup). Its auto-ignition temperature is about 465 °C and the lower and upper explosive limits in air are 2.5 and 13% v/v (see Section 17). It is miscible in all proportions with water, but Inoue in ref. 39 notes that the acetone must be considerably diluted if the flash point is to be brought to a relatively safe level.

Acetone vapour density is twice that of air and, if in sufficient quantity, the vapour can flow a considerable distance to an ignition source and then 'flash back'.

Fire-exposed containers of acetone can explode when heated and, if feasible, should be cooled with flooding amounts of water applied from as far as possible until well after the fire has been extinguished.

Small fires may be fought using dry powder, carbon dioxide or vaporizing liquid extinguishers. Considerable dilution with water (see above) may be effective. For larger fires, alcohol-resistant foam, water spray or fog can be used. A water jet risks scattering the burning liquid despite its water miscibility. Immediate withdrawal is imperative if the fire causes discolouration of a storage tank or in the case of rising sound from a venting safety device. Firefighters should keep upwind of the fire and avoid breathing acetone vapour and possibly toxic combustion products.

Ref. 44 notes that where a fire is fed by an uncontrolled flow of combustible liquid the decision on how or if to fight it will depend on the size and type of fire anticipated and must be carefully considered. This may call for special engineering judgement, particularly in large-scale applications. If the fire is uncontrollable or if containers are exposed to direct flame then the area should be evacuated to a radius of 1,500 feet.[3]

Bretherick[45] notes that in the absence of other ignition sources, fires in plant used to recover acetone from air by use of active carbon are due to the bulk-surface effect of oxidative heating when the air flow is too low for effective cooling.

[Notwithstanding the foregoing review of various authorities noted in Sections 12 and 13, it is essential that managers and others responsible for the planning, implementation and overseeing of personnel and plant safety should be familiar with the legal constraints and official guidelines applicable to them and that they liaise with their local emergency services in the planning of plant and storage facilities and in the preparation of contingency plans for dealing with fires and other emergencies. Managers should regularly monitor staff knowledge of, and ability to implement, emergency procedures and should ensure that equipment provided for use in emergencies is regularly inspected and maintained.]

14. HAZARDOUS REACTIONS[3,11,45,48–50]

Bretherick includes the following information in ref. 45.

Interaction of acetone with bromoform and chloroform in the presence of solid potassium hydroxide or other bases (calcium hydroxide being mentioned for chloroform) is violently exothermic. Bromoform is reported to react thus even in the presence of diluting solvents.

Acetone–isoprene mixtures can become peroxidised.

Acetone is oxidized with explosive violence if brought in contact with mixed nitric and sulphuric acids ('nitrating acid'), particularly under confinement.

Violent reaction may occur with bromine trifluoride.

Sudden and violent reaction may occur during bromination of acetone to bromoacetone should a large excess of bromine be present.

A cold, sealed tube containing nitrosyl chloride, platinum wire and traces of acetone exploded violently on being allowed to warm up.

Acetone ignites and then explodes on reaction with nitrosyl perchlorate and with nitryl perchlorate. Acetone ignites on contact with chromyl chloride, with chromium trioxide and with fuming nitric acid.

Acetone and hydrogen peroxide readily form explosive dimeric and trimeric cyclic peroxides, particularly during evaporation of the mixture. Mixtures of acetone, hydrogen peroxide and nitric acid can overheat and explode violently unless conditions and concentrations are carefully controlled.

Peroxomonosulphuric acid reacts with acetone to produce explosive acetone peroxide.

Acetone liquid or vapour, in contact with solid potassium t-butoxide, may ignite within a few minutes unless such as an excess of solvent enables the heat to be dissipated.

Precipitated thiotrithiazyl perchlorate exploded on being washed with acetone.

OSHA[3] notes that reaction with potassium hypobromite results in explosion; that there is violent reaction with sulphur dichloride and that explosion is possible with sodium hypoiodite, with hexachloromelamine and with trichloromelamine. With 1,1,1-trichloroethane, there is exothermic condensation in the presence of a basic catalyst.

Sax[48] notes incompatibility with boron trifluoride, sodium hypobromite and chromium(II) oxide.

Streng[49] notes that the addition of acetone to a mixture of dioxygen difluoride absorbed in 'dry ice' (solid carbon dioxide) resulted in sparking accompanied by an explosion.

Certain active chlorine compounds may react with acetone to form toxic chloroketones.[11]

Reaction of hydrogen sulphide with acetone in the presence of hydrochloric acid produces a noxious material (possibly thioacetone) known to have caused fainting, nausea and vomiting up to $\frac{3}{4}$ km from the laboratory source.[50]

15. EMERGENCY MEASURES IN THE CASE OF ACCIDENTAL SPILLAGE[3,47,51]

All ignition sources should be eliminated and any leak stopped if this can be done without risk. Water spray may be used to reduce vapours.[3] Very small spills may be mopped up with plenty of water and run to waste (a sewer) diluting greatly with running water,[47] or else such spillage may be absorbed on paper, sand or vermiculite and allowed to evaporate in a fume cupboard with a flame proof extractor until the whole of the ducting is clear of vapour.

For spillages up to *ca.* 25 l, the liquid may be absorbed onto sand or vermiculite, followed by transfer in suitable containers [*e.g.*, covered buckets] to a safe, open area where atmospheric evaporation can take place. The site of the spillage should be washed thoroughly with water and biodegradable detergent.[51] The washings should be discharged to a sewer and not to a 'soakaway' nor a land drain. Larger spillages should be bunded far ahead of the spill for later disposal.[3] Protective wear appropriate to the degree of spillage should be worn and only necessary personnel allowed to enter the hazard area.

Warren[51] notes that "*ideally* all hydrocarbons and related flammable organic chemicals should be burned in an incinerator with an afterburner."

16. FIRST AID

General

The casualty should be removed from danger of further exposure into fresh air. Rescuers should ensure their own safety from inhalation and skin contamination where necessary. Conscious casualties should be asked for information about what has happened. Bear in mind that the casualty may lose consciousness at any time. Where necessary continued observation and care should be ensured. First aiders should take care not to become contaminated. Note that acetone is highly flammable.

Inhalation

The conscious casualty should be kept at rest. The unconscious casualty should be placed in the recovery position and an open airway maintained. If breathing or heart beat stops, resuscitation should be commenced immediately. Medical aid should be obtained or the casualty removed to hospital immediately.

Skin Contamination

Remove contaminated clothing. Wash with copious amounts of water, or soap and water for at least 15 minutes. If necessary, seek medical aid. In the case of persistent skin irritation refer for medical advice.

Eye Contamination

Irrigate with copious amounts of water for at least ten minutes but avoid further damage to the eye from use of excessive force. Seek medical aid. Remember the possible presence of contact lenses which may be affected by some solvents and may impede decontamination of the eye.

Ingestion

If the lips or mouth are contaminated rinse thoroughly with water. Do not induce vomiting. Give further supportive treatment as for inhalation. In the case of ingestion of significant amounts wash out mouth with water. Obtain immediate medical attention. Contact a hospital or poisons centre at once for advice. Remember that symptoms may develop many hours after exposure so continued care and observation may be necessary.

In all cases note information on the nature of exposure, and give this to ambulance or medical personnel. Information for doctors is provided in some technical literature issued by manufacturers.

17. PHYSICO-CHEMICAL PROPERTIES

17.1 General[38,40,52]

Acetone is a colourless, volatile, highly flammable liquid with a characteristic odour.[38,40] Merck[38] describes it as having a pungent, sweetish taste.

Weast[52] records that acetone is miscible with a wide range of organic solvents, including benzene, diethyl ether, ethanol, butanol and chloroform. Miscibility was determined at 20 °C, admixing equal volumes of each solvent.

Acetone is miscible in all proportions with water.

17.2 Melting Point[52]

−95.35 °C.

17.3 Boiling Point[52,53]

Weast[52] records the boiling point of acetone as:

56.2 °C.

Acetone will form azeotropes with a number of other compounds. Details are listed below of some binary ones[53]:

Wt % acetone	Second component	Wt % second component	Boiling pt. azeotrope (°C)
59	Hexane	41	49.8
51.4	1-Hexene	48.6	50.1
61	Isopropyl ether	39	54.2

17.4 Density/Specific Gravity[38,40,52]

The specific gravity of acetone at 20 °C referred to water at 4 °C is given by Weast[52] as 0.7899. The specific gravity at 25 °C referred to water also at 25 °C is given by Merck[38] as 0.788.

The critical density of acetone is given by Marsden and Mann[40] as 0.268 g/cm³. From this the calculated critical volume is 3.731 cm³/g.

17.5 Vapour Pressure[52]

A selection of vapour pressures of acetone below atmospheric pressure are listed by Weast[52]:

Temperature in °C	Vapour pressure in mmHg
−59.4	1
−31.1	10
−9.4	40
7.7	100
39.5	400

Weast[52] also gives the following list of temperatures corresponding with elevated vapour pressures:

Temperature in °C	Vapour pressure in atm
56.5	1
78.6	2
113.0	5
144.5	10
181.0	20
214.5	40

The critical temperature and critical pressure of acetone are recorded as 235.5 °C, and 47 atm [*ca.* 35,700 mmHg] respectively.[52]

17.6 Vapour Density[44]

2.0 (relative to air = 1).

17.7 Flash Point[44]

− 20 °C, closed cup test.

17.8 Explosive Limits[44]

The limits of flammability of acetone in air at normal atmospheric temperature and pressure are given by the NFPA[44] as:

Lower limit: 2.5% v/v
Upper limit: 13% v/v.

The NFPA[44] give the ignition temperature of acetone as 465 °C. However, they point out that different test conditions (and also different definitions of "ignition temperature") can result in widely varying values being quoted. They therefore recommend that any value should be treated as an approximation.

17.9 Viscosity[52]

The viscosity of liquid acetone is recorded by Weast[52] as follows:

Temperature in °C	Viscosity in cp
− 92.5	2.148
− 80.0	1.487
− 59.6	0.932
− 42.5	0.695
− 30.0	0.575
− 20.9	0.510
− 13.0	0.470
− 10.0	0.450
0	0.399
15	0.337
30	0.295
41	0.280

Weast also gives the viscosity of acetone vapour[52]:

Temperature in °C	Viscosity in μp
100	93.1
119.0	99.1
190.4	118.6
247.7	133.4
306.4	148.1

17.10 Concentration Conversion Factors

At 25 °C and 760 Torr (1 atm)

1 mg/m³ = 0.42 ppm, and
1 ppm = 2.38 mg/m³.

18. TOXICITY

18.1 General[4,39,54,55]

Acetone is considered one of the least toxic solvents used in industry.[39,54] Inhalation of acetone vapour though can cause headache, restlessness and fatigue, leading to narcosis and unconsciousness at high concentrations.[4] Vomiting and haematemesis may occur and there is often a latent period of several hours.[4] ILO notes that, in some cases of over-exposure, albumin and red and white cells in the urine indicate the possibility of kidney damage and, in others, the high levels of urobilin and the early appearance of bilirubin reported presume liver damage. The longer the exposure, the lower the respiratory rate and pulse, these changes being roughly proportionate to the acetone concentration. Various toxic effects have been noted at chronic inhalational exposures down to 500 ppm, with subjective symptoms and blood changes being very slight at 250 ppm.[39] Acetone is one of the solvents known to have been abused in 'glue-sniffing'.[4] Prolonged or repeated skin contact may defat the skin and produce dermatitis.[54] Liquid acetone is moderately irritating to the eyes. The vapour has produced only slight eye irritation at or below 1,000 ppm.[54]

Ruth[55] gives the odour threshold of acetone as ranging from about 47.5 mg/m³ [20 ppm] to 1,613.9 mg/m³ [680 ppm] with irritation at 474.7 mg/m³ [200 ppm].

18.2 Toxicokinetics[10,54,56–58]

Krasavage *et al.* in ref. 54 note that acetone is readily absorbed by all routes of administration and that its high water solubility ensures widespread distribution in the tissues.

Sax[10] notes work on acetone metabolism in male beagle dogs and in man where exposure was by inhalation to 100, 500 and 1,000 ppm for two or four hours. In both species the half-life in blood was three hours and there was direct proportionality between the concentrations of acetone in blood and breath and the magnitude of the exposure, with man absorbing more than the dog, given comparable doses. In humans, there was no direct proportionality found between urinary acetone and the exposure limit observed.

Krasavage *et al.* in ref. 54 cite a study in which workers were exposed to 833 ppm of acetone vapour for three hours, twice daily, with a one hour break between exposures. At the end of the day the expired air had levels of 190 µg/l of acetone. Sixteen hours later this had decreased to 32 µg/l and the level returned to background values over the weekend, suggesting that repeated exposure to high concentrations of acetone may lead to some accumulation during the work week.

Price and Rittenberg in ref. 56 state that large doses of ingested acetone are predominantly excreted unchanged in expired air whereas small doses (up to 7 mg/kg) are largely oxidized to carbon dioxide.

Krasavage *et al.* in ref. 54 note that following dosage of rats with ^{14}C acetone, radio-labelled carbon was found in cholesterol, hepatic glycogen, various amino acids and in carcass protein. These authors also note that acetone is converted to lactate in mice and, when rats and mice were exposed to 30 mg/l of acetone in air, and rabbits and guinea pigs were exposed to 72 mg/l in air for two hours, increased levels of acetone, acetoacetic acid and β-hydroxybutyric acid were found in the blood and urine immediately after exposure and 24 hours later.[54]

A study by Casazza *et al.*[57] involved the intraperitoneal injection of 5 µmol of acetone per gram body weight into rats previously fed with 1% (v/v) acetone in their drinking water. This resulted in the appearance in blood serum of 16 ± 2 nmol of 1,2-propanediol/ml and 8 ± 1 nmol of 2,3-butanediol/ml. Neither of these two compounds was found in the serum after acetone or saline injection without the prior addition of acetone to drinking water, nor in the serum of animals injected with saline after having been maintained on drinking water with 1% acetone. The authors suggest that acetone both acts to induce a critical enzyme or enzymes and serves as a precursor for the production of 1,2-propanediol and that chronic acetone feeding plays a role in 2,3-butanediol production in the rat.

Krasavage *et al.* in ref. 54 note a study showing that isopropanol is oxidized extensively to acetone *in vivo* by laboratory animals. On the other hand, Lewis *et al.*[58] investigated the possibility of acetone being metabolized to isopropanol both in normal rats and in rats with alloxan-induced diabetes. Dosages of 1, 2, and 4 g/kg were given orally. All treated animals had increased levels of blood acetone and measurable levels of isopropanol. Untreated animals had no detectable isopropanol. In diabetic animals, isopropanol production increased significantly only at the highest acetone dose given, whereas in non-diabetic animals, there was a sixfold increase (significant at the 0.10 level) in blood levels of isopropanol when the acetone dose was doubled from 1 to 2 g/kg but with no difference on further doubling the dose to 4 g/kg.

Krasavage *et al.* in ref. 54 observe that many studies have shown that the hepatic microsomal mixed function oxidase system is induced by acetone. It enhances the rate of microsomal aniline-*p*-hydroxylase activity. Inhaled acetone enhanced ethoxycoumarin *o*-demethylase activity and cytochrome P450 content of the liver in rats but had no effect on NADPH cytochrome P450 reductase activity.[54]

Krasavage *et al.* further note that pretreatment with acetone for six days has potentiated acute ethanol toxicity in rats and that combined treatment with acetone and ethanol for six days potentiated acetone depression of total liver microsomal protein and cytochrome P450 and b5, prolonged the pharmacologic effect of hexobarbital, and decreased hexobarbital concentrations in rats recovering from narcosis. Acetone pretreatment has also potentiated chlorinated hydrocarbon toxicity.

18.3 Human Acute Toxicity[10,39,54,59,60]

ACGIH[59] and Krasavage *et al.* in ref. 54 cite a paper reporting the ingestion of 200 ml of acetone by a 42-year old man who became stuporous a half hour later and developed flushed cheeks and shallow respiration. Later, he relapsed into a coma but regained consciousness after about 12 hours when given supportive therapy. Subsequently hyperglycaemia of unknown cause was diagnosed and it has been suggested that this may have been due to the acetone ingestion.

Sax[10] notes that acetone has a similar effect in humans to that of ethanol for equal blood levels but that acetone has greater anaesthetic potency and that no ill effect was reported after an ingestion of 20 ml of acetone. Sax further notes that in acute cases there is a latent period which may be followed by restlessness and vomiting, then haematemesis and progressive collapse with stupor.

Sax[10] and Krasavage *et al.* in ref. 54 observe that human epidermis exposed for 90 minutes to undiluted liquid acetone produced cellular damage in the stratum corneum and stratum spinosum, there being marked intracellular oedema of keratinized cells and vacuolization of spinous cells. The structural alterations were minimal by 72 hours after treatment.

Krasavage *et al.* in ref. 54 cite an occurrence where eight men entered a pit wherein they were exposed to a mixture of 12,000 ppm acetone and 50 ppm 1,1,1-trichloroethane. They showed signs of throat and eye irritation, limb weakness, headache, dizziness, lightheadedness and a feeling of general malaise. Levels of acetone in urine collected 90 minutes after exposure were elevated and it was considered that the symptoms were due to acetone intoxication.

These authors and ACGIH[59] also note a case of acute acetone poisoning in a 10-year old boy who wore a hip cast set with a mixture of 90% acetone, 9% pentane and 1% methyl salicylate. Eight hours after the cast was in place the boy became restless and complained of headache. Four hours later he vomited (positive benzidine test for blood) and collapsed. The cast was removed, after which the boy became comatose. The skin that had been covered by the cast appeared normal. Recovery was complete after four days. The relative amounts of acetone inhaled or percutaneously absorbed were not known.[54,59]

ILO[39] records that exposure at 9,300 ppm could not be tolerated by a man for more than five minutes because of throat irritation, while at 2,000 ppm a slight narcotic effect was observed.

A study is noted by Krasavage *et al.* in ref. 54 wherein groups of five students were exposed to 0, 100, 250, 500 or 1,000 ppm acetone vapour for six hours. Irritation of eyes, nose and throat was noted at 500 and 1,000 ppm. The odour of acetone was detected at 100 ppm and acclimatization occurred rapidly. In a study by Nelson *et al.*[60], 500 ppm for 5–8 minutes irritated the eyes, nose and throat of the majority of about ten persons of mixed sexes, while the highest concentration which the majority of them estimated would be satisfactory for an eight hour exposure was 200 ppm. ACGIH[59] comment on the shortness of exposure in this study.

Krasavage *et al.* in ref. 54 record the reporting of hepatorenal lesions in two men and two women acutely exposed to acetone [exposure levels not stated]. One person had inhaled vapour and the others had ingested the material. Clinical manifestation of liver injury was noted in all four workers. Renal lesions were detected in two of them.

18.4 Animal Acute Toxicity[2,6,10,54,59]

ACGIH[59] notes that acetone concentrations of 52,200 ppm produced narcosis in rats and a concentration of 126,600 ppm for one hour was fatal.

Krasavage *et al.* in ref. 54 observe that oral LD_{50}s have been reported for rats in the range 8.5–10.7 ml/kg. For mice, LD_{50}s ranging from 4–8 g/kg are reported, and an oral LD_{50} of 5.3 g/kg is reported for rabbits. A minimum lethal oral dose of 8 g/kg is reported for dogs.

Also noted are studies giving the following minimum lethal concentrations for rats and mice exposed to acetone vapours, *viz.* rats – 16,000 ppm for four hours; mice – 46,000 ppm for one hour.[54] Other studies quoted give lethal concentrations for rats as 42,000 ppm and for guinea pigs as 40,000 ppm (with slight irritation noted at 20,000 ppm and none at 10,000 ppm) [durations unstated]. Also quoted by Krasavage *et al.*[54] are lethal concentrations for rats of 126,000 ppm over two hours; for guinea pigs 50,000 ppm over $\frac{3}{4}$ hour and for cats 21,000 ppm over three hours. The anaesthetic concentration of acetone for male and female mice is quoted as 99 mg/l [*ca.* 41,680 ppm].[54]

In an open Draize test, 395 mg acetone applied to the skin of rabbits produced mild skin irritation. The dermal LD_{50} was 20 g/kg.[2,54] Sax[10] notes a study in which groups of four white rabbits were tested with covered applications of acetone on abraded skin and groups of four male Hartley guinea pigs were similarly tested. For both species, the LD_{50} was reported as greater than 9.4 ml/kg (greater than 7.4 g/kg).

Browning[6] states that the preliminary symptoms of acute intoxication are irritative, *viz.* salivation, lacrymation, giddiness, ataxia, twitchings and convulsions. This author also cites a report that irritation of the nasal mucosa causes a temporary cessation of breathing as a protective reflex. Other reports noted by Browning are that of there being a specific irritative effect on the respiratory centre and that of a fall in blood pressure after intravenous and intramuscular injection which some researchers have regarded as being primarily due to a decrease in cardiac output.

Browning[6] notes a study wherein direct application of acetone to rabbits' eyes caused marked oedema of the conjunctiva. It has been since suggested that this injury is due to dehydration of the sclera followed by gelatinous flocculation and opacity. Other workers, noted by Krasavage *et al.* in ref. 54, reported that acetone produced moderate corneal injury to rabbit eyes.

18.5 Human Sub-acute and Chronic Toxicity Including Reproductive Effects[3,39,43,54,59,61–63]

General

ILO[39] notes that cases of chronic poisoning from prolonged inhalational exposure to low concentrations of acetone are rare. However, repeated exposures to low concentrations have given rise to complaints of headache, drowsiness, vertigo, irritation of the throat and coughing. Although acetone itself is considered to have relatively low chronic toxicity, it may enhance the hepatotoxicity of chlorinated solvents such as 1,1-dichloroethylene and 1,1,2-trichloroethane.[61]

Workers exposed to 500 ppm acetone vapour for six hours per day for six days experienced irritation of mucous membranes, an unpleasant smell, heavy eyes, overnight headache and general weakness accompanied by haematologic changes such as an increase of leucocyte and eosinophil counts and decrease in the phagocytic activity of neutrophils. Recovery took several days, but with 250 ppm exposure the subjective symptoms and blood changes were very slight.[3,39]

ACGIH[59] notes an examination of workers engaged in acetone production and subject to 700 ppm in air for three hours a day for 7–15 years. Inflammation of the respiratory tract, stomach and duodenum were reported plus attacks of giddiness and loss of strength. There had been no mention of other solvent exposures.

ACGIH[59] and Krasavage *et al.* in ref. 54 note a study of workers exposed for eight hours a day to a varying concentration of acetone from their work in cleaning filter press plates containing a thick syrup of cellulose acetate dissolved in acetone. For three hours per shift acetone concentrations ranged from 155–6,596 ppm and, at other times, from 25–904 ppm with time-weighted daily averages ranging from 950–1,060 ppm. Eye irritation was noted by seven of nine workers. Less frequently reported symptoms included headache, lightheadedness, nasal irritation and throat irritation. The symptoms were originally reported as being intermittent, transient and occurring at acetone concentrations in air considerably in excess of 1,000 ppm. However, a review of the raw data indicated that at least four instances of eye irritation occurred at exposure levels below 1,000 ppm.

Arena[62] states that acetone taken orally in doses of 15–20 g daily for several days usually produces no ill effects other than slight drowsiness.

Prolonged contact of acetone with the skin can irritate and lead to dryness, cracking and dermatitis.[61]

Reproductive Effects in Man

Ref. 43 notes a Russian report suggesting that long-term exposure of women to about 30–40 mg/m³ [approx. 13–17 ppm] and about 200–300 mg/m³ [approx. 85–125 ppm] produced an increased incidence of pregnancy complications and reduced foetal weight, apparently accompanied by maternal metabolic disorders. No teratogenicity was reported. [The significance of these findings is open to question since the original paper (ref. 63) gives insufficient detail, such as the nature of the control group, for substantiation.] (A concomitant animal study also reported in ref. 63 is noted in Section 18.6.)

Carcinogenicity to Man

No reports on carcinogenic effects of acetone were found. Refs. 61 and 59 state there is no evidence of carcinogenesis due to acetone.

18.6 Animal   Sub-acute   and   Chronic   Toxicity   Including   Reproductive Effects[2,10,54,59,63,64]

General

Sax[10] notes several studies of rats exposed to acetone vapour by inhalation:

Rats exposed to 25–37 mg/l [approx. 10,500–15,600 ppm] for eight hours/day for 28 days showed a decrease in serum albumin of 20.33%, with increases in α-globulins of 38%, β-globulins of 20.45% and total lipids of 37.55%. There was a slight increase in β-lipoproteins.

Sax[10] also notes a study wherein juvenile baboons showed slow response time when exposed by inhalation to 500 ppm acetone for 24 hours/day for seven days.

RTECS[2] notes an oral LD_{50} of 7,400 mg/kg for the rat; ACGIH[59] notes a study wherein the oral LD_{50} for the rat was found to be 10.7 ml/kg [8,460 mg/kg]. RTECS[2] also notes an inhalational LCLo for the mouse of 110,000 mg/m³ [approx. 46,000 ppm] for 62 months.

Reproductive Effects in Animals

RTECS[2] notes a mammalian study in which females [species unstated] were subjected to 31,500 μg/m³ [approx. 13.3 ppm] by inhalation for 24 hours/day through days 1–13 of pregnancy. This regimen was reported to cause increased post-implantation mortality. [The original paper (ref. 63) gives insufficient detail for substantiation.]

An inhalational study with an acetone concentration initially averaging 3.65 mg/l, increased stepwise to 75.7 mg/l over 11 weeks, where rats were exposed over this period for 40 min/day, three days/week, caused an apparent reduction in large ovarian follicles but not the smaller maturing ones after eleven weeks but not seven weeks, as a non-specific habituation response. [The response would probably be reversible; detail in the paper was inadequate for further substantiation or evaluation].[64]

Krasavage *et al.* in ref. 54 report a study where no evidence of teratogenicity was found when 39 mg or 78 mg acetone was injected into the yolk sacs of fertile chick eggs prior to incubation.

Carcinogenicity to Animals

No reports were seen implicating acetone as a carcinogen. Krasavage *et al.* in ref. 54 note a study in which acetone did not produce tumours when applied to the skin of mice three times per week for one year. [Acetone has often been used as a vehicle in skin carcinogenicity tests with no indication of positive effects.]

18.7 Mutagenicity[65-67]

A concentration of 4.76% of acetone in the aqueous medium was found to be the most effective in inducing mitotic aneuploidy in strain D61.M of *Saccharomyces cerevisiae*.[65]

A concentration of 40 mg/ml of liquid gave a positive result in chromosomal aberration tests *in vitro*, using a Chinese hamster fibroblast cell line. In allied studies a dose (maximum) of 10 mg/plate was negative in *Salmonella*/microsome (Ames) tests.[66]

Acetone was negative in the Ames reversion test using strains TA1535, TA100, TA1538, TA98, TA1537 and TA97 of *S. typhimurium* and was negative in a DNA repair test using various strains of *Escherichia coli*.[67]

18.8 Summary

[Acetone is of generally low acute and cumulative toxicity. Acute over-exposure may cause temporary headache and narcosis, with high vapour concentrations irritating the respiratory tract, while the liquid is irritating to eyes and prolonged or repeated skin contact may cause defatting leading to mild dermatitis. Long term severe or acute gross over-exposure may additionally cause liver or kidney damage, perhaps with reversible haematological changes. No indications of carcinogenicity have been traced, and a report suggesting adverse effects on pregnancy, but not teratogenicity, has not been substantiated. Absorbed acetone is fairly rapidly excreted unchanged, with some oxidation to carbon dioxide or incorporation into the general carbon pool, with some evidence of partial accumulation unchanged during daily exposure but not with longer intervals. Acute or chronic exposure to inhaled vapour concentrations of 500 ppm in air causes definite systemic toxic effects and respiratory tract irritation, while the effects of 200–250 ppm are slight and generally tolerable.]

19. MEDICAL / HEALTH SURVEILLANCE

A decision on the need for, and content of, medical surveillance should be based on an assessment of the possibility and extent of exposure in the work operation. In addition, medical examination may be directed to identifying any pre-existing or newly arising condition in the individual workers which might either be aggravated by subsequent exposure, or might confuse any subsequent medical assessment in the event of excessive exposure or an illness not related to exposure. A particular aspect to be considered is the identification of sensitive subjects not adequately protected by the control limit in operation. A professional medical judgement may be required on continuance of employment in the specified process. The following information is relevant in the case of this solvent.

Pre-employment Medical Examination

Particular attention in the history and pre-employment medical examination should focus on the skin and conjunctivae, respiratory, liver and kidney functions, and the nervous system.

The employment of diabetics, persons with liver disease, persons who consume substantial amounts of alcohol, those with overt skin diseases and those with anaemia should be assessed by a physician.

Periodic Medical Examination

If repeated medical examinations are judged to be necessary related to the environmental and skin contact exposure then the same aspects as in the pre-employment examination should be considered. The possibility of exposure outside work should not be overlooked.

Acetone determination in blood, urine, and exhaled air have been suggested as exposure indicators but whereas acetone concentrations in blood and exhaled air are closely correlated with exposure, no correlation has been reported between urinary acetone and exposure levels.

It should be noted that observation over a long period has not demonstrated any prolonged health effect following occupational exposure to acetone, nor are there any known delayed effects after a single exposure.

20. OCCUPATIONAL EXPOSURE LIMITS

[The Committee felt that a time-weighted average limit of 250 ppm and a short term limit of 500 ppm were appropriate for acetone. The Committee accepts that because of practical difficulties it may take some time to attain these levels.]

REFERENCES

1. Council of Europe. 'Dangerous Chemical Substances and Proposals Concerning their Labelling'. ('The Yellow Book'.) 4th ed. Maisonneuvre, 1978.
2. National Institute for Occupational Safety & Health. 'Registry of Toxic Effects of Chemical Substances (RTECS)'. DHHS (NIOSH) Publication No.84-101-6. US Department of Health & Human Services, April, 1986.
3. Occupational Safety & Health Administration. Material safety data sheets from the Occupational Health Services database, OSHA, Washington, D.C., 1986.
4. Reynolds, J.E.F. and A.B. Prasad, Eds. 'Martindale, The Extra Pharmacopoeia'. 28th ed. The Pharmaceutical Press, 1982.
5. Verschueren, K. 'Handbook of Environmental Data on Organic Chemicals'. 2nd ed. Van Nostrand Reinhold, 1983.
6. Browning, E. 'Toxicity and Metabolism of Industrial Solvents'. Elsevier, 1965.
7. Graedel, T.E. 'Chemical Compounds in the Atmosphere'. Academic Press, 1978.
8. Sexton, K. and H. Westberg. Environmental Science & Technology 14; No.3(1980):329. 'Ambient hydrocarbon and ozone measurements downwind of a large automotive painting plant'.

9. Fielding, M. *et al.* 'Organic Micropollutants in Drinking Water'. Technical Report TR 159. WRC Environmental Protection, Water Research Centre, UK, 1981.
10. Sax, N.I., Ed.-in-Chief. Dangerous Properties of Industrial Materials Report 4; No.3(1984):9. 'Chemical review: acetone'. Van Nostrand Reinhold.
11. Grayson, M., Exec. Ed. 'Kirk-Othmer Encyclopedia of Chemical Technology'. 3rd ed. John Wiley & Sons, 1979.
12. Grasselli, J.G. and W.M. Ritchey, Eds. 'Atlas of Spectral Data and Physical Constants for Organic Compounds'. 2nd ed. CRC Press, 1975.
13. National Institute for Occupational Safety & Health. 'NIOSH Manual of Analytical Methods'. 3rd ed. DHHS (NIOSH) Publication No.84-100. US Department of Health & Human Services, 1984.
14. Ciupe, R. *Revista de Chimie (Bucharest)*, **32**, No.6(1981):584.
15. Ciupe, R. *et al. Revue Roumaine de Chimie*, **30**, No.11-12(1985):1053.
16. Mal'tseva, G.A. and Y.N. Talakin. *Gigiena i Sanitariya*, **50**, No.9(1985):55.
17. Diaz-Rueda, J. *et al. Applied Spectroscopy*, **31**, No.4(1977):298.
18. Kalab, P. *Collection of Czechoslovak Chemical Communications*, **47**, No.9(1982):2491.
19. Campbell, D.N. and R.H. Moore. *American Industrial Hygiene Association Journal*, **40**, No.10(1979):904.
20. Hagemann, R. *et al. Analusis*, **6**, No.9(1978):401.
21. Gage, J.C. *et al. Annals of Occupational Hygiene*, **20**, No.2(1977):127.
22. Pinigina, I.A. and Y.B. Yatsukovich. *Gigiena i Sanitariya*, **49**, No.2(1984):46.
23. Stentsel, I.I. *et al. Khimicheskaya Tekhnologiya (Kiev)*, 5(1982):53.
24. Stentsel, I.I. and V.V. Tishchuk. *Zavodskaya Laboratoriyat*, **50**, No.5(1984):1.
25. King, M.V. *et al. American Industrial Hygiene Association Journal*, **44**, No.8(1983):615.
26. Saito, T. *et al. Bunseki Kagaku*, **32**, No.1(1983):33.
27. Tiunov, L.A. *et al. Gigiena i Sanitariya*, **49**, No.2(1984):58. 'Device for taking samples of exhaled air from laboratory animals in toxicological experiments'.
28. Fredericks, G.E. and R.S. Scott. *Review of Scientific Instruments*, **46**, No.6(1975):749.
29. Smith, N.B. *Clinical Chemistry*, **30**, No.10(1984):1672.
30. Vickers, T.G. and R.A. Owen. *Biochemical Society Transactions*, **12**, No.6(1984):1130.
31. Holm, S. and E. Lundgren. *Analytical Biochemistry*, **136**, No.1(1984):157.
32. Lopez-Soriano, F.J. and J.M. Argiles. *Journal of Chromatographic Science*, **23**, No.3(1985):120.
33. Bock, J.L. *Clinical Chemistry*, **28**, No.9(1982):1873.
34. Grayson, M., Exec. Ed. 'Kirk-Othmer Concise Encyclopedia of Chemical Technology'. 3rd ed. John Wiley & Sons, 1985.
35. Lowenheim, F.A. and M.K. Moran. 'Faith, Keyes and Clark's Industrial Chemicals'. 4th ed. John Wiley & Sons, 1975.
36. Commission of the European Communities. 'Environmental Chemicals Data and Information Network (ECDIN)' Databank. Commission of the European Communities Joint Research Centre, ECDIN Group, I-21020 Ispra (Varese), Italy, 1986.
37. Hawley, G.G., Ed. 'The Condensed Chemical Dictionary'. 10th ed. Van Nostrand Reinhold, 1981.
38. Windholz, M. *et al.*, Eds. 'The Merck Index'. 10th ed. Merck & Co., 1983.
39. International Labour Office. 'Encyclopaedia of Occupational Health and Safety'. 3rd rev. ed. ILO, 1983.
40. Marsden, C. and S. Mann, Eds. 'Solvents Guide'. 2nd ed. rev. Cleaver-Hume Press, 1963.
41. Sax, N.I., Ed.-in-Chief. Dangerous Properties of Industrial Materials Report 1; No.4(1981):25. 'Acetone'. Van Nostrand Reinhold.
42. National Institute for Occupational Safety & Health. 'NIOSH Recommended standard for occupational exposure to ketones'. US Department of Health, Education & Welfare, 1978.
43. The Royal Society of Chemistry. Laboratory Hazards Data Sheet No.21:'Acetone', 1984.
44. National Fire Protection Association. 'Fire Protection Guide on Hazardous Materials'. 9th ed. NFPA, Massachusetts, USA, 1986.

45. Bretherick, L. 'Handbook of Reactive Chemical Hazards'. 3rd ed. Butterworths, 1985.
46. Canada Safety Council. Occupational Safety and Health Data Sheet No. F-1, rev. 'Acetone'. CSC, Canada, 1982.
47. Bretherick, L., Ed. 'Hazards in the Chemical Laboratory'. 4th ed. Royal Society of Chemistry, UK, 1986.
48. Sax, N.I. 'Dangerous Properties of Industrial Materials'. 6th ed. Van Nostrand Reinhold, 1984.
49. Streng, A.G. *Chemical Reviews*, 63(1963):607. 'The oxygen fluorides'.
50. Farrar, W.V. *Chemistry & Industry*, 29(18 July 1964):1310. 'A persistent odour'.
51. Warren, P.J., Ed. 'Dangerous Chemicals Emergency Spillage Guide'. 1st ed. Jensons (Scientific) Ltd., Leighton Buzzard, UK, 1985.
52. Weast, R.C., Ed.-in-Chief. 'CRC Handbook of Chemistry and Physics' ('The Rubber Handbook'). 67th ed. CRC Press, 1986-1987.
53. Horsley, L.H. 'Azeotropic Data III'. Advances in Chemistry Series No.116, American Chemical Society, 1973.
54. Clayton, G.D. and F.E. Clayton, Eds. 'Patty's Industrial Hygiene and Toxicology'. 3rd ed. rev. Wiley Interscience, 1982.
55. Ruth, J.H. *American Industrial Hygiene Association Journal*, 47(1986):A-142. 'Odor thresholds and irritation levels of several chemical substances: a review'.
56. Price, T.D. and D. Rittenberg. *Journal of Biological Chemistry*, 185(1950):449. 'The metabolism of acetone - I. Gross aspects of catabolism and excretion'.
57. Casazza, J.P. et al. *Journal of Biological Chemistry*, **259**, No.1(1984):231. 'The metabolism of acetone in rat'.
58. Lewis, G.D. *et al. Journal of Forensic Sciences*, **29**, No.2(1984):541. 'Metabolism of acetone to isopropyl alcohol in rats and humans'.
59. American Conference of Governmental Industrial Hygienists. 'Documentation of the threshold limit values for chemical substances in the workroom environment'. ACGIH, 1986.
60. Nelson, K.W. *et al. Journal of Industrial Hygiene & Toxicology*, 25(1943):282. 'Sensory response to certain industrial solvent vapors'.
61. Safety Practitioner 3 No.2(1985):6. 'Hazard Data Bank, Sheet No. 62: Acetone'.
62. Arena, J.M. 'Poisoning'. 3rd ed. Charles C. Thomas, 1974.
63. Nizyaeva, I.V. *Gigiena Truda I Professional'nye Zabolevaniia (Labour Hygiene & Occupational Diseases)*, **26**, No.6(1982):24. 'On hygienic assessment of acetone'.
64. Berliner, E.G. *Nekotorye Voprosy Eksperimental'noi Promyshlennoi Toksikologii*, (1977):93. 'Effect of organic solvents on morphofunctional state of rat ovaries'.
65. Zimmerman, F.K. *Annals of the New York Academy of Sciences*, 407(1983):186. 'Mutagenicity screening with fungal systems'.
66. Ishidate, M., jr. *et al. Food & Chemical Toxicology*, **22**, No.8(1984):623. 'Primary mutagenicity screening of food additives currently used in Japan'.
67. De Flora, S. *et al. Mutation Research*, 133(1984):161. 'Genotoxic activity and potency of 135 compounds in the Ames reversion test and in a bacterial DNA-repair test'.

CONTENTS

1. CHEMICAL ABSTRACTS NAME

carbon disulfide
(10th Collective Index)

2. SYNONYMS AND TRADE NAMES[1-4]

carbon bisulfide
carbon disulphide
carbon sulfide
dithiocarbonic anhydride
kohlendisulfid (German)
kooldisulfide (Dutch)
koolstofdisulphide (Dutch)
NCI-CO4591
OHS04280
P022
schwefelkohlenstoff (German)
solfuro di carbonio (Italian)
sulfure de carbone (French)
sulphocarbonic anhydride
UN 1131
Weeviltox
wegla dwusiarczek (Polish)
zwavelkoolstof (Dutch)

3. CHEMICAL ABSTRACTS SERVICES REGISTRY NUMBER

75-15-0

4. NIOSH NUMBER

FF6650000

5. CHEMICAL FORMULA

CS_2
(Molecular weight 76.14)

6. STRUCTURAL FORMULA

$$S = C = S$$

7. OCCURRENCE[5,6]

Carbon disulphide is present in the atmosphere to the extent of several hundred ppt. Emission sources include animal waste, fish processing, plastics and refuse combustion, petroleum, rubber, synthetic fibre and starch manufacture, turbines, natural gas and volcanoes.[5] Carbon disulphide has been found to be emitted from salt marshes along with other sulphur compounds.[6] Biogenic sulphur compounds derive from plant decomposition and through the action of sulphate reducing bacteria. All sulphur species are eventually oxidized to sulphur dioxide and sulphates so the concentration of atmospheric sulphur dioxide gives a general indication of the original total concentration of other sulphur compounds. Also, as all gaseous sulphur compounds except for carbonyl sulphide and carbon disulphide are oxidized rapidly in the troposphere it has been suggested that these two compounds are responsible for the presence of sulphur in the stratosphere.[6]

Carbon disulphide reacts with atomic oxygen by addition followed by cleavage to form carbon monoxide and sulphur monoxide, which latter would rapidly be oxidized to sulphur dioxide. However, the reaction of carbon disulphide with hydroxyl radical will be favoured in the atmosphere, possible pathways being

$$CS_2 + HO^{\cdot} \rightarrow CS + HOS^{\cdot}$$

(with the carbon monosulphide presumably being then oxidized to sulphur monoxide and thence to sulphur dioxide), and

$$CS_2 + HO^{\cdot} \rightarrow COS + SH^{\cdot}$$

the carbon disulphide thus serving as a source of carbonyl sulphide for which a long atmospheric lifetime has been suggested.[5]

8. COMMERCIAL AND INDUSTRIAL CARBON DISULPHIDE[7-9]

Pure carbon disulphide is clear and colourless with a mild ethereal odour. Minor impurities impart a disagreeable sulphurous odour and the liquid may be faintly yellow.[7,8] The impurities are sulphur compounds.[7]

Timmerman in ref. 9 gives the following examples of specifications for carbon disulphide.

Technical – industry (typical):

Specific gravity	(by pycnometer) 1.270–1.272 (15 °C/4 °C)
Residue %	(dry at 60 °C) 0.002 max.
Colour	(APHA Pt-Co 500 std) < 20
Foreign sulphide	(by lead acetate, no colour) Pass

Technical – US Federal:

Specific gravity	(pycnometer) 1.262–1.267 (20 °C/20 °C)
Residue	(dry at 60 °C) 10 mg/100 ml
Colour	Special test
Boiling range	(ACS distillation) 45.5–47.5 °C
Foreign sulphide	No discolour - Cu
Water	No turbidity

USP and ACS reagent quality:

Residue %	(dry at 60 °C) 0.002 max.
Colour	(APHA Pt-Co 500 std) 10 max.
Boiling range	1 °C including 46.3 °C
Foreign sulphide	Hg drop, no colour
Sulphite and sulphate	(I_2–$BaCl_2$) Pass
Water %	(Karl Fischer) 0.05 max.

9. SPECTROSCOPIC DATA[10]

Infrared, Raman, ultraviolet and mass spectral data for carbon disulphide have been tabulated.[10]

10. MEASUREMENT TECHNIQUES[11–25]

The method recommended for the determination of carbon disulphide in workshop atmospheres is the NIOSH[11] activated charcoal trapping method for collection and concentration, followed by solvent extraction of the charcoal and a gas chromatographic (gc) analysis of the extract.

The collection tube is 7 cm long with a 4 mm internal diameter. It contains a total of 150 mg of activated charcoal, (20–40 mesh) divided into a front section of 100 mg and a rear section of 50 mg separated by a small plug of urethane foam. The collection tube is preceded by a drying tube containing 270 mg anhydrous sodium sulphate. Air is sampled at a flow-rate of between 0.01 and 0.2 l per minute, by means of a small pump, the higher rate being used for ten minutes if the expected concentration is of the order of 10 ppm. The maximum volume of sample recommended is 25 l. The entire apparatus is portable so that it may be carried in a pocket with the sampling tube in the breathing zone (normally a coat lapel). The apparatus can also be static.

Toluene is used to extract the carbon disulphide, which is separated on a 2 m × 6 mm glass column packed with 5% OV-17 on Gas Chrom Q (80–100 mesh) at a temperature of 30 °C and using flame photometric detection in the sulphur mode.

The method has been used to determine carbon disulphide in working atmospheres of the viscose rayon industry and in carbon disulphide production facilities. For full details of the method see ref. 11.

The sampling method uses a small portable apparatus with no liquids and is therefore easy to handle and maintain. The analytical method is relatively specific with the additional advantage of flexibility of operating conditions.

Air samples are normally taken by means of suction using an electric pump capable of operating at low flow-rates (*i.e.* 10–300 ml per minute). If higher flow-rates are used, breakthrough can occur (*i.e.*, the carbon disulphide is not completely adsorbed and some of it passes through the adsorbent). It is also convenient either to establish the trapping device in a static position, or to have it attached to the user in such a way that the breathing zone is sampled (this usually means attachment to the worker's lapel).

It should be noted that suitable personal samplers may be commercially available.

The most popular accurate measurement technique following solvent extraction or thermal desorption of carbon disulphide is gc. There is a detection method for carbon disulphide in air issued by the H M Factory Inspectorate of the Department of the Environment.[12] This involves trapping of carbon disulphide in diethylamine and cupric acetate solution after the removal of interfering hydrogen sulphide by passage through lead acetate. The intensity of the resulting yellow copper diethyldithiocarbamate is assessed against standards and the method is used up to 40 ppm carbon disulphide.

A similar method to the recommended one involves trapping the carbon disulphide from air on 1.2 g activated charcoal contained in a 15 cm × 8 mm tube.[13] The carbon disulphide, together with any hydrogen sulphide, is extracted using a mixture of benzene and methanol (19:1) and gc is performed at 70 °C on a 1.83 m × 6.35 mm column packed with 5% OV-17 on Gas Chrom Q. Nitrogen is used as carrier gas and detection is made using a flame photometric detector in the sulphur mode.

Activated charcoal contained in personal dosimeters was extracted with benzene prior to gc analysis on a 9 ft × 0.125 in column packed with 18% OV-1 on Chromosorb W (80–100 mesh).[14] The chromatography was performed at 150 °C using helium as carrier gas and with thermal conductivity detection. This was part of a study of the effects of carbon disulphide on workers exposed in a Belgian viscose plant and, the metabolite, 2-thiothiazolidine-4-carboxylic acid, was determined in urine by a high-performance liquid chromatographic method using a 25 cm × 4.6 mm column packed with LiChrosorb RP-18, the metabolite being detected at 273 mm.

Acetonitrile has been used to extract carbon disulphide from activated charcoal.[15] After gc separation, detection was made using a photo-ionization system with an ion-chamber temperature of 210 °C. This method can detect 0.1 ppm carbon disulphide in air.

Apart from solvent extraction, the alternative method of obtaining carbon disulphide from a solid trapping medium is by thermal desorption. This is seldom recommended when activated charcoal is the trapping medium but is preferred if certain gc stationary phases are used *e.g.*, Tenax-GC. Carbon disulphide, hydrogen sulphide and other sulphides in air were trapped in a 21.3 cm × 4 mm glass tube containing 0.8 g molecular sieve 5A (60–80 mesh) in the front 9 cm and 0.2 g activated Tenax-GC (60–80 mesh) at the rear end.[16] The sulphides were desorbed by heating and sweeping them into a cold PTFE loop and then on to the 10 ft × 0.125 in gc column packed with Chromosil 330. Chromatography was performed starting at 1 °C and programming up to 100 °C to give good separation of the sulphides; detection was made by a flame photometric detector. The detection limit was between 8.8 and 20 pg as sulphur, depending on the sulphide.

In a study of sulphides in an urban atmosphere, one litre samples were trapped on a column of 1,2,3-tris-(2-cyanoethoxy)propane (TCEP) on Shimalite cooled with liquid oxygen.[17] This column was heated to 70 °C and sulphides were swept onto a similar column for gc analysis using flame ionization detection. The method was used for carbon disulphide concentrations of 0.1–75 ppb.

An oxygen-cooled 15 cm tube containing 25% TCEP on a silanised support was also used to trap carbon disulphide from air passed through it at 4.17 ml per second.[18] The tube was heated and the desorbed carbon disulphide was swept on to a similar column for gc analysis at 65 °C and using thermionic detection.

Colorimetric methods for determining carbon disulphide, because of their relative insensitivity, have been superseded by gc ones. As in ref. 12, the formation of copper diethyldithiocarbamate is the usual product for colour measurement following trapping of carbon disulphide. Silica gel treated with a mixture of ethanolic piperazine and ethanolic cupric acetate has been used as a trapping and reaction medium, the resulting copper diethyldithiocarbamate being extracted and matched with standards.[19] The method was used down to concentrations of 3.5 mg carbon disulphide per cubic metre of air [approx. 1 ppm].

Carbon disulphide from industrial atmospheres was adsorbed on activated charcoal, extracted into aqueous solution and reacted with pyrrolidine and subsequently converted to copper diethyldithiocarbamate with a copper reagent.[20] The cupric ions were then measured either potentiometrically or by atomic absorption spectrometry, thus producing an indirect method of determining the carbon disulphide.

In a titrimetric method, carbon disulphide, hydrogen sulphide and carbonyl sulphide in air were passed at 20 l per hour through a 10% ethylenediamine solution containing tributyl tin as an antioxidant.[21] After making the solution alkaline, it was titrated with 2-(hydroxymercuri)-benzoic acid, using dithizone as indicator. Before trapping, and by using separate samples, hydrogen sulphide may be removed by passage through 40% potassium hydroxide solution and activated charcoal could be used to adsorb both hydrogen sulphide and carbon disulphide, thus producing a method for quantitatively determining all three sulphides.

In another titrimetric method, carbon disulphide in air was adsorbed on to activated charcoal with prior removal of hydrogen sulphide by passage through cotton wool impregnated with lead acetate.[22] The carbon disulphide was extracted with ethanolic potassium hydroxide solution and after addition of acetate buffer a mixture of sodium azide, potassium iodide and iodine was added. After 15 minutes, unconsumed iodine was titrated with sodium arsenite solution. The method was used in the range 16–160 mg carbon disulphide per cubic metre of air [approx. 5–50 ppm].

A novel photometric method has been used to determine carbon disulphide and toluene in air.[23] The gas mixture is led into a detection chamber where it is ionized by means of a hydrogen discharge tube emitting radiation at 100–170 nm and the ion currents are measured by a photo-ionization detector provided with a lithium fluoride window and a sensitive picoammeter. The method is suitable for organic air pollutants with photo-ionization potentials in the 7–13 eV range.

There has been some interest in the determination of the metabolites of carbon disulphide in urine.[14] (See above.)

Blood as well as urine has been examined for the metabolites, dithiocarbamates and thiothiazolidones.[24] The sample was made acidic with 1.5M sulphuric acid, heated to 60 °C and the headspace sampled for gc analysis. This was performed on two different columns, *viz*. 15% SE-30 on Aeropak and 0.3% SE-30 plus 0.5% phosphoric acid on Carbopack HT-100. Electron capture detection was used with the former column, producing a more precise result and flame photometric detection in the sulphur mode was used with the latter column, producing a more selective result.

The urine of workers exposed to carbon disulphide in the air of a rayon plant was tested for the presence of metabolites using the catalytic effect the metabolites have on the reaction between sodium azide and iodine.[25] The test was applied to concentrations of metabolites greater than 20 ppm in the urine.

11. CONDITIONS UNDER WHICH CARBON DISULPHIDE IS PUT ON THE MARKET

Production[9,26–29]

Beauchamp *et al*. in ref. 26 note that carbon disulphide was first discovered in 1796 when a mixture of charcoal and iron pyrites was heated at elevated temperatures. Timmerman in ref. 9 states that the material has been an important industrial chemical since the nineteenth century with commercial manufacture beginning about 1880.

Until the 1950s the traditional method of manufacture was the reaction of charcoal with sulphur vapour at elevated temperatures.[27] Since then, hydrocarbons have been replacing charcoal as the carbon source.[27] Beauchamp *et al*. in ref. 26 quote 1979 data showing that by then all US and the majority of Western European production utilised the catalysed reaction of sulphur vapour and methane gas. Timmerman in ref. 9 notes that, in the USA, hydrocarbon feedstock has replaced charcoal since about 1965.

In the charcoal-sulphur processes, either fuel-fired retorts or electric furnaces have been used. Low-ash sulphur is essential, especially in the retort method, in order to limit the rate of accumulation of non-volatile residues. Preferred are uniformly sized lumps of low-ash, hardwood charcoal that have been prepared under carefully controlled conditions, including charring at temperatures of 400–500 °C. Retorts operate at 750–900 °C and at slightly above atmospheric pressure. The reaction is endothermic, there is little scope for heat economy, and external heat has to provide many times the theoretical energy input of 1,950 kJ

(466 kcal)/kg of carbon disulphide, (based on carbon and sulphur fed into the reactor at 25 °C and carbon disulphide produced at 750 °C). Internal electric heating at the reaction zone supplies an efficient energy source and several electric furnace designs for this purpose have been proposed.[9]

Carbon disulphide gases from either type of furnace pass into sulphur separators and then to coolers or water condensers.[28] The uncondensed gases, which include some carbon disulphide, are fed into absorbers where this residual carbon disulphide is removed by countercurrently flowing mineral oil. The washed gases, mostly hydrogen sulphide, pass into storage tanks. The absorbed carbon disulphide, having been freed from mineral oil by stripping, is combined with the liquefied carbon disulphide from the condensers and the whole is then passed to continuous distillation units in which the product is freed from unreacted sulphur in a hot water still. Subsequent fractionation removes hydrogen sulphide which goes to storage from which it may be passed to a packed tower to be burned under reducing conditions in the presence of a catalyst, such as bauxite, in order to recover sulphur. The fractionated carbon disulphide may be washed with 5% caustic soda in a tower packed with porcelain rings and condensed to give a 99.99% product.

An example of the material and utility requirements given for the production of one tonne of carbon disulphide is: charcoal 320 kg, sulphur 920 kg, electricity 1,650 kWh (5,950 MJ) (or equivalent fuel).[28]

Timmerman in ref. 9 states that where methane (as purified natural gas) is the carbon source, advantage can be taken of large, continuous processes with low operating cost and a high efficiency compared with solid carbon methods. In an example given by Lowenheim and Moran,[28] molten sulphur is vaporized in a furnace filled with high-chrome-steel coils and is mixed with high purity (99 + %) methane. The mixed gases are then reacted in a vertical, high-chrome steel or refractory reactor, filled with an activated alumina or clay catalyst, at 675 °C and a pressure of 200–275 kPa (*ca.* 2–2.7 atm). Carbon disulphide and hydrogen sulphide are formed along with small amounts of light mercaptans and heavy di- and poly-sulphides. Usually about 5–10% excess methane is required to give a sulphur conversion of 90–95%.

$$CH_4 + 4S \rightarrow CS_2 + 2H_2S$$

Unreacted sulphur is removed by contact with liquid sulphur. The desulphurised gases are then fed to an absorption column where carbon disulphide is removed by mineral oil, from which the product is then stripped and sent to steel distillation columns. Overhead from the absorption column, about 90% hydrogen sulphide, goes to a sulphur recovery unit.

An example of material requirements for the production of one tonne of carbon disulphide is methane 345 m^3 (SC); sulphur 925 kg.[28]

Technical grade carbon disulphide is of 99% purity.

Timmerman in ref. 9 notes the use of other catalysts such as silica gel, bauxite, various metals, oxides, sulphides and salts. Other hydrocarbon sources noted by Timmerman include ethane, propylene and 1-butene; light hydrocarbon gases are favoured as they result in fewer by-products.

Containers for carbon disulphide range through bottles, cans, drums, rail tank-cars, road tankers and barges (refs. 28 and Timmerman in 9). Transportation of carbon disulphide by air is forbidden.[28]

EC production of carbon disulphide was 250 kilotonnes in 1980. This decreased to 150 kilotonnes in 1983.[29] In the US, 1981 production had dropped to 172 kilotonnes[29] from 229 kilotonnes in 1977 and about 347 kilotonnes in 1974 (Timmerman in ref. 9). Factors affecting the uptake of carbon disulphide are considered below.

Uses[8.9.26.28.30–34]

Timmerman in ref. 9 notes that in the United States carbon tetrachloride production by chlorination of carbon disulphide accounted for less than 10% of the latter's consumption until about 1960. After then, this use increased considerably due to the rapid rise in demand for carbon tetrachloride as an intermediate for fluorocarbon aerosol propellants and refrigerants. In 1974 this use of carbon disulphide amounted to 32% of the US production.

$$(\text{Overall reaction: } CS_2 + 2Cl_2 = CCl_4 + 2S)$$

The other major use of carbon disulphide is in the manufacture of rayon and cellulose film.[28] Both these products face competition from new fibres and films. In addition, chlorination of short-chain hydrocarbons offers a competing route for the production of carbon tetrachloride.[28]

Other considerable uses of carbon disulphide are in the manufacture of rubber vulcanizers, flotation chemicals and pesticides.[26] Representative are the xanthates, the mono- and dithiocarbamates and their derived disulphides (*e.g.*, tetramethyl-thiuram disulphide) and the disulphide from 2-mercaptobenzothiazole which last is prepared from carbon disulphide, aniline and sulphur and is used as a vulcanizer.[26]

Solvent uses for carbon disulphide have been reported for fats, lipids, resins, rubbers, sulphur monochloride and white phosphorus.[30] Other solvent applications are associated with processes involving hydrocarbon conversion, reforming, extraction and separation. Removal of wax plugs occurring in petroleum wells during drilling and maintenance operations by purging with carbon disulphide has been effective. Among other uses for carbon disulphide are or have been use as a grain fumigant, in fresh fruit preservation, in electroplating, as a corrosion inhibitor, as a sulphiding agent[9.30] and in analysis.

To reduce its flammability, carbon disulphide may be diluted with carbon tetrachloride (commonly one part carbon disulphide to four parts carbon tetrachloride).[31] For grain fumigation, such diluted material was usually used.[32]

In the USA, carbon disulphide, carbon tetrachloride and ethylene dichloride were three materials banned by the Environmental Protection Agency as fumigants as from 30 June 1986.[33] [This use has also been discontinued in most EC countries.]

Timmerman in ref. 8 (published 1985) gives the largest single application of carbon disulphide (about one third of production) as being in the manufacture of viscose rayon. About 31% is used in carbon tetrachloride production and about 13% in cellulose film manufacture. Another source[34] in 1986 gives the US breakdown as rayon 35%; carbon tetrachloride, 30%; cellulose film and other regenerated cellulosics, 10%; rubber chemicals, 10%; other, 15%, with total US capacity being 700 Mlb/year (approx. 320 kilotonnes/year).

12. STORAGE, HANDLING AND USE PRECAUTIONS[2,9,35–41]

Carbon disulphide is highly flammable, has a boiling point of 46.25 °C, a flash point of −30 °C (closed cup), an auto-ignition temperature variously quoted between 90 °C and 125 °C (*e.g.*, refs. 2, 35, 36), and explosive limits in air of 1.3% and 50.0% v/v. Its specific gravity is 1.26 and its solubility in water is 0.294% at 20 °C.[37] (See Section 17.) Carbon disulphide is very toxic and is a much more powerful anaesthetic than chloroform.[38]

The foregoing require that carefully controlled conditions are maintained in storage, process and drying plant in order to ensure that personnel are not exposed to fire, explosion or toxic hazards.

[As with other flammable solvents the organization of storage facilities has to take into account two separate requirements. These are that the stored material be protected from a fire elsewhere on the premises and that the premises be protected from a fire involving the stored materials. On an industrial scale, sufficient physical separation and water spray curtains are often used effectively. On a laboratory scale, flammable solvents need to be stored in fire resistant, thermally insulated cupboards.

It is important to appreciate that a non-combustible barrier is not synonymous with a fire barrier. A fire barrier has to protect against both conducted and radiant heat. For example, a metal cupboard is useless on both these counts.]

In the case of carbon disulphide the liquid level in a storage tank must be blanketed with water or inert gas (nitrogen) at all times (Timmerman in ref. 9). If water blanketed, the entire tank is preferably completely submerged in a below-ground pit containing water. If above ground, the tank should be over a bunded area containing water at least equal in volume to that of the tank (refs. 39 and Timmerman in 9). The latter notes that storage and handling equipment is generally of conventional carbon steel construction.

As carbon disulphide is not corrosive to metals, use may be made of iron, aluminium or copper as well as of mild steel for plant and containers.[40] Copper will discolour badly due to the formation of a surface coat of sulphide but is not thereafter attacked.[40] Note though that Timmerman in ref. 9 states that copper and copper alloys can be corroded by impure carbon disulphide and that equipment using these materials should be used with caution. Above about 250 °C, carbon disulphide liquid and vapour become increasingly reactive with iron, steel and other common metals. For handling and processing at elevated temperatures, high-chromium stainless steel such as types 316, 309 and 310 may be suitable, while glass and ceramic materials are resistant to carbon disulphide at all temperatures. Teflon may also be used as a container material.[37] The ILO recommends that not more than 200 tonne quantities should be stored at individual plants that are within 500 metres of each other.[41] Storage areas should be bunded.

Timmerman in ref. 9 states that all parts of a system including piping, valves and movable containers must be electrically earthed and firmly bonded by good electrical conductors to eliminate the possibility of static charge build-up and spark discharge. Carbon disulphide may be safely transferred from an enclosed container by displacement with water or nitrogen. In the workplace it should be kept away from heat, sparks and flames. Adequate ventilation should be provided.

[Personnel should not be allowed to enter areas containing potentially hazardous concentrations of carbon disulphide unless all relevant legislation and official guidelines have been strictly adhered to. Suitable self-contained respiratory protective equipment should be used.] Such exposed personnel should be able to communicate with each other and with those (*i.e.*, more than one person) monitoring the work from outside the hazardous area. Monitors should be suitably equipped to effect rescue (possibly with lifelines to exposed personnel) and should themselves ensure, so far as is possible, that one of their number remains outside the hazard area in the event of an emergency.

NIOSH[39] instruct as follows regarding control of airborne carbon disulphide and storage of the material.

> Engineering controls shall be used when needed to keep carbon disulphide concentrations below the recommended limits. Local exhaust ventilation may also be effective, used alone or in combination with process enclosure. Spark-proof ventilation systems shall be designed to prevent recirculation of air in the workroom, to keep concentrations of carbon disulfide below the recommended occupational exposure limit, and to remove carbon disulfide from the breathing zones of workers. Ventilation systems shall be subject to regular preventive maintenance and cleaning to ensure effectiveness, which shall be verified by periodic airflow measurement. Make-up air shall be provided to workrooms in which exhaust ventilation is operating.

Drums of liquid carbon disulfide shall not be stored in direct sunlight or near a source of heat. The storage area should be fire resistant, cool, and either open or well ventilated at floor level. The storage area shall be equipped with an adequate supply of portable fire extinguishers and automatic sprinklers.

N.B. the caveat in Section 13 (final paragraph).

13. FIRE HAZARDS[2,4,26,35,36,42]

Carbon disulphide is a highly flammable liquid of boiling point 46.25 °C, specific gravity 1.26 and about 0.3% water solubility. It is dangerous when exposed to heat, flame, sparks or friction and presents a severe explosion hazard. The flash point is −30 °C (closed cup). The auto-ignition temperature is variously quoted between 90 and 125 °C (*e.g.*, refs. 2, 35, 36). The lower and upper explosive limits in air are 1.3% v/v and 50.0% v/v (see Section 17). The vapour density is approximately 2.6 times that of air and, if in sufficient quantity, the vapour can travel to a remote ignition source and 'flash back'. Rust can exacerbate the tendency to inflame, and examples are quoted by Bretherick.[42] For example, disposal of two litres into a rusted iron sewer caused an explosion. Ignition occurred when a bottle of carbon disulphide fell and broke behind a rusted steel cupboard. The vapour or liquid is known to ignite on contact with steam pipes, particularly if these are rusted.

Discharge of a quartz lamp used in flash photolysis has been known to cause ignition of carbon disulphide vapour in air. Care is also necessary with the vapour and other UV sources.[42]

Oxidation of carbon disulphide in air can produce carbonyl sulphide,[26] sulphur dioxide and carbon monoxide.

Fire-exposed containers of carbon disulphide can explode when heated and, if feasible, should be cooled with flooding amounts of water from as far as possible until well after the fire is extinguished.

The US National Fire Protection Association states[36] that foam is ineffective in fighting carbon disulphide fires. Fires may however be fought using suitable dry chemical, carbon dioxide or other inert gas. Cooling and blanketing with water spray is effective, in the case of fires in metal containers or tanks, by helping to prevent re-ignition by hot surfaces. Water used in fire-fighting should be bunded for later disposal. Immediate withdrawal is imperative if the fire causes discolouration of a storage tank or in the case of rising sound from a venting safety device. Fire-fighters should keep upwind of a fire and avoid breathing the vapours of carbon disulphide and its toxic combustion products.

Ref. 36 notes that where a fire is fed by an uncontrolled flow of combustible liquid the decision on how or if to fight it will depend on the size and type of fire anticipated and must be carefully considered. This may call for special engineering judgement, particularly in large-scale applications.

[Notwithstanding the foregoing review of various authorities noted in Sections 12 and 13, it is essential that managers and others responsible for the planning, implementation and overseeing of personnel and plant safety should be familiar with the legal constraints and official guidelines applicable to them and that they liaise with their local emergency services in the planning of plant and storage facilities and in the preparation of contingency plans for dealing with fires and other emergencies. Managers should regularly monitor staff knowledge of, and ability to implement, emergency procedures and should ensure that equipment provided for use in emergencies is regularly inspected and maintained.]

14. HAZARDOUS REACTIONS[2,42]

Carbon disulphide will decompose explosively to its elements with mercury fulminate initiation.[42]

Carbon disulphide and aqueous solutions of metal azides interact to produce metal azidodithioformates, most of which are explosive with varying degrees of power and with varying sensitivity to shock or heat.[42]

The bis- or tris-complexes of phenyl copper with triphenylphosphine react violently and exothermally with carbon disulphide, even at 0 °C.[42]

An explosion occurred when liquid chlorine was added to carbon disulphide in an iron cylinder owing to the iron-catalysed chlorination of carbon disulphide to carbon tetrachloride.[42]

Carbon disulphide ignites in contact with fluorine at ambient temperature.[42]

Aluminium powder ignites in carbon disulphide vapour.[42] Zinc powder and carbon disulphide interact with incandescence.[42] Mixtures of potassium–sodium alloy with carbon disulphide are powerful explosives being more sensitive to shock than mercury fulminate.[42]

A solution of permanganic acid (or its explosive anhydride, dimanganese heptoxide) produced by the interaction of permanganates and sulphuric acid, will explode on contact with carbon disulphide.[42]

A demonstration of combustion of carbon disulphide in nitrogen oxide (both endothermic compounds) exploded violently.[42]

Liquid mixtures of dinitrogen tetraoxide and carbon disulphide proposed for use as explosives are stable up to 200 °C but can be detonated by mercury fulminate, as can the vapours by sparking.[42]

Sax[2] notes that carbon disulphide reacts violently with aluminium, potassium and zinc; chlorine and fluorine; azides, (including those of the alkali metals and lead); chlorine monoxide; nitric oxide; dinitrogen tetraoxide, ethylene diamine, ethylene imine; sulphuric acid and permanganate mixtures. It can react vigorously with oxidizing materials and when heated to decomposition [in air] emits highly toxic fumes of SO_x.

15. EMERGENCY MEASURES IN THE CASE OF ACCIDENTAL SPILLAGE[43-45]

Warren[43] states that for first action it is essential to eliminate all possible sources of ignition; to instruct others to keep well away from the spillage area; to open windows if possible and to close workroom doors on leaving the area – the aim being to ventilate the area but to isolate the material.

If the spillage is not controllable on-site, the building should be evacuated by sounding the [flame-proof] fire alarm and the fire brigade should be called and the water authorities informed, stating the extent of the incident and the chemical involved.

If the spillage is controllable and small enough (*e.g.*, less than five litres) then the spillage should be absorbed on dry sand which should then be placed in such as a bucket with a lid and then removed to the open air. The sand can then be left in a metal tray to enable the carbon disulphide to evaporate.

In the case of large spills (*ca.* 5 l–50 l), the spillage should be covered with sand, vermiculite or ashes and then covered with water. This should then be transferred under water in buckets to an outside, safe, open area and emptied into a steel tray (*not* spilled on the ground). Taking extreme care, the material can be ignited from a distance with an excelsior train. Note, combustion products may include carbon monoxide and carbonyl sulphide with the sulphur oxides formed.

Where a spill or release is large enough to have an environmental impact, amelioration of the situation might be achievable by pumping or vacuuming off any bottom layer of undissolved material and using peat or carbon to absorb a soluble fraction. Disposal by allowing to stand open to atmosphere should not be allowed. Countermeasure materials suggested are flameproof pumps or vacuum apparatus; peat (*e.g.*, from horticultural suppliers); carbon (from water treatment plants, sugar refineries and their suppliers).

There is industrial fouling potential in that carbon disulphide's volatility may cause explosions if the material is present in boiler feed or cooling water (from which it should be removed by aeration). Its taste-imparting characteristics may ruin food processing waters.[44]

OSHA,[45] in addition, states that persons not wearing protective equipment and clothing should be restricted from spillage or leakage areas until cleanup is complete. Small quantities may be absorbed on paper towels from which the carbon disulphide may be allowed to evaporate in a safe place (*e.g.*, a fume hood whose ducts must finally be completely cleared of the vapour). The paper may then be burned at a suitably safe location. Large quantities can be reclaimed or collected and atomized in a suitable combustion chamber equipped with an appropriate effluent gas cleaning device. Carbon disulphide should not be allowed to enter a confined space, such as a sewer, because of the explosion risk.

[Every effort should be made to prevent carbon disulphide entering a sewage system. 35 mg/l of the material will cause 75% inhibition of nitrification in a sewage treatment works. Furthermore, carbon disulphide will there react with ethylenediamine (used in several industrial processes) to form ethylenethiourea (2-imidazolidinethione) which causes nitrification inhibition at 1.2 mg/l. Ethylene-thiourea is a known carcinogen and teratogen. It is persistent enough then to be able to enter a potable water recovery system.]

16. FIRST AID

General

The casualty should be removed from the danger of further exposure into fresh air. Rescuers should ensure their own safety from inhalation and skin contamination where necessary. Conscious casualties should be asked for information about what has happened. Bear in mind that the casualty may lose consciousness at any time. Where necessary, continued observation and care should be ensured. First aiders should take care not to become contaminated. Note that carbon disulphide is highly flammable and oxidizes to toxic and irritant sulphur dioxide.

Inhalation

Exposure to 1,000 ppm carbon disulphide for one hour is dangerous to life.

The conscious casualty should be kept at rest. The unconscious casualty should be placed in the recovery position and an open airway maintained. If breathing or heart beat stops, resuscitation should be commenced immediately. Medical aid should be obtained or the casualty removed to hospital immediately.

Skin contamination

Remove contaminated clothing. Wash with copious amounts of water, or soap and water, for at least 15 minutes. If necessary, seek medical aid. In case of persistent skin irritation refer for medical advice.

Eye contamination

Irrigate with copious amounts of water for at least ten minutes but avoid further damage to the eye from use of excessive force. Seek medical aid. Remember the possible presence of contact lenses, which may be affected by some solvents and may impede decontamination of the eye.

Ingestion

An oral dose to an adult of about 15 ml of carbon disulphide has proved lethal.

If the lips or mouth are contaminated rinse thoroughly with water. Do not induce vomiting. Give further supportive treatment as for inhalation. In the case of ingestion of significant amounts wash out mouth with water. Obtain immediate medical attention. Contact a hospital or poisons centre at once for advice. Remember that symptoms may develop many hours after exposure so continued care and observation may be necessary.

In all cases note information on the nature of exposure, and give this to ambulance or medical personnel. Information for doctors is provided in some technical literature issued by manufacturers. Antidotes are available for retention at the place of work for use by persons competent to treat casualties.

17. PHYSICO-CHEMICAL PROPERTIES

17.1 General[37,46,47]

Carbon disulphide is a colourless, flammable liquid.[46] When extremely pure its smell is reported to be sweet, pleasing and ethereal, but commercial grades are foul smelling.[37] It is highly refractive.[37] The Merck Index[37] states that it will decompose if left standing for a long time. Mellor[47] states that carbon disulphide, when exposed for many months to sunlight in the absence of air, decomposes, forming sulphur and a brown sulphide C_nS_n.

Carbon disulphide is miscible with anhydrous methanol, ethanol, diethyl ether, chloroform, carbon tetrachloride, benzene and oils[37] The solubility of carbon disulphide in water is recorded as 0.22 g/100 cm³ at 22 °C, and 0.14 g/100 cm³ at 50 °C.[46] The solubility of water in carbon disulphide is quoted as less than 0.005%.[37]

17.2 Melting Point[48]

Dreisbach[48] records the freezing point as

$-111.53\,°C.$

17.3 Boiling Point[48,49]

Dreisbach[48] gives the boiling point of carbon disulphide as

$46.25\,°C.$

Carbon disulphide forms azeotropes with a number of other compounds. Binary ones formed with water and also with some organic solvents are as follows[49]:

Wt% carbon disulphide	Second component	Wt% second component	Boiling pt azeotrope (°C)
67	Acetone	33	39.25
35	Dichloromethane	65	35.7
71	Methanol	29	39.8
97.2	Water	2.8	42.6

17.4 Density/Specific Gravity[37,48]

The specific gravity of carbon disulphide (referred to water at 4 °C) is given by Merck[37] as follows:

Temperature in °C	Specific gravity
0	1.29272
15	1.27055
20	1.2632
30	1.24817

The critical density of carbon disulphide is reported as 0.3679 g/ml.[48] From this the calculated critical volume is 2.718 ml/g.

17.5 Vapour Pressure[46,48]

The variation of vapour pressure of carbon disulphide with temperature and below atmospheric pressure is given by Weast[46] as:

Temperature in °C	Vapour pressure in mmHg
− 73.8	1
− 44.7	10
− 22.5	40
− 5.1	100
28.0	400

Weast also quotes the following values, for pressures at and above atmospheric pressure[46]:

Temperature in °C	Pressure in atm
46.5	1
69.1	2
104.8	5
136.3	10
175.5	20
222.8	40
256.0	60

The critical temperature and critical pressure of carbon disulphide are reported as 273.0 °C and 57,760 mmHg [76 atm], respectively[48].

17.6 Vapour Density[37]

2.67 (relative to air = 1).

17.7 Flash Point[48]

− 30.0 °C, closed cup test.

17.8 Explosive Limits[2,35,36]

The limits of flammability of carbon disulphide in air at normal atmospheric temperature and pressure are given by the National Fire Protection Association[36] as:

Lower limit: 1.3% v/v
Upper limit: 50.0% v/v.

The NFPA[36] give the ignition temperature of carbon disulphide as 90 °C. However, they point out that different test conditions (and also different definitions of "ignition temperature") can result in widely varying values being quoted. They therefore recommend that any value should be treated as an approximation. Examples of other values quoted are 100 °C[35] and 125 °C[2].

17.9 Viscosity [46,48]

The viscosity of liquid carbon disulphide at a range of temperatures is reported[46] as follows:

Temperature in °C	Viscosity in cp
−13	0.514
−10	0.495
0	0.436
5	0.380
20	0.363
40	0.330

The viscosity of carbon disulphide vapour is also given for a range of temperatures[46]:

Temperature in °C	Viscosity in μp
0	91.1
14.2	96.4
114.3	130.3
190.2	156.1
309.8	196.6

Dreisbach[48] quotes the kinematic viscosity (= absolute viscosity/density) of carbon disulphide as:

Temperature in °C	Kinematic viscosity in cSt
0	0.3453
15	0.3032
20	0.2916
35	0.2660

17.10 Concentration Conversion Factors

At 25 °C and 760 Torr (1 atm),

$1 \text{ mg/m}^3 = 0.32$ ppm, and
$1 \text{ ppm} = 3.11 \text{ mg/m}^3$.

18. TOXICITY

18.1 General[50-52]

Carbon disulphide is a toxic, low-boiling point (46 °C) liquid. Patty in ref.[50] notes that the predominant effect of high concentrations of vapour is narcosis. Death may follow from respiratory failure. Less severe exposures may result in headache, giddiness, respiratory disturbances, precordial distress and gastrointestinal disturbances. Chronic carbon disulphide poisoning gives rise to a number of subjective complaints such as fatigue, loss of memory, insomnia, listlessness, headache, excessive irritability, melancholia, vertigo, weakness, loss of appetite, gastrointestinal disturbances and impairment of sexual functions. Visual disturbances, loss of reflexes, hallucinations, mania or chronic dementia may occur and lung irritation has been reported.[50] Ingestion of carbon disulphide is very dangerous, half an ounce (*ca.* 15 ml) having proved fatal, with death from respiratory paralysis within a few hours.[51]. The odour threshold of carbon disulphide varies considerably. Ruth[52] quotes reported low and high values of 0.0243 and 23.1000 mg/m³ [0.0078 ppm and 7.4 ppm] respectively with the odour variously described as disagreeable or sweet.

18.2 Toxicokinetics[26,27,29,30,53,54]

With occupational exposure, inhalation is the major route by which carbon disulphide enters the body. Up to 90% of absorbed carbon disulphide is metabolized and the remainder is eliminated unchanged by various routes. Equilibrium between inhaled and exhaled carbon disulphide is usually attained within the first two hours of exposure at which point retention declines to about 15–45% of inhaled vapours. Retention may be greater in those not chronically exposed to the material[26,27].

Exhalation is the primary route of excretion of unmetabolized material. A triphasic pulmonary excretion lasting up to 97 hours has been described and various investigations have estimated that 10–30% of total body burden of carbon disulphide is expired from the lung while excretion unchanged by the kidney accounts for less than 1% of absorbed material. The remainder is excreted in the form of metabolites[26,30].

Carbon disulphide not exhaled is distributed in the human body by the bloodstream with twice as much taken up by erythrocytes as by the plasma. The partition coefficient for carbon disulphide from air to blood is about 2.8 and from blood to organs about 100 thus accounting for the rapid disappearance of the material from the blood though traces have nevertheless been found in the blood of exposed workers 80 hours after exposure has ceased. A Chinese study[26] has shown the presence of carbon disulphide in the milk of nursing mothers occupationally exposed to carbon disulphide and, depending on level and duration of exposure, averages of up to 12.3 µg/100 ml were measured.[26]

Carbon disulphide reacts with amino acids and peptides yielding thiocarbamates and thiazolinones. This binding and the solubility in lipids and fats accounts for the affinity of carbon disulphide for all tissues and organs, distribution being greatest to lipid-rich tissues and organs such as the brain and liver, accumulation in these two sites being probably as thiocarbamates.[27,30]

It has been found that, *in vitro*, rat liver microsomes oxidize carbon disulphide to carbonyl sulphide and thence to carbon dioxide utilizing cytochrome P450.[27] Metabolism of carbonyl sulphide to carbon dioxide has also been accomplished utilizing cytosolic carbonic anhydrase.[26] A proposed pathway for such oxidation to carbon dioxide quoted by Beauchamp *et al.* is

$$CS_2 \rightarrow COS + S \rightarrow CO_2 + S$$

Thus the concomitant formation of carbonyl sulphide and atomic sulphur is also proposed.[26]

Several *in vitro* studies have indicated the potential for atomic sulphur, formed during the oxidative desulphuration stages proposed for the metabolism of carbon disulphide to carbon dioxide, to bind covalently to many cellular macromolecules. It has been found that the binding to liver microsomes of radioactivity from ^{35}S carbon disulphide is NADPH dependent and also increases in microsomes from phenobarbital pretreated rats. Carbon monoxide decreases the microsomal binding and so further implicates the catalysing of reactive metabolite formation by mixed function oxidase. ^{35}S binding has been found to be significantly greater than that of the radioactivity from ^{14}C carbon disulphide, thus implying the binding of sulphur that was free of its carbon atom. Other work has provided evidence of the reaction of atomic sulphur with protein cysteine residues.[26]

Acute CNS toxicity may in part be accounted for by formation of dithiocarbamates as also may be the peripheral neurotoxicity. Nonenzymatic reaction of carbon disulphide with free amino groups indicates possible reaction with macromolecules such as enzymes, structural proteins, polypeptides and nucleic acids. Also, disturbance of lipid metabolism, catecholamine metabolism and vitamin metabolism (vitamin B_6 and nicotinic acid) have been postulated. Dithiocarbamates can chelate several polyvalent inorganic ions such as copper and zinc and so may inactivate numerous essential enzymes whose activity is dependent on such ions.[26,29] On the other hand the formation of reactive sulphur might inhibit the microsomal mono-oxygenase system and so disturb the metabolism of other endogenous or exogenous compounds.[26,29]

In connection with enzyme inhibition, Cavanagh[53] notes that an increased sensitivity of brain tissue to monoamine oxidase inhibitors has been shown in carbon disulphide intoxication and that this is due to inhibition in the brain of dopamine β-hydroxylase in exposed animals. This enzyme converts the catecholamine, dopamine, to noradrenaline. The inhibition of the enzyme occurs through the conversion of carbon disulphide to diethylthiocarbamate which last can chelate the cupric ions essential to the proper functioning of the enzyme. Magos[54] notes that dopamine in the brain is mainly concentrated in the corpus striatum, a part of the extrapyrimidal system, and that 'any change in the level or turnover of dopamine has the utmost clinical consequences'.

Beauchamp *et al.*[26] point out that it remains to be clarified as to the relative contributions, if any, made by the various metabolic pathways to the development of the acute and chronic toxicity of carbon disulphide. They also point out that the majority of metabolism studies had utilised single, high-dose exposures to carbon disulphide and that the contribution of metabolic routes to the overall disposition of carbon disulphide under conditions of single or repeated low-dose exposures needed to be determined.[26]

It was found that a metabolite present in the urine of animals exposed to carbon disulphide catalyses the reaction of iodine with sodium azide[30]

$$2NaN_3 + I_2 \rightarrow 3N_2 + 2NaI$$

Subsequently, C–SH and C–S groups were found to catalyse the reaction which has been adapted as a quick, inexpensive method of assessing a person's exposure level to carbon disulphide. The test though is not sensitive enough for detection of exposures of less than 60 mg/m^3 [*ca.* 19 ppm] and diet may affect the results of the test.[29]

High levels of thiourea and mercaptothiazolinone have been identified in humans following carbon disulphide exposures.[27] It has also been indicated that carbon disulphide and carbonyl sulphide could be conjugated with endogenous glutathione to yield 2-thiothiazolidine-4-carboxylic acid and 2-oxythiazolidine-4-carboxylic acid respectively. In humans, the major metabolites are thiourea, 2-thio-5-thiazolidone and 2-thiothiazolidine-4-carboxylic acid.[27]

It has been proposed that urinary excretion of 2-thiothiazolidine-4-carboxylic acid, when normalised to urinary creatinine content, can be used as a biological indicator of dose. Increased amounts of this metabolite were detected in the urine of a worker who had been exposed to approximately 2.5 ppm carbon disulphide. This metabolite has only a short half-life however and measurement can only be performed on urine sampled at the end of the work-shift (see Section 19).

It should be noted that tetraethylthiuramdisulphide (TETD, Disulfiram, Antabuse), a substance used in the treatment of alcoholics, is metabolized in a way that liberates carbon disulphide *via* diethyldithiocarbamate. Alcoholics so treated are thus exposed to carbon disulphide and its metabolites, and the diethyldithiocarbamates present in their urine have been measured using the iodine-azide reaction. These results have led to the use of TETD as a test for the evaluation of the metabolic rate of sulphur compounds in workers (the 'Antabuse test').[30] The susceptibility to carbon disulphide can be explored by measuring the amount of diethyldithiocarbamate excreted within four hours of oral administration of TETD. It has been recommended that only persons with excretion of over 150 mg/g creatinine should be engaged in occupational exposure to carbon disulphide.[29]

18.3 Human Acute Toxicity[4,35,41,45,50,51,55-66]

Ingestion

Reports of poisoning by ingestion are uncommon. Half an ounce of carbon disulphide is stated to have been fatal on at least three occasions.[51] The victims exhibited spasmodic tremor, prostration, dyspnoea, cyanosis, peripheral vascular collapse, hypothermia, mydriasis, convulsions, coma and, in a few hours, death from respiratory paralysis. However, in a case of attempted suicide, the ingestion of two ounces of carbon disulphide failed to kill.[51]

Hayes[55] cites the case of the accidental ingestion of about 5 ml by a 42 year-old woman who for some years had used carbon disulphide to control insects in warehouses. Hayes in his quotation continues,

> 'She induced vomiting by sticking her finger down her throat, but about 5 hours after ingestion she felt numbness of the lips and nausea, and she vomited again. Twelve hours after ingestion, a hot feeling in the upper abdomen and intermittent agitation led to her hospitalization. Examination revealed accentuated tendon reflexes, bilateral positive Babinski reaction, and hyperesthesia in addition to the agitation. By 16 hours after ingestion, transient sinus tachycardia and sharp P waves appeared in the electrocardiogram, and there were some abnormalities of the electroencephalogram from the second through the sixth day of illness. Except for dizziness while walking, the patient seemed normal by the second day of hospitalization. However, about midnight of the fifth day she came to a nurse's station carrying a grapefruit and expressing very confused ideas about it and about God. The same illusion and delusion returned several times for about a week. The only other finding at this time was some loss of memory. Following discharge, the patient remained normal for 8 days and then for 2 days experienced an intermittent return of agitation, illusions, and delusions. By a little over 2 months after ingestion she was entirely healthy.'

In another case, reported in 1886, a patient who had swallowed a glass of carbon disulphide was seen by a physician about an hour later and found to be "sensible and wanting to go in the backyard to defecate". Physical examination showed "normal pupils, slow, long stertorous respirations and a small rapid pulse varying from 150 to 160 a minute". He was given an emetic, became unconscious soon after, and died about two hours after drinking the poison. "Postmortem examination revealed a quarter pint of fluid in the stomach along with minute hemorrhages of the gastric mucosa."[56]

Skin contact

Thienes and Haley[57] note that contact of liquid carbon disulphide with the skin or mucous membrane is highly irritating. The burning sensation is followed by numbness of the part, blistering is common and deep burns may result. Adjacent nerve fibres degenerate. ACGIH[35] notes a report that toxic quantities of carbon disulphide vapour can be absorbed through the skin.

Fielder *et al.*[58] note a series of experiments wherein the rate of absorption of carbon disulphide from dilute aqueous solutions of increasing concentration (0.33–1.67 g/l) was investigated. Volunteers immersed their arms in the solutions at 21 °C for one hour. The rate of absorption, as calculated from the amount of carbon disulphide lost from solution, increased from 232 to 789 μg/cm^2/hour with increasing concentration. Lower values (21–96 μg/cm^2/hour) were derived from the carbon disulphide content of the volunteers' exhaled breath. However, only brief details of the experimental procedure were given, allowing the possibility of other factors enhancing the former, higher range of values. The lower range of values (from exhaled breath analysis) might have been depressed by the assumption that as much as 50% of absorbed material would appear here. Fielder *et al.*[58] conclude that the results suggest that appreciable absorption may occur through the skin. From the higher range of values it was estimated that immersion of one hand for one hour in a washing bath solution containing 0.1 mg/l carbon disulphide at a viscose rayon plant would result in the absorption of 17.5 mg of carbon disulphide.

Davidson and Feinleib[56] quote a 1936 report wherein it was believed that carbon disulphide was the severest skin irritant of the organic solvents. A study was made of the blisters of certain viscose rayon workers (doffers) and the lesions were reproduced in animal experimentation. Conclusions reached were that finger blisters were caused by carbon disulphide coming from the "feed-wheel drip"; that the blisters often resembled second and third degree burns and that skin contact with carbon disulphide caused degenerative changes in local peripheral nerves.

Eye contact

Above vapour concentrations of 100–1,000 ppm, conjunctivitis, pain and blurred vision may occur. Direct contact with the liquid produces severe irritation.[4] Bretherick[59] states that both the vapour and the liquid irritate the eyes, though OSHA[45] notes that carbon disulphide [presumably as vapour] is not known to be an eye irritant. Houston[60] states that eye contact causes severe irritation and redness and that corneal burns may occur.

Grant[61] notes that irritation of the eye and keratitis epithelialis have been erroneously attributed to carbon disulphide in occasional reports from such industries as viscose rayon manufacture in which the actual cause, hydrogen sulphide, is also liberated.

An ancillary effect is that on contact lenses which should be removed immediately.[62] OSHA[45] goes further and states that contact lenses should not be worn when working with carbon disulphide.

Inhalation

The ILO[41] notes that carbon disulphide is primarily a neurotoxic poison and so the symptoms indicating damage to the central and peripheral nervous system are the most important. Concentrations of 0.5–0.7 mg/l (160–230 ppm) were reported as causing no acute symptoms in man. Concentrations of 1–1.2 mg/l (320–390 ppm) were bearable for several hours with the appearance of headaches and unpleasant feelings after eight hours of exposure. At 3.6 mg/l (1,150 ppm) giddiness set in, and at 6.4–10 mg/l (2,000–3,200 ppm) light intoxication, paraesthesia and irregular breathing occurred within $\frac{1}{2}$–1 hour. A concentration of 15 mg/l (4,800 ppm) has proved lethal after 30 minutes. At higher concentrations, unconsciousness has occurred after several inhalations.[41]

Exposure of human volunteers for six hours to inhaled carbon disulphide vapour concentrations down to 10 ppm caused reversible inhibition of drug metabolizing enzymes.[63]

The predominant effect of high concentrations is narcosis.[50] Unconsciousness, frequently rather deep, with extinction of cornea and tendon reflexes, occurs after only a short time. Death, should it occur, sets in by a blockage of the respiratory centre.[41]

Where consciousness is regained, motor agitation and disorientation follow. Psychic disturbances as well as permanent damage to the central and peripheral nervous systems are often late sequelae following recovery.[41]

Browning[64] reports that most cases of gassing by carbon disulphide, especially in Great Britain, have occurred during the cleaning of tanks that have contained the chemical. Headache, vomiting, general weakness, paralysis of the legs, abdominal pain and loss of consciousness for a short period were the main symptoms. In two cases there was mental confusion.[64] Psychotic cases, sometimes with permanent dementia developing, are now much rarer than were at one time recorded (*e.g.*, in 1899) following short severe exposures, and prognosis is good if there is rapid removal from exposure.[64] Trevethick[65] in noting symptoms of reddening of the face, euphoria, restlessness and muscular cramps as well as those of stupor, unconsciousness and death, also states that during recovery, symptoms of narcosis and muscular cramps may persist for a long time.

In 1979 a railway tank-car leaking carbon disulphide caught fire at Charlottesville, Virginia, USA. The fire was extinguished and the carbon disulphide transferred to another tank-car. Spyker *et al.*[66] detail the symptoms found in twenty seven individuals. These were mainly police and firefighters who had been exposed to the chemical and who were later given standard questionnaires covering description of exposure, smoking and general health.

Spirometry, single-breath carbon monoxide diffusing capacity and arterial blood gases were measured. Measurements were made in all four from the cohort having shortness of breath and chest pain and in seven others who appeared clinically to be the most severely ill. Symptoms of all twenty seven subjects together with measurements made were presented as follows:

> Complaints of the subjects are summarized in Table 1. Slow vital capacity and the partial pressure of arterial oxygen (pO_2) were lower at the day of exposure than when measured 9 d later (Table 2). No significant changes were observed in forced vital capacity, forced expired volume in 1 s, forced expired volume from 25 to 75% of vital capacity, or in diffusing capacity.
>
> Airborne carbon disulfide levels of 20 ppm were observed at a site outside of Charlottesville during transfer of CS_2 from the leaking tanker to an intact railroad tank car. No measurements were made in Charlottesville. [Most of the effects observed were probably attributable to carbon disulphide rather than to combustion products.]

Table 1 *Symptoms of 27 patients exposed to carbon disulfide*

Symptom	Number of subjects	
Headache:	16	(59%)
Slight	8	(30%)
Moderate	6	(21%)
Severe	2	(7%)
Nausea	14	(52%)
Vomiting	1	(4%)
Burning of throat, lips or skin	11	(40%)
Dizziness	16	(59%)
Shortness of breath or chest pain	4	(15%)
Impotence	2	(7%)

Table 2 *Changes in pO_2 and vital capacity following exposure to carbon disulfide**

		Result mean ± SD			
Test	N	Aug 4(Day 0)	Aug 13(Day 9)	t value	p value
pO_2(mmHg)	9	75.9 ± 6.2	83.8 ± 5.8	3.11	0.02
Slow vital capacity(l)	11	4.72 ± 0.46	5.09 ± 0.41	2.97	0.02

*pO_2: Partial pressure of O_2 in arterial blood. N: Number of patients. SD: Standard Deviation. t: Student's test for paired variables. p: Associated p value.

It was noted that headache and dizziness (59%), indicative of CNS effects, were more frequent than direct irritant effects (40%). Further, it was stated that the changes observed in slow vital capacity and arterial pO_2 could be explained by transient inflammation in small airways causing mild ventilation perfusion-abnormalities and mild air trapping. Such would be consistent with an earlier clinical observer's noting that workers often had a chronic cough which he attributed to the bronchial irritant effect of carbon disulphide fumes.

It was stated that the study demonstrated that subtle changes in pulmonary function occurring in the patients were transient and that residual, permanent lung damage appeared unlikely. Further, none of the patients evaluated appears to have sustained injury from the level of exposure experienced lasting beyond the first few days.

18.4 Animal Acute Toxicity[58,67,68]

WHO[67] notes that high concentrations of carbon disulphide act as a narcotic, and hepatic injury may also occur. Neurological symptoms prevail in acute cases and many biochemical changes have been demonstrated such as inhibition of microsomal mixed-function oxidases and destruction of cytochrome P450 in the liver, disturbances in catecholamine metabolism and different enzymatic changes.[67]

Rats dosed orally with 1.0 ml/kg (1.26 g/kg) of carbon disulphide produced no overt signs of toxicity over a twenty four hour observation period. Apart from some congestion and haemorrhages in the lungs, no lesions were noted at autopsy. There was no sign of any hepatotoxic effect unless there had been pretreatment with phenobarbital whereupon hepatotoxicity could be severe.[58]

Carbon disulphide is a marked irritant to rabbit skin. Severe inflammation with blistering after 2–3 days was produced when cotton-wool plugs soaked in carbon disulphide were fixed to the tips of rabbits ears.[58]

RTECS[68] includes:

oral	– rat LD_{50}	3,188 mg/kg
oral	– mouse LD_{50}	2,780 mg/kg
oral	– rabbit LD_{50}	2,550 mg/kg
oral	– guinea pig LD_{50}	2,125 mg/kg

18.5 Human Sub-acute and Chronic Toxicity Including Reproductive Effects[26.27.30.45.51.53.57.64.69–72]

Thienes and Haley[57] note that chronic carbon disulphide poisoning results in headache, malaise, fatigue, muscle weakness, anorexia, motor disturbances of the bowel, anaemia, disturbance of cardiac rhythm, loss of weight, polyuria, menstrual disorders, visual disturbances and paraesthesias. Neuritis is frequent and in severe chronic poisoning there is often hepatic degeneration with jaundice. Psychic changes are frequent, daytime somnolence and nocturnal insomnia are common, and partial blindness may occur. Peptic ulcers are found in over 15% of cases of carbon disulphide intoxication and achlorhydria has been reported.

Cavanagh in ref. 53 notes that the mental changes recorded include strange and terrifying dreams, distortions of perception, brobdingnagian and lilliputian hallucinations, outbursts of mania and periods of deep depression. Browning[64] notes a British report of 1899 that the windows of a vulcanising room had to be barred to keep men from leaping out during attacks of mania. Cavanagh notes that these mental effects occur rapidly after relatively short exposure to fairly high concentrations of carbon disulphide but recovery is rapid once exposure is stopped.

WHO[30] quotes the following classification of the different syndromes as suggested by Nesswetha and Nesswetha in 1967, *viz.*

- psychoses characterized by manic and depressive symptomatology and disorientation;

- polyneuropathy of the lower extremities, with diminished or completely absent Achilles and patellar tendon reflexes, sensory disturbances in a glove-stocking distribution, diminished faradic and galvanic excitability, and decrease of the motor and sensory conduction velocity in the peripheral nerves;

- disturbances of the gastrointestinal tract in the form of chronic, hyper- and hypoacidic gastritis and duodenal ulceration;

- myopathy of the calf muscles;

- neurasthenic syndrome with disturbances in the autonomous nervous system;

- optic neuritis;

- atherosclerotic vasculoencephalopathy; the principal forms being bulbar-paralytic, hemiplegic, or extrapyramidal.

The typical mental deterioration has been called an organic psycho-syndrome, which may be due to general cerebral atherosclerosis, to direct toxic action upon the brain cells, or to both.

WHO[30] also notes that the pattern of carbon disulphide poisoning has changed with improvements in hygienic standards in industry. Gosselin *et al.*[51] observe similarly, stating that the classical picture of carbon disulphide intoxication, *viz.* encephalopathy, psychosis, polyneuritis, is increasingly less frequent and that more recent intoxications develop slowly over many years of exposure and tend to be milder in form. They further note that victims exhibit marked cerebral vascular damage on cerebral arteriography or at autopsy. Neurological findings are associated with pyramidal, extrapyramidal and pseudobulbar tracts and concomitant findings often include arterial hypertension, elevation of serum cholesterol (but not of cholesterol esters), and renal damage. They note also a report that women appear to be more sensitive than men to the neurotoxic effects of carbon disulphide.[51]

Cavanagh in ref. 53 notes that in contrast with the mental symptoms of carbon disulphide intoxication, the signs and symptoms of neuropathy are slow to appear and develop, and that recovery is correspondingly slow if the neuropathy has been in any way severe. The earliest clinical signs are distal sensory abnormalities coupled with numbness and mild anaesthesia. Electrophysiological methods show that nerve conduction velocity slows before the onset of definite physical signs in exposed individuals but these electrophysiological changes are completely though slowly reversible if the exposure is stopped. Cavanagh continues that, nevertheless, marked signs of distal neuropathy and glove and stocking anaesthesia will develop in established cases of intoxication. Neither ataxia nor Rombergism occurs and the cranial nerves are not affected although ophthalmological inspection shows the 'blind spot' to be significantly enlarged in an appreciable proportion of cases.

As regards neurological disturbances there is some evidence of a synergistic effect of alcohol and carbon disulphide. Age also appears to be a factor here and, with viscose rayon workers, simultaneous exposure to hydrogen sulphide may be a critical factor in producing these effects.[26]

Prolonged exposures to vapour concentrations as low as 0.1 mg/l [*ca.* 30 ppm] have resulted in severe intoxications.[51] Santodonato *et al.*[27] note that occupational exposure, primarily in the rayon industry, is associated with coronary heart disease, encephalopathy and neuropathy through vascular degeneration, and that ocular, auditory and vestibular functions are impaired by muscular and neurological atrophy following chronic exposure.

Attention has been given to reported retinopathy in workers exposed to carbon disulphide as a means of early diagnosis of the effects of exposure.[26] However, the implications of some studies are debatable in that their results are not consistently found. Beauchamp *et al.*[26] state that in practice one can only conclude that (as given in ref. 45) one of the manifestations of long-term exposure may be eye damage with blind spots, narrowing of vision and decreased ability to see in the dark. Beauchamp *et al.*[26] note though that whether such eye damage occurs at exposure levels below 20 ppm (TWA, eight hour) is open to question.

Beauchamp *et al.*[26] note that hearing loss, particularly to high frequencies, and impairment of vestibular function are well-recognised features of chronic carbon disulphide intoxication. Audiometric examinations and electronystagmography are useful in clinical assessment. Vestibular function may also be studied by electroencephalography.

With the much improved ventilation now present in viscose rayon plants, the encephalopathy, neuropathy and hypertension associated with carbon disulphide's properties as a vasotropic poison have given way to coronary heart disease (CHD) as a principle concern.[26] Finnish workers have observed this correlation between CHD and carbon disulphide exposure even where exposures ranged from 10 to 30 ppm. In an eight-year follow-up, a twofold excess death rate in relation to a matched reference group was noted, but coronary mortality had decreased following the institution of protective measures and a reduction of exposure to less than 10 ppm. Later work on the same cohort made provision for the additional risk factors of more advanced age and raised blood pressure. While age and hypertension were the predominant factors determining CHD, the effect of carbon disulphide exposure alone was found to be statistically significant with a relative risk of 2.3.

Beauchamp et al.[26] note the question as to whether or not reductions in overall carbon disulphide concentrations to 10 ppm, or lower, will serve to eliminate the excess risk from CHD, contrasting (a) the suggestion that in Finland (where there is an existing high incidence of CHD) occupational contact with carbon disulphide not only increased the incidence but worsened the prognosis, with (b) a note that in Japan, where there is a notably low incidence of CHD, there was no evidence that carbon disulphide exposure had an effect on myocardial infarction, angina pectoris, coronary electrocardiograms or hypertension. The implication here is that the coronary effects of carbon disulphide are only manifested in the presence of other factors which themselves predispose to the development of CHD. Beauchamp *et al.*[26] conclude that at 10 ppm or lower, the question of risk of CHD remains to be resolved.

More recently, Sweetnam *et al.*,[69] following up an earlier cohort study of viscose rayon workers, found that there was significantly higher mortality from all causes for spinners (the group most exposed to carbon disulphide) compared with the least exposed group, for those of ages 45 to 64. This excess mortality was largely accounted for by ischaemic heart disease (IHD) for which the spinners had a standardised mortality ratio of 172 (p < 0.001). When the analysis was repeated covering all ages, the trend of higher mortality with increasing exposure became less strong with only the trend for IHD remaining significant (p < 0.05).

Over the age of 65 though [when exposure would cease] there was no trend of total mortality with exposure and IHD mortality tended to decline with increasing exposure. Sweetnam *et al.* state that this finding suggests that any hazard may cease when exposure ceases and hence that the effect is some type of reversible, direct cardiotoxic or thrombotic effect. This they observe to be in contrast to the main hypothesis put forward to explain the excess mortality from IHD, which is that carbon disulphide promotes atherosclerosis with the consequent expectancy of IHD mortality remaining high after cessation of exposure to carbon disulphide. These authors also note the decrease in coronary mortality in Finnish workers following the institution of a preventative programme and the conclusion that had in consequence been drawn that the "cardiotoxic effect of carbon disulphide is reversible." Sweetnam *et al.* note earlier studies resulting in negative findings that had been conducted and also observe that the few morbidity studies had provided little convincing evidence, either that IHD is more prevalent among workers exposed to carbon disulphide, or that levels of coronary risk factors are higher. Sweetnam *et al.* in conclusion observe that these negative findings are consistent with the hypothesis that carbon disulphide may act to trigger a clinical event in a man who is at high risk and that "perhaps morbidity studies should instead be looking at the effect of carbon disulphide on factors that are related to thrombosis, on the autonomic innervation of the cardiovascular system, and on the cardiac muscle and enzymes."

Carcinogenicity to Man

Santodonato *et al.*[27] report that limited data are available regarding the carcinogenicity of carbon disulphide in humans. Reference is made to a 10 year closed-cohort study, commencing in 1964, of 6,678 workers in a rubber and tyre manufacturing plant in the USA, and where workers had shown excess mortality from several different cancers.[27,70]

The solvent exposure histories of a 20% age-stratified random sample of the cohort were compared with those of cohort members (aged 40–84 at commencement of the study) who died during 1964–1973 from stomach cancer, respiratory system cancer, prostate cancer, lymphosarcoma or lymphatic leukaemia. Work history records were used to identify workers *authorized* to use specific solvents and it is recognized that an unknown number of workers may not actually have been exposed and other confounding exposures were not identified. Only workers with an accumulation of not less than one year of such 'exposure' were considered as exposed. Carbon disulphide was one of 25 solvents under consideration. Of the various cancers, lymphatic leukaemia was strongly related to carbon disulphide use, with an odds-ratio of 8.9 (p = 0.0003), and lymphosarcoma showed similar but weaker association (odds-ratio 5.6, p < 0.01). The authors of ref. 70 point out that carbon disulphide has not been shown to be the solvent which caused these cancers directly, though another report suggests that it may cause bone marrow hyperplasia. Note was made that confounding occupational or non-occupational exposures were not identified and that relatively small numbers of cancer cases (less than ten per solvent for lymphatic leukaemia or lymphosarcoma) were involved. The findings require cautious interpretation.[70]

Reproductive Effects in Man

The previous mentioned problem of simultaneous exposure to hydrogen sulphide may here too be a factor in conflicting findings relating to the effect of carbon disulphide on male and female reproductive functions.[26] Conflicting reports have appeared in the literature regarding the effect of carbon disulphide on the reproductive functions of women. Some studies have associated occupational exposures ranging from 18 ppm to less than 3.2 ppm with irregularities of the menstrual cycle, increased menstrual flow and pain and a slight increase in miscarriages, while other studies under similar conditions have failed to elicit similar results. In many of these studies though, the possible complicating effect of hydrogen sulphide was not adequately evaluated.[26] (See also Section 18.6.)

As regards the effect of carbon disulphide exposure on male reproductive function, a study[26,71] was made on 33 young male workers (average age 22 years) at an artificial fibre factory and who presented a clinical image of chronic carbon disulphide poisoning. A significant degree of hypospermia and abnormal sperm morphology was reported. The average exposure of the subjects was 21 months and during the last year various plant malfunctions had raised their exposure to carbon disulphide to average values of 40–80 mg/m³ [*ca.* 13–26 ppm] with peaks of 780 mg/m³ [ca. 250 ppm]. However Beauchamp *et al.*[26] note that data on actual exposures were not presented and there was no obvious correlation between length of employment and the frequency of germ cell disorders.

Santodonato *et al.*[27] note a 1983 report that reproductive capacity may be affected by altering gonadotropin and testosterone levels.

General

A caveat regarding exposure data is sounded by Vanhoorne and Grosjean,[72] who claim that a review of 94 papers, including 32 mentioned in TLV documentation, shows measurement methodology to have been poorly described in many studies on exposure in the viscose industry. In their own studies in two Belgian viscose plants they found that in a viscose rayon factory, workers performing different tasks in the same spinning room experienced average daily carbon disulphide exposures varying by a factor of over five times between individuals. A single individual's exposure might vary during the day by a factor of three times and noticeable day to day variation was also observed. Variation was found between the exposure experienced by two spinners performing the same work on the same day and during some activities there were considerable peak exposures. In another (a rayon staple) factory where it was claimed that the exposure in the spinning room was about 20 ppm [*ca.* 62 mg/m³], it was found that average daily exposure could be much higher for various workers and amount to almost 390 mg/m³ [*ca.* 125 ppm] for a spinning head washer, which task could also be subject to short term exposure of up to 900 mg/m³ [*ca.* 290 ppm].

18.6 Animal Sub-acute and Chronic Toxicity Including Reproductive Effects[26,27,68,73-76]

General

With oral or intraperitoneal or inhalational administration of carbon disulphide to rats to evaluate the effect on the liver, a marked decline in cytochrome P450 and cytochrome P450 dependent mono-oxygenase activity measured *in vitro* is detectable. With phenobarbital treated rats the effects are even greater and there is also moderate to severe centrilobular hydropic degeneration of the liver.[73] Studies with ^{35}S and ^{14}C labelled carbon disulphide have been noted in Section 18.2 (*v.s.*).

O'Donoghue[74] states that neurotoxicity studies have been conducted to confirm findings made in case studies of man and to explore mechanisms. Rats, mice, rabbits, cats and dogs have been intoxicated with carbon disulphide, with rats appearing the most resistant and dogs the most sensitive to neurotoxicity.[74]

In one study rats were exposed to approximately 190 ppm of carbon disulphide, six hours/day and five days/week for six months without showing clinical or electromyographic evidence of neurotoxicity. At 385 ppm or greater concentrations of carbon disulphide, there were signs of neuropathy but the onset of this was altered by diet, especially supplementation with copper, zinc and vitamin B_6. A study in which rats were exposed to 48 ppm carbon disulphide for five days/week for 40 weeks produced no motor impairment but a neuropathy was produced after 18 weeks at 385 ppm or eight weeks at 770 ppm.[74]

In another study, levels of exposure that did not produce clinical evidence of neuropathy in rats did produce reduction in peripheral nerve conduction velocity. Rats exposed to *ca.* 290 ppm carbon disulphide for five hours/day, six days/week for six months had reduced conduction velocity in the sciatic and tibial nerves. Recovery was incomplete six months after the end of a twelve month exposure period.

Experimental studies have reported histological lesions in the brain, particularly the basal ganglia; alteration of selected reflexes; rigidity and tremor; ataxia and behavioural changes in dogs exposed to 400 ppm carbon disulphide eight hours/day, five days/week for 10–15 weeks.[26,74] Marked vascular damage and neuronal degeneration were found in cats exposed to 2,560–3,210 ppm for 0.5–3 hours/day totalling 19–74 hours over a period of 24–92 days.[26,74] O'Donoghue[74] notes that experimental studies make it clear that carbon disulphide can directly affect the central and peripheral nervous system, resulting in axonal damage which fits the pattern of a central-peripheral distal axonopathy.

Carbon disulphide and hydrogen sulphide mixtures

Beauchamp *et al.*[26] note that although simultaneous exposure of humans to carbon disulphide and hydrogen sulphide is usual, there is little information on simultaneous exposure of laboratory animals.

Histological lesions in the liver, kidney and testes were noted in rabbits exposed to 300 ppm carbon disulphide and 100 ppm hydrogen sulphide combined inhalational exposure [duration not stated].[26] They also occurred in the same organs with hydrogen sulphide alone but were more severe with the combined gases.

The two gases also have additive effects on blood parameters in rabbits (carbon disulphide 300 ppm + hydrogen sulphide 100 ppm inhalational). Elevations in blood cholinesterase, plasma aspartate aminotransferase and urinary copro-porphyrin were noted in rats inhalationally exposed for six months to a carbon disulphide (0.032 ppm) and hydrogen sulphide (0.032 ppm) mixture. Either chemical alone produced either no or lesser alterations in these enzymes. The forgoing suggest either an additive or synergistic toxic effect of the two chemicals.[26]

Carcinogenicity to Animals

Santodonato *et al.*[27] writing in 1985 stated that pertinent data on the carcinogenicity of carbon disulphide in animals had not been located in the available literature. Beauchamp *et al.*[26] noted that the results of chronic bioassays sponsored by the NCI in the 1970s, in which carbon disulphide had been administered to rats and mice by gavage, had been considered inadequate for the evaluation of carcinogenicity of carbon disulphide because of the poor survival of both species. Santodonato *et al.*[27] noted that two studies in the USA were underway at the time of their writing.

Reproductive Effects in Animals

A report reviewed in ref. 75 notes that carbon disulphide treatment resulted in significantly increased prenatal mortality but not malformations in rabbits at a dose of 25 mg/kg which itself was not toxic to dams. Significant maternal toxicity and increased maternal liver weight were observed at 75 and 150 mg/kg. In rats, with a regimen of 0, 100, 200, 400 or 600 mg/kg, all doses produced significantly reduced gestational body weight gain in the dams compared with the control group, but only 200 mg/kg and above produced foetal toxicity in the form of reduced foetal weight.

A review in ref. 76 notes work wherein exposure to 10.0 mg/m³ [3.2 ppm] or 0.03 mg/m³ [0.01 ppm] of inhaled carbon disulphide throughout gestation did not produce congenital malformations or functional biochemical changes in the neonate but that 10 mg/m³ affected the postnatal development causing impairment of viability and retardation of morphological and sensory development. Behavioural deviations occurred with both exposure levels. The changes were more pronounced in early postnatal life. Lipid metabolism is also adversely affected in the offspring of albino rats.

The same review[76] also notes that when female rats and rabbits were exposed to 20 ppm and 40 ppm of carbon disulphide vapour for three weeks before impregnation and up to 30 days after gestation, there was no evidence of mutagenic or teratogenetic abnormalities.

RTECS gives the following[68]:

oral – rat, TDLo – foetotoxicity, except death - 2 g/kg for 6th–15th days of pregnancy

inhalation – rat, TCLo – pre-implantation mortality – 200 mg/m³ [64 ppm] for 24 hrs for 1st–21st days of pregnancy

inhalation – rat, TCLo – eye/ear; effects on newborn (viability index) – 10 mg/m³ [3.2 ppm] for 8 hrs for 1st–22nd days of pregnancy

inhalation – rat, TCLo – effects on newborn (growth statistics) – 100 mg/m³ [32 ppm] for 8 hrs for 1st–22nd days of pregnancy

inhalation – rat, TCLo – foetal death; craniofacial; homeostasis – 100 mg/m³ [32 ppm] for 8 hrs for 1st–21st days of pregnancy

inhalation – rat, TCLo – behavioural – 30 µg/m³ [0.01 ppm] for 8 hrs for 1st–22nd days of pregnancy

inhalation – rat, TCLo – spermatogenesis; prostate and associated organs – 600 ppm [1,750 mg/m³] for 6 hrs for 50 days pre-mating

inhalation – mouse, TCLo – pre-implantation mortality; litter size – 2,000 mg/m³ [640 ppm] for 2 hrs for 1st–21st day of pregnancy

oral – rabbit, TDLo – post-implantation mortality – 350 mg/kg for 6th–19th day of pregnancy

oral – rabbit, TDLo – developmental abnormalities - 2,100 mg/kg for 6th–19th day of pregnancy.

18.7 Mutagenicity[26,27]

Carbon disulphide at concentrations of 30–5,128 µg/ml was not mutagenic in *Salmonella typhimurium* strains TA98, TA100, TA1535 and TA1537 in the absence of systems for metabolic activation, in the presence of induced rat liver homogenate (S-9) and in the presence of induced hamster liver S-9.[26,27]

In another study, carbon disulphide at 0.1–5 µl/ml failed to induce unscheduled DNA synthesis (UDS) in human fibroblasts (WI-38) in the absence of metabolic activation. In the presence of mouse liver S-9, a small but statistically significant amount of UDS appeared to occur, though a benzo(*a*)pyrene positive control failed to induce UDS here.[26]

Other investigations found carbon disulphide to lack mutagenic potential for *S. Typhimurium* and *E. coli*.

Carbon disulphide at relatively low concentrations (in terms of OELs) of 20 and 40 ppm failed to produce sex-linked recessive lethals in *Drosophila*, failed to produce cytogenetic abnormalities in rat bone marrow and had no significant effects in a rat dominant lethal study.[26]

Other, difficult to evaluate, reports have suggested that carbon disulphide may have genotoxic properties. For example, the material was reported to produce cytogenetic abnormalities in primary cultures of human blood leucocytes, and rats exposed to 3.3 ppm carbon disulphide and hydrogen sulphide mixtures are reported to have increased incidence of aneuploid cells in bone marrow. Lack of technical detail and other factors make evaluation of these reports difficult though.[26]

Beauchamp *et al.*[26] conclude that apparent positive results obtained in various studies should be regarded as inconclusive pending corroboration from further work.

18.8 Summary

[Carbon disulphide is of high acute and cumulative toxicity by inhalation, ingestion or skin penetration. Effects of acute over-exposure may include narcosis, headache, giddiness, and respiratory and digestive disturbances. Additional effects of severe or chronic over-exposure may include visual, nervous system, cardiac, mental and reproductive damage, together with other less specific effects. The liquid is irritating to skin, strongly so with prolonged or repeated contact, and severely so to eyes and mucous membranes, while high vapour concentrations irritate the eyes and respiratory tract. Problems may arise from use of contact lenses. The threshold vapour concentration for eye irritation is about 100–1,000 ppm, for acute toxic effects about 300 ppm, and for cumulative or chronic toxic effects about 10–30 ppm or less. No reliable evidence of carcinogenicity or mutagenicity has been found.]

19. MEDICAL / HEALTH SURVEILLANCE

A decision on the need for, and content of, medical surveillance should be based on an assessment of the possibility and extent of exposure in the work operation. In addition, medical examination may be directed to identifying any pre-existing or newly arising condition in the individual workers which might either be aggravated by subsequent exposure, or might confuse any subsequent medical assessment in the event of excessive exposure or an illness not related to exposure. A particular aspect to be considered is the identification of sensitive subjects not adequately protected by the control limit in operation. A professional medical judgement is required on continuance of employment in the specified process. The following information is relevant in the case of this solvent.

Pre-employment Medical Examination

Pre-employment medical examination should include a full physical assessment; particular attention to those systems known to be main targets of carbon disulphide toxic action (*i.e.*, the cardiovascular and nervous systems) has been advocated but exclusion of workers on such grounds is challenged. Identification of a coronary heart disease risk factor at initial or later medical examination offers the opportunity of advice on removing or reducing the risk factor.

Urinalysis and electrocardiography have however been advocated by some authorities with a view to identifying pre-existing conditions which might increase the risk of adverse effect or could confuse later assessment in relation to exposure which might occur. For work where the use of personal respiratory protective equipment will arise, assurance of fitness to wear this will be required.

Periodic Medical Examination

Further medical supervision during work with carbon disulphide will be related to the effectiveness of hygiene control of exposure. Normally an annual review of health would be adequate but examination may be advisable after exposure above a specific level, or in the event of skin or eye contamination. A medical decision may be required on further employment on carbon disulphide processes.

Blood cholesterol assay and electrocardiography may be undertaken but decision on removal from work with carbon disulphide requires careful consideration of all factors by the supervising occupational physician.

Biological monitoring either on a spot-check basis or on a planned sample system may be administered. The iodine-azide test rests on the detection of organo-sulphur metabolites present in urine. It has been shown to have poor sensitivity particularly at low exposures (below 15 ppm) and in any case is non-specific. Recent work shows that the urinary concentration of 2-thiothiazolidine-4-carboxylic acid (TTCA) is quantitatively related to the uptake of carbon disulphide. End of shift urine TTCA concentration indicates uptake over the whole working shift.

In the event of skin or eye contamination or accidental excessive inhalational exposure an immediate medical assessment may need to be followed by further reviews at appropriate intervals. Decision on continuance of work on the process would follow.

20. OCCUPATIONAL EXPOSURE LIMITS

[The Committee felt that a time-weighted average limit of 10 ppm and a short term limit of 20 ppm were appropriate for carbon disulphide. This substance may produce serious toxic effects by absorption through the skin (see Introduction).]

REFERENCES

1. Council of Europe. 'Dangerous Chemical Substances and Proposals Concerning their Labelling'. ('The Yellow Book'.) 4th ed. Maisonneuvre, 1978.
2. Sax, N.I. 'Dangerous Properties of Industrial Materials'. 6th ed. Van Nostrand Reinhold, 1984.
3. Sax, N.I., Ed.-in-Chief. Dangerous Properties of Industrial Materials Report 3 No.4(1983):16. 'International toxicity update: carbon disulphide'. Van Nostrand Reinhold.

4.   Occupational Safety & Health Administration. Material safety data sheets from the Occupational Health Services database, OSHA, Washington D.C., 1985.
5.   Graedel, T.E. 'Chemical Compounds in the Atmosphere'. Academic Press, 1978.
6.   Aneja, V.P. *et al. Journal of the Air Pollution Control Association*, **32,** No.8(1982):803. 'Biogenic sulfur compounds and the global sulfur cycle'.
7.   Hawley, G.G., Ed. 'The Condensed Chemical Dictionary'. 10th ed. Van Nostrand Reinhold, 1981.
8.   Grayson, M., Exec. Ed. 'Kirk-Othmer Concise Encyclopedia of Chemical Technology'. 3rd ed. John Wiley & Sons, 1985.
9.   Grayson, M., Exec. Ed. 'Kirk-Othmer Encyclopedia of Chemical Technology'. 3rd ed. John Wiley & Sons, 1979.
10. Grasselli, J.G. and W.M. Ritchey, Eds. 'Atlas of Spectral Data and Physical Constants for Organic Compounds'. 2nd ed. CRC Press, 1975.
11. National Institute for Occupational Safety & Health. 'NIOSH Manual of Analytical Methods'. 3rd ed. DHHS (NIOSH) Publication No.84-100. US Department of Health & Human Services, 1984.
12. H.M. Factory Inspectorate Booklet No.6. 'Methods for detection of toxic substances in air'. Department of the Environment, UK, 1974.
13. Vincent, R. *et al. Analusis*, **13,** No.9(1985):415.
14. Rosier, J. *et al. International Archives of Occupational & Environmental Health*, **51,** No.2(1982):159.
15. Smith, D.B. and L.A. Krause. *American Industrial Hygiene Association Journal*, **39,** No.12(1978):939.16. Steudler, P.A. and W. Kijowski. *Analytical Chemistry*, **56,** No.8(1984):1432.
17. Hoshika, Y. *et al. Japan Analyst*, **23,** No. 11(1974):1393.
18. Saulova, E. and M. Pankova. *Khimiya i Industriya (Sofia)*, **52,** No.4(1980):164.
19. Osin, I.A. *et al. Zavodskaya Laboratoriya*, **40,** No.10(1974):1199.
20. Kneebone, B.M. and H. Freiser. *Analytical Chemistry*, **47,** No.6(1975):942.
21. Wronski, M. and L. Walendziak. *Chemia Analityczna*, **21,** No.5(1976):1147.
22. Waldman, M. and M. Vanecek. *Annals of Occupational Hygiene*, **25,** No.1(1982):5.
23. Yasouka, T. *et al. Proceedings of the Faculty of Science of Tokai University*, 15(1979):81.
24. Herber, R.F.M. and H. Poppe. *Journal of Chromatography*, **118,** No.1(1976):23.
25. Boillat, M.A. *Sozial - und Praeventivmedizin*, **25,** No.4(1980):203.26. Beauchamp, R.O., jr. *et al. CRC Critical Reviews in Toxicology*, **11,** No.3(1983):169. (Ed. L. Golberg.) 'A critical review of the literature on carbon disulfide toxicity'. CRC Press.
27. Santodonato, J. *et al.* 'Monograph on human exposure to chemicals in the workplace: carbon disulphide'. Report No.SRC-TR-84-986. National Cancer Institute, USA, 1985.
28. Lowenheim, F.A. and M.K. Moran. 'Faith, Keyes and Clark's Industrial Chemicals'. 4th ed. John Wiley & Sons, 1975.
29. Commission of the European Communities. 'Environmental Chemicals Data and Information Network (ECDIN)' Databank. Commission of the European Communities Joint Research Centre, ECDIN Group, I-21020 Ispra (Varese), Italy, 1986.
30. World Health Organization. 'Environmental Health Criteria 10: Carbon disulfide'. UNEP/WHO, Geneva, 1979.
31. Osol, A., Ed. 'Remington's Pharmaceutical Sciences'. 16th ed. Mack Publishing Co., 1980.
32. Berg, G.L., Ed. 'Farm Chemicals Handbook '83'. Meister Publishing Co., Willoughby, Ohio, USA, 1983.
33. *Chemical Marketing Reporter*, 228(1985):4. 'Fumigants are removed from market by EPA'. (Abstract).
34. *Chemical Marketing Reporter*, 229 No.2(1986):46. 'Chemical profile: carbon disulphide'. (Abstract).
35. American Conference of Governmental Industrial Hygienists. 'Documentation of the threshold limit values for chemical substances in the workroom environment'. ACGIH, 1986.
36. National Fire Protection Association. 'Fire Protection Guide on Hazardous Materials'. 9th ed. NFPA, Massachusetts, USA, 1986.

37. Windholz, M. *et al.*, Eds. 'The Merck Index'. 10th ed. Merck & Co., 1983.
38. Sax, N.I. 'Dangerous Properties of Industrial Materials'. 5th ed. Van Nostrand Reinhold, 1979.
39. National Institute for Occupational Safety & Health. 'NIOSH Recommended standard for occupational exposure to carbon disulphide'. US Department of Health, Education & Welfare, 1977.
40. Marsden, C. and S. Mann, Eds. 'Solvents Guide'. 2nd ed. rev. Cleaver-Hume Press, 1963.
41. International Labour Office. 'Encyclopaedia of Occupational Health and Safety'. 3rd rev. ed. ILO, 1983.
42. Bretherick, L. 'Handbook of Reactive Chemical Hazards'. 3rd ed. Butterworths, 1985.
43. Warren, P.J., Ed. 'Dangerous Chemicals Emergency Spillage Guide'. 1st ed. Jensons (Scientific) Ltd., Leighton Buzzard, UK, 1985.
44. Sax, N.I., Ed.-in-Chief. Dangerous Properties of Industrial Materials. Report 3 No.5(1983):84. 'Carbon disulfide'. Van Nostrand Reinhold.
45. National Institute for Occupational Safety & Health/Occupational Safety & Health Administration. 'NIOSH/OSHA Occupational health guideline for carbon disulphide'. DHHS (NIOSH) Publication. US Department of Health & Human Services/US Department of Labor, 1978.
46. Weast, R.C., Ed.-in-Chief. 'CRC Handbook of Chemistry and Physics' ('The Rubber Handbook'). 67th ed. CRC Press, 1986-1987.
47. Mellor, J.W. 'A Comprehensive Treatise on Inorganic and Theoretical Chemistry'. Vol.6. Longmans, Green & Co., 1925.
48. Dreisbach, R.R. 'Physical Properties of Chemical Compounds', Vol.3. Advances in Chemistry Series No.29, American Chemical Society, 1961.
49. Horsley, L.H. 'Azeotropic Data III'. Advances in Chemistry Series No.116, American Chemical Society, 1973.
50. Fassett, D.W. and D.D. Irish, Eds. 'Patty's Industrial Hygiene and Toxicology', Vol.II: 'Toxicology'. 2nd rev. ed. Wiley Interscience, 1962.
51. Gosselin, R.E. *et al.* 'Clinical Toxicology of Commercial Products'. 4th ed. Williams and Wilkins, 1976.
52. Ruth, J.H. *American Industrial Hygiene Association Journal*, 47(1986):A-142. 'Odor thresholds and irritation levels of several chemical substances: a review'.
53. World Health Organization/Nordic Council of Ministers. Environmental Health Series 5(1985). 'Organic solvents and the central nervous system'. Report on a joint WHO/Nordic Council of Ministers Working Group.
54. Magos, L. *Reviews on Environmental Health*, **2**, No.1(1975):65. 'The clinical and experimental aspects of carbon disulphide intoxication'.
55. Hayes, W.J., jr. 'Pesticides studied in man'. Williams & Wilkins, 1982.
56. Davidson, M. and M. Feinleib. *American Heart Journal*, **83**, No.1(1972):100. 'Carbon disulfide poisoning: a review'.
57. Thienes, C.H. and T.J. Haley. 'Clinical Toxicology'. 5th ed. Lea & Febiger, 1972.
58. Fielder, R.J. *et al.* 'Toxicity review 3: carbon disulphide'. Health & Safety Executive, Her Majesty's Stationery Office, London, 1981.
59. Bretherick, L., Ed. 'Hazards in the Chemical Laboratory'. 3rd ed. Royal Society of Chemistry, 1981.
60. Houston, A., Ed. 'Dangerous Chemicals Emergency First Aid Guide'. 2nd ed. Wolters Samson (United Kingdom), 1986.
61. Grant, W.M. 'Toxicology of the Eye'. 2nd ed. Charles C. Thomas, Illinois, USA, 1974.
62. Keith, L.H. and D.B. Walters, Eds. 'Compendium of safety data sheets for research and industrial chemicals'. VCH Publishers, 1985.
63. Mack, T. *et al.* *Biochemical Pharmacology*, 23(1974):607. 'Inhibition of oxidative N-demethylation in man by low doses of inhaled carbon disulphide'.
64. Browning, E. 'Toxicity and Metabolism of Industrial Solvents'. Elsevier, 1965.
65. Trevethick, R.A. 'Environmental and industrial health hazards: a practical guide'. William Heinemann Medical Books, London, 1973.

66. Spyker, D.A. *et al. Journal of Toxicology, Clinical Toxicology*, **19**, No.1(1982):87. 'Health effects of acute carbon disulphide exposure'.

67. World Health Organization. Technical Report Series No. 664: 'Recommended health-based limits in occupational exposure to selected organic solvents'. WHO, Geneva, 1981.

68. National Institute for Occupational Safety & Health. 'Registry of Toxic Effects of Chemical Substances (RTECS)'. DHHS (NIOSH) Publication No.84-101-6. US Department of Health & Human Services, April, 1986.

69. Sweetnam, P.M. *et al. British Journal of Industrial Medicine*, 44(1987):220. 'Exposure to carbon disulphide and ischaemic heart disease in a viscose rayon factory'.

70. Wilcosky, T.C. *et al. American Industrial Hygiene Association Journal*, **45**, No.12(1984):809. 'Cancer mortality and solvent exposures in the rubber industry'.

71. Lancranjan, J. and H.I. Popescu. *Medicina del Lavoro*, **60**, No.10(1969):567. 'Changes of the gonadic function in chronic carbon disulphide poisoning'.

72. Vanhoorne, M. and R. Grosjean. *Giornale Italiano di Medicini del Lavoro*, 6(1984):95. 'Exposure data in the viscose industry; Achilles' heel of carbon disulphide epidemiology?'

73. Neal, R.A. 'Thiono-sulfur Compounds'. In 'Bioactivation of Foreign Compounds' (1985):519. Ed. M.W. Anders. Academic Press.

74. O'Donoghue, J.L. 'Neurotoxicity of industrial and commercial chemicals', Vol.2. CRC Press, Florida, 1985.

75. Price, C.J. *et al. Toxicologist*, 4(1984):86. 'Developmental toxicity of carbon disulphide in rabbits and rats'. (Abstract.)

76. Council on Scientific Affairs Report. *Journal of the American Medical Association*, **253**, No.23(1985):3431. 'Effects of toxic chemicals on the reproductive system'.

Diethyl Ether

CONTENTS

1. CHEMICAL ABSTRACTS NAME

ethane, 1,1'-oxybis-
(10th Collective Index)

2. SYNONYMS AND TRADE NAMES[1-6]

aether
anaesthetic ether
anesthesia ether
anesthetic ether
diaethylaether (German)
diethyl ether
diethylether (German, Dutch)
diethyl oxide
dwuetylowy eter (Polish)
etere etilico (Italian)
ether
ether éthylique (French)
ethoxyethane
ethyl ether
ethyl oxide
OHS08980
1,1'-oxybisethane
1,1'-oxybis-ethane
oxyde de diéthyle (French)
oxyde d'éthyle (French)
pronarcol
solvent ether
sulfuric ether
sulphuric ether
U117
UN 1155

3. CHEMICAL ABSTRACTS SERVICES REGISTRY NUMBER

60-29-7

4. NIOSH NUMBER

KI5775000

5. CHEMICAL FORMULA

$C_4H_{10}O$
(Molecular weight 74.12)

6. STRUCTURAL FORMULA

$$CH_3-CH_2-O-CH_2-CH_3$$

7. OCCURRENCE[7]

Graedel[7] found no data available on the atmospheric concentrations of ethers though diethyl ether has been detected in ambient air presumably due to its industrial and solvent usages.

Oxygen atoms and hydroxyl radicals can abstract hydrogen from aliphatic ethers. Likely products appear to be aldehydes and ketones.[7] Graedel[7] gives reaction with dimethyl ether as an example, *viz.*

$$CH_3OCH_3 \xrightarrow{HO^{\cdot}} CH_3OCH_2{}^{\cdot} \xrightarrow{O_2} CH_3OCH_2O_2{}^{\cdot} \xrightarrow{NO} CH_3OCH_2O^{\cdot}$$

$$CH_3OCH_2O^{\cdot} \rightarrow HCHO + CH_3O^{\cdot} \xrightarrow{O_2} HCHO$$

Graedel[7] notes that the oxygen atom in the ethers makes the C–H bonds weaker than in the corresponding hydrocarbon. The ethers are thus much more reactive and have estimated tropospheric lifetimes of about a day.

8. COMMERCIAL AND INDUSTRIAL DIETHYL ETHER[8-10]

Marsden and Mann[8] state that diethyl ether is produced in several different grades ranging from technical material for industrial use to specially purified material for anaesthetic use and for spectroscopic analysis. They note also that the main impurities in industrial material are water, ethanol, and denaturant from the spirit [ethanol] used in manufacture. Lesser impurities that may be present are acetaldehyde, vinyl alcohol, acetone, acetic acid and peroxides formed during storage. (See Section 12 regarding peroxide hazards.)

Marsden and Mann[8] give the following example of a specification. (Higher purity standards also exist.)

B.S. 579:1957 grade.

Colour	Colourless
Specific gravity ($d_{15.5}^{15.5}$)	0.719–0.725
Specific gravity (d_{20}^{20})	0.714–0.720
Specific gravity (d_{25}^{25})	0.709–0.715

Distillation range	95% (min.) below 36 °C
Water	1.0% w/w (max.)
Residue	0.0005% (max.)
Acidity as H_2SO_4	0.002% (max.)
Peroxides	none
Sulphur compounds	none

Ref. 9 notes the availability of USP, ACS absolute and industrial or technical (97.5%) grades of diethyl ether.

Keeley in ref. 10 quotes typical specifications for various grades of diethyl ether, *viz.*

Technical refined grade.

Colour, max.	water-white
Density (d_{25}^{25})	0.710–0.713
Acidity, as acetic acid, wt%	0.0025
Peroxides, max. wt%	passes USP test – no colour with potassium iodide reagent
Aldehydes, max. wt%	passes USP test – no turbidity with alkaline mercuric chloride– potassium iodide reagent
Alcohol, max. %	0.5% v/v
Non-volatile matter, max. wt%	0.002
Water, max. wt%	0.3
Odour	non-residual
Net container contents:	
19 l (5 [US] gal) drum	13.6 kg
208 l (55 [US] gal) drum	147 kg

The technical grade can also be shipped [in USA] in insulated tank cars of 4,000, 6,000, 8,000, and 10,000 [US] gal capacity.

Ref. 10 also quotes US specifications for purer grades.

9. SPECTROSCOPIC DATA[11]

Infrared, Raman, ultraviolet, both ^{1}H and ^{13}C NMR, and mass spectral data for diethyl ether have been tabulated.[11]

10. MEASUREMENT TECHNIQUES[12,13]

The method recommended for the determination of diethyl ether in workshop atmospheres is the NIOSH[12] activated charcoal trapping method for collection and concentration, followed by solvent extraction of the charcoal and a gas chromatographic (gc) analysis of the extract.

The collection tube is 7 cm long with a 4 mm internal diameter. It contains a total of 150 mg of activated charcoal (20–40 mesh) divided into a front section of 100 mg and a rear section of 50 mg separated by a small plug of urethane foam. Air is sampled at a flow-rate of 0.01–0.2 l per minute for a sample size of 0.25 to 3 l by means of a small pump. The entire apparatus is portable so that it may be carried in a pocket with the sampling tube in the breathing zone (normally a coat lapel). The apparatus can also be static. Ethyl acetate is used to extract the diethyl ether, which is separated on a 1.2 m x 6 mm stainless steel column packed with Porapak Q (50–80 mesh) using flame ionization detection. For full details of the method see ref. 12.

The sampling method uses a small portable apparatus with no liquids and is therefore easy to handle and maintain. The analytical method is relatively specific with the additional advantage of flexibility of operating conditions.

Air samples are normally taken by means of suction using an electric pump capable of operating at low flow-rates (10–300 ml per minute). If higher flow-rates are used, breakthrough can occur (*i.e.*, the diethyl ether is not completely adsorbed and some of it passes through the adsorbent). It is also convenient either to establish the trapping device in a static position or to have it attached to the user in such a way that the breathing zone is sampled (this usually means attachment to the worker's lapel).

It should be noted that suitable personal samplers may be commercially available.

There appears to be very little interest in the determination of diethyl ether in air. There has been a study of general organic air contamination near an American waste disposal site and although many organics were found these did not include diethyl ether.[13] The method involved trapping on Tenax-GC, thermal desorption and a gc-ms analysis.

11. CONDITIONS UNDER WHICH DIETHYL ETHER IS PUT ON THE MARKET

Production[9,10,14]

Ref. 10 notes that most diethyl ether is now produced *via* the vapour-phase hydration of ethylene to ethanol over a supported phosphoric acid catalyst. The process is flexible enough to enable the relative proportions of ethanol and diethyl ether produced to be varied to some extent and any additional diethyl ether required can be prepared, in greater than 95% yield, by the vapour phase dehydration of ethanol in a fixed-bed reactor using an alumina catalyst.

Ref. 9 describing syntheses of ethanol details the catalytic hydration. Ninety seven percent ethylene (with impurities of methane, ethane, acetylene and higher olefins) is compressed to 1,000 psig and mixed with a recycled stream containing 85% ethylene. This combined ethylene stream is mixed with a stream of water (0.6 mole water per mole of ethylene), then vaporized (by heat exchange with the product leaving the reactor) and finally heated in a gas-fired furnace to the required 300 °C reaction temperature.

In the reactor, the downflow feed of reactants passes over a catalyst made by impregnating diatomaceous earth with phosphoric acid (85–90% by weight), and whose maximum activity is maintained by injections of phosphoric acid into the top of the reactor at predetermined intervals. Conversion per pass is only 4.2%. Outflow from the reactor is partially condensed by heat exchange and dilute caustic soda solution is added to neutralize any phosphoric acid vaporized from the catalyst. The product stream is further condensed; liquid and vapour are separated in a high pressure separator; the recycle gas is cooled and scrubbed with water and then, before being fed back, is partly bled off in order to avoid methane and ethane build-up in the reactor feed.

Liquid phases from separator and scrubber are concentrated in a stripping column from which water is removed as bottoms. Overheads contain products plus acetaldehyde (from hydration of acetylene). This last is reduced to ethanol by mixing the vapour stream with hydrogen and passing through another reactor containing a nickel hydrogenation catalyst. Condensation in a product heat exchanger and separation in a product accumulator enables excess hydrogen to be separated and recycled. Liquid products (diethyl ether and ethanol) are then separated and purified by conventional means.

Diethyl ether may be produced via the low temperature reaction of ethylene and sulphuric acid which yields a mixture of ethyl sulphates. On hydrolysis ethanol is formed and this can react with ethyl sulphate to form diethyl ether.[9]

The ethylene is passed into the bottom of an absorber, countercurrent to 90% sulphuric acid. Unabsorbed gases come off the top and are used as fuel. A mixture of mono- and diethyl sulphates is discharged from the bottom together with a small amount of free acid and water. The exothermic reaction is controlled by the rate of feed of the two reactants.

The mixed esters are passed to a hydrolyser with a measured volume of water. Hydrolysis takes 1.5–2 hours at 70 °C. In a stripping column, steam carries over ethanol, diethyl ether and small amounts of acid while dilute sulphuric acid is taken from the bottom for recovery by vacuum evaporation.

The vapours from the stripping column are scrubbed with countercurrent sodium hydroxide solution to neutralize entrained acid and then passed to an ether column where the diethyl ether is removed by live steam. Ether yield is about 5–10% by weight of product [ethanol] but the method can be modified to produce diethyl ether entirely.

An older method for producing diethyl ether is to heat 95% ethanol with 96% sulphuric acid in a lead-lined steel reaction kettle, using about three parts of acid per one part of ethanol. Reaction is initiated at 125–140 °C by means of a steam jacket or internal steam coils. Vaporized alcohol is passed into the reactor at such rate that, with suitable steam-heating adjustment, the temperature is maintained at approximately 127 °C.

In a scrubber, countercurrent dilute sodium hydroxide solution removes entrained sulphuric acid together with any acidic reaction products. From this aqueous phase, on its being passed to the lower section of a fractionating column, alcohol and ether are removed.

The vapours from the top of the scrubber (ether, alcohol and water) are separated in a continuous fractionating column, ether vapours being taken from the top *via* a reflux condenser maintained at 34 °C. Material boiling at above 34 °C is returned to the column while the ether is condensed and passed to storage. This technical grade material contains very small amounts of alcohol, water, aldehydes, peroxides and other impurities. Redistillation and dehydration, followed by an alkali or charcoal treatment enables more refined grades to be obtained such as that required for anaesthesia.

The yield of technical diethyl ether is 94–95% on the ethanol processed. The process is practically continuous and may be run for months before recharging with sulphuric acid. The formation of tarry products and reduction of some acid to sulphur dioxide necessitate periodic recharging.[9]

EC production of diethyl ether in 1984 was 50 kilotonnes. EC consumption in 1984 was 45 kilotonnes.[14] Between 1970 and 1975 the US production fell from 47.6 kilotonnes to 12.5 kilotonnes.[10]

Containers for diethyl ether range from glass bottles through cans, tins and drums to tankers.[9] Legislation may limit the quantities transportable by various modes and any labelling requirements must be adhered to.

Uses[10,14,15]

In the USA, large quantities of diethyl ether were used in World War II and in the Vietnam War as a solvent in the manufacture of smokeless powder. Other uses for diethyl ether in the chemical industry are as a solvent or extractant for waxes, oils, perfumes, resins, dyes, gums and alkaloids. It is used as an extractant to recover acetic acid from dilute aqueous solutions; as a denaturant for ethanol; as a starting fuel for diesel engines; as an entrainer for the dehydration of ethanol and isopropyl alcohol; as an anhydrous inert reaction medium and, [to a lessening extent] as a general anaesthetic [10].

Diethyl ether is also used as a reactant in chemical syntheses. It may be chlorinated to yield 2,2'-dichlorodiethyl ether[14] and, even though itself synthesised from ethylene, may be used as a commercial source of ethylene in plants not having access to petroleum refinery gases. Diethyl ether is more readily transportable than ethylene and, either alone or denatured with ethanol gives a high yield of very pure ethylene when passed over alumina at approximately 343°C.[10]

For 1974, ref. 10 gives the estimated usage breakdown [presumably in the USA] as

Solvents and smokeless powder	65%
Chemical synthesis	25%
Anaesthetic and other medicinal uses	3%
Miscellaneous uses	7%

In the EC, the Council of Ministers has agreed to allow alcohol and diethyl ether to replace lead in petrol from 1988. The lead is to be eliminated progressively and these oxygenated organics can be used to increase the octane value of petrol.[15]

12. STORAGE, HANDLING AND USE PRECAUTIONS[6,8,10,14,16-21]

Diethyl ether is a volatile dangerously flammable liquid with a vapour density about $2\frac{1}{2}$ times that of air. It is also liable to peroxidise on standing in contact with air, especially in sunlight, and this may well constitute an additional hazard with some uses of the material (such as distillation or evaporation) and in such cases the peroxides should be removed or be inhibited from forming.

Formation of peroxide in stored diethyl ether may be prevented by the presence of 0.05 ppm of sodium diethyldithio carbamate which probably deactivates traces of metals which catalyse peroxidation. Alternatives are 1 ppm pyrogallol or larger proportions (5–20 ppm) of other inhibitors. Peroxides in ether may be detected by the starch-iodine test. Some of the methods described for their removal are percolation through anion exchange resin, or activated alumina, which leaves the diethyl ether dry, or by shaking with aqueous ferrous sulphate or sodium sulphite solutions.[16] Ref. 6 recommends that bottles con taining only a little diethyl ether should not be kept longer than three months.

Diethyl ether is not corrosive to metals and *e.g.*, mild steel, copper or aluminium is suitable for plant and containers.[8] Special regulations govern the transport and storage of diethyl ether. Local legislation may exist, governing its storage and industrial use, especially in built-up areas.[8] [In the USA] the area in which it is handled should be considered a Class I hazardous location as defined by the [US] National Electrical Code.[10] Martindale[17] notes a report that diethyl ether, even in low concentrations, caused softening of PVC bottles and was associated with loss by permeation. ECDIN[14] notes that diethyl ether attacks some forms of plastics, rubber and coatings.

Keeley in ref. 10 notes that special containers have been developed to prevent deterioration of anaesthetic ether before use. Their effectiveness usually depends on the presence of the lower oxidation state of a polyvalent metal. For example, the insides of tin-plate containers can be electroplated with copper containing a small amount of cuprous oxide. Stannous oxide may also be used in tin container linings. As an alternative to special containers, the diethyl ether may be stabilized by the addition of iron wire or certain other metals and alloys or various organic compounds[10] [*e.g.*, *vide supra*]. Ref. 18 notes that diethyl ether may be supplied mixed with 3% ethanol to retard oxidation.

Ref. 19 notes US regulations that anaesthetic diethyl ether must be preserved in tight containers of not more than 3 kg capacity and not be used for anaesthesia if more than 24 hours have elapsed since its removal from the original container. It may though be transported in larger containers and repackaged as above, provided the material meets the requirements of the USP at the time of repackaging.

Care must be taken with storage, process and drying plant to ensure that hazard-free conditions exist regarding fire and ex plosion risks and exposure of personnel to diethyl ether. All tools used in making connections or repairs to containers or other equipment should be of the non-sparking type. Tankers should be earthed and bonded before any connection is made to them and should always be unloaded through dome connections rather than through bottom outlets. Natural gas or a positive suction pump can be used to remove diethyl ether from a tanker. Air pressure should never be used for this purpose.[10]

Precautions should be taken when storing ether-contaminated material in refrigerators, so that an explosion cannot be caused by sparking at the thermostat or light switch.

[As with other flammable solvents the organization of storage facilities has to take into account two separate requirements. These are that the stored material be protected from a fire elsewhere on the premises and that the premises be protected from a fire involving the stored materials. On an industrial scale, sufficient physical separation and water spray curtains are often used effectively. On a laboratory scale, flammable solvents need to be stored in fire resistant, thermally insulated cupboards.

It is important to appreciate that a non-combustible barrier is not synonymous with a fire barrier. A fire barrier has to protect against both conducted and radiant heat. For example, a metal cupboard is useless on both these counts.]

Sax,[20] for general storage procedure, advises that detached outside storage be preferred. There should be isolation from combustible materials and direct sunlight should be avoided. Exposed lots should be moved to a well ventilated area. With large quantity storage, an automatic sprinkler system and total flooding carbon dioxide systems should be installed.

OSHA[21] states that employees should be provided with, and required to use, impervious clothing, gloves, eight inch (minimum) face shields and other protective clothing necessary to prevent repeated or prolonged skin contact with liquid diethyl ether. Splash-proof safety goggles should be used in situations where eye-splashes are possible. OSHA[21] also states that suitable respiratory protection is required for personnel exposed to vapour concentrations above 400 ppm. [Personnel should not be allowed to enter areas containing potentially hazardous concentrations of diethyl ether unless all relevant legislation and official guidelines have been strictly adhered to. Suitable respiratory protective equipment should be used.] Such exposed personnel should be able to communicate with each other and with those (*i.e.*, more than one person) monitoring the work from outside the hazard area. Monitors should be suitably equipped to effect rescue (possibly with lifelines to exposed personnel) and should themselves ensure, so far as is possible, that one of their number remains outside the hazard area in the event of an emergency.

Adequate ventilation should be provided in order to keep personnel exposures from being above the recommended limits. For many operations in which diethyl ether is used or liberated, general dilution ventilation is recommended together with local exhaust ventilation and process enclosure where either or both of these last are feasible. Ref. 21 notes examples.

N.B. the caveat in Section 13 (final paragraph).

13. FIRE HAZARDS[4,6,10,16,20-23]

As given in Section 17, diethyl ether has a boiling point of 34.5 °C, a specific gravity of 0.7 and a water solubility of about 7%. It has a flash point of -45 °C (closed cup) and an auto-ignition temperature of 170 °C. Its lower and upper explosive limits in air are 1.85% and 48% v/v. It is highly flammable and is highly volatile at ambient temperatures. Its vapour is about $2\frac{1}{2}$ times denser than air and can, if in sufficient quantity, accumulate at low levels, flow a considerable distance to an ignition source and 'flash back'. It is easily ignited by flames, sparks and, given its low auto-ignition temperature, even hot surfaces. As a non-conductor it can generate static electric charges that may result in ignition or vapour explosion. It forms unstable peroxides in the presence of air, especially in sunlight. Bretherick[16] notes that the hydroperoxide is not particularly explosive but that, on standing and evaporation, polymeric 1-oxyperoxides are formed which are dangerously explosive even below 100 °C. These may explode spontaneously especially if concentrated by distillation or if deposited *e.g.*, under the caps of bottles of diethyl ether from which repeated evaporation has taken place.[21]

Fire-exposed containers of diethyl ether can explode when heated and, if feasible, should be cooled with flooding amounts of water from as far as possible until well after the fire is extinguished.

For fire-fighting, foam, carbon dioxide, dry powder or vaporizing liquid are recommended. Water may be ineffective as an extinguishant and may spread a burning spillage. Note that the danger of reignition is high. Immediate withdrawal is imperative in the event of rising sound from a venting safety device or of any discolouration of a storage tank due to the fire. Personnel should keep upwind and avoid breathing diethyl ether vapour and possibly toxic combustion products. The area downwind should be evacuated to avoid danger from leaks.

Ref. 23 notes that where a fire is fed by an uncontrolled flow of combustible liquid the decision on how or if to fight it will depend on the size and type of fire anticipated and must be carefully considered. This may call for special engineering judgement, particularly in large-scale applications. For uncontrollable fires, a radius of 1,500 feet should be evacuated.[4]

[Notwithstanding the foregoing review of various authorities noted in Sections 12 and 13, it is essential that managers and others responsible for the planning, implementation and overseeing of personnel and plant safety should be familiar with the legal constraints and official guidelines applicable to them and that they liaise with their local emergency services in the planning of plant and storage facilities and in the preparation of contingency plans for dealing with fires and other emergencies. Managers should regularly monitor staff knowledge of, and ability to implement, emergency procedures and should ensure that equipment provided for use in emergencies is regularly inspected and maintained.]

14. HAZARDOUS REACTIONS[3,4,16]

Bretherick[16] reports the following hazardous reactions.

Boron triazide detonated in contact with diethyl ether vapour at -35 °C (probably initiated by the heat of co-ordination to oxygen).

Bromine trifluoride explodes on contact with diethyl ether.

Bromine pentafluoride may cause fire or explosion on contact with hydrogen-containing materials such as diethyl ether.

A case is mentioned where, shortly after the addition of bromine to diethyl ether, the solution erupted violently (or exploded softly). Photocatalytic bromination may have been involved.

Chlorine has caused ignition of diethyl ether on contact. Exposure of an ethereal solution of chlorine caused a mild, photocatalysed explosion.

Diethyl ether ignites in contact with gaseous iodine heptafluoride.

Silver perchlorate exploded violently upon being crushed after crystallization from diethyl ether, whereas it previously had been considered stable since it melts without decomposition. However a similar incident with silver perchlorate not previously in contact with organic materials is also reported.

It is noted that the explosions sometimes observed when anhydrous perchloric acid is in contact with diethyl ether are probably due to the formation of ethyl perchlorate by scission of the ether, or possibly due to formation of diethyloxonium perchlorate.

Diethyl ether evolves gas with nitrosyl perchlorate and then explodes after a few seconds delay.

Sharp explosions and ignition were observed with the action of nitryl perchlorate on diethyl ether.

Diethyl ether ignites on contact with chromyl chloride.

Liquid nitryl hypofluorite ('fluorine nitrate') explodes immediately on contact with diethyl ether. (It may also be sensitive in the vapour or the solid state.)

Anhydrous permanganic acid ignited diethyl ether explosively. The solution of acid (or its anhydride) produced by the interaction of permanganates and sulphuric acid will explode in contact with diethyl ether.

It is noted that it is dangerous to extract nitric acid solutions with diethyl ether or other organic solvents without first removing the excess acid, either by neutralization or by ion-exchange. Examples are given of violent reactions and explosions which the above procedure should obviate.

On evaporation of an ethereal solution of hydrogen peroxide a drop of the residue on a platinum spatula exploded weakly on exposure to flame. When 1–2 g of the sample was stirred with a non-fire polished glass rod, an extremely violent detonation occurred.

Peroxodisulphuric acid in uncontrolled contact with diethyl ether may cause explosion.

Washing the (incompletely characterized) solid iodine(VII) oxide with diethyl ether has occasionally led to explosive decomposition.

Simultaneous contact of sodium peroxide with water and diethyl ether causes ignition.

Contact of diethyl ether with ozonized oxygen produces some explosive diethyl peroxide.

Addition of liquid air to diethyl ether in a dish caused a violent explosion after a short delay. Previous demonstrations had been uneventful though it was known that such mixtures were impact- and friction-sensitive.

Precipitated thiotrithiazyl perchlorate exploded on washing with diethyl ether.

A solution of sulphuryl chloride in diethyl ether vigorously decomposed, evolving hydrogen chloride. Peroxides accelerate this reaction hence the need for the (careful) use of peroxide-free diethyl ether.

An explosion of great violence occurred on the evaporation of an ethereal extract of sulphur. It was shown that the evaporation of wet, peroxidised diethyl ether gave a mildly explosive residue which became violently explosive on addition of sulphur.

Solvent diethyl ether in the crystals of uranyl nitrate (six moles of water may be replaced by two of the ether) causes mild detonations when the salt is disturbed. Solutions of the nitrate in diethyl ether should not be exposed to sunlight due to the possibility of explosions.

Explosions occurred during the extraction of fats and waxes from peat soils by the use of diethyl ether, and also when heating the extract at 100 °C. Comment is made that although the latter event is not surprising (the diethyl ether contained 230 ppm of peroxides), the former is unusual.

Peroxide-free diethyl ether was used to extract wood pulp. The extracts were then stored for three weeks before being concentrated; during or after which processing, explosions occurred. It was noted that, during the three week interval, the ether and/or the extracted terpenes would be expected to form peroxides but that no attempt had been made to test for or remove them prior to distillation.

Sax[3] adds that diethyl ether can also react vigorously with acetyl peroxide, bromoazide, chlorine trifluoride, chromic anhydride, nitrile chloride, potassium peroxide, triethyl- or trimethyl aluminium in the presence of air, lithium aluminium hydride (OSHA[4] gives lithium aluminium hydride which has carbon dioxide as impurity).

15. EMERGENCY MEASURES IN THE CASE OF ACCIDENTAL SPILLAGE[4,20,21,24]

OSHA[4,21] instruct that persons not wearing protective equipment (see Section 12) should be restricted from areas of spills and leaks until clean-up of the diethyl ether spillage is complete. In the event of spillage all ignition sources should be removed or extinguished and the area should be ventilated. A leak may be stopped if this can be done without risk. Small quantities may be absorbed on paper towels, sand or vermiculite[24] which should be placed in a covered container (*e.g.*, a bucket with lid) if to be disposed of elsewhere. The absorbed ether may then be evaporated in a safe place, (such as a fume cupboard, but ensuring that all ductwork is eventually free of vapour,[21]) or else removed to a safe open area for evaporation on such as a metal tray.[24] It should not be disposed of on open ground as groundwater contamination may occur. If combustible, the absorbent may then be burnt in a suitable location away from other combustible materials.[21] Large quantities may, if practicable, be collected, dissolved in an alcohol of greater molecular weight than butanol and atomized in a suitable combustion chamber. This latter method may be used as a general method for the disposal of waste diethyl ether which itself should never be allowed to enter a confined space such as a drain or sewer, because of the risk of an explosion.[21] For large spills, the area ahead should be bunded for later disposal.[4]

Warren[24] states that for such as laboratory fires the aim should be to ventilate the area but to isolate the material if possible by closing the doors of the spillage room when leaving it. If the spillage cannot be controlled on site then the building should be evacuated by sounding the fire alarm and the fire brigade should be called.

Major spillage or leakage situations that have an environmental effect are considered by Sax in ref. 20. The air, fire and water authorities should be notified and the civil defence warned of a possible explosion. The area should be evacuated and entered from upwind. Spark resistant equipment should be used.

An attempt should be made to contain any slick, *e.g.*, in booms. Skimming may present a fire hazard. This might be lessened by the use of peat or absorbent foams or by carbon treatment. Diethyl ether contamination of shores or beaches should not be ignited. Booms might be obtained at major ports or from water authorities.

[Note that diethyl ether that has been in contact with air for any protracted period may have become peroxidised. See Section 12 for methods of removing peroxides.]

16. FIRST AID

General

The casualty should be removed from the danger of further exposure into the fresh air. Rescuers should ensure their own safety from inhalation and skin contamination where necessary. Conscious casualties should be asked for information about what has happened. Bear in mind that the casualty may lose consciousness at any time. Where necessary continued observation and care should be ensured. First aiders should take care not to become contaminated. Note that diethyl ether is highly flammable.

Inhalation

The conscious casualty should be kept at rest. The unconscious casualty should be placed in the recovery position and an open airway maintained. If breathing or heart beat stops, resuscitation should be commenced immediately. Medical aid should be obtained or the casualty removed to hospital immediately.

Diethyl ether has well known anaesthetic and respiratory irritation effects but there are no known delayed effects after a single exposure.

Skin contamination

Remove contaminated clothing. Wash with copious amounts of water, or soap and water, for at least 15 minutes. If necessary, seek medical aid. In the case of persistent skin irritation refer for medical advice.

Eye contamination

Irrigate with copious amounts of water for at least ten minutes but avoid further damage to the eye from use of excessive force. Seek medical aid. Remember that contact lenses may be worn and that these may be affected by some solvents and may impede decontamination of the eye.

Ingestion

If the lips or mouth are contaminated rinse thoroughly with water. Do not induce vomiting. Give further supportive treatment as for inhalation. In the case of ingestion of significant amounts wash out the mouth with water. Obtain immediate medical attention. Contact a hospital or poisons centre at once for advice. Remember that symptoms may develop many hours after exposure so continued care and observation may be necessary.

In all cases note information on the nature of exposure and give this to ambulance or medical personnel. Information for doctors is provided in some technical literature issued by manufacturers. Antidotes may be available for retention at the place of work for use by persons competent to treat casualties.

17. PHYSICO-CHEMICAL PROPERTIES

17.1 General[8,23,25,26]

Diethyl ether is a very volatile, highly flammable, colourless liquid with a characteristic, sweetish, pungent odour and a burning taste.[23,26] It is a good insulator.[26] Diethyl ether can form explosive peroxides when exposed to air and light[26] (see Sections 12 and 13).

Weast describes diethyl ether as miscible with a wide range of organic solvents, including acetone, ethanol, benzene, chloroform, butanol and 3-heptanol. Miscibility was determined at 20 °C, admixing equal volumes of each solvent.[25]

The Merck Index states that diethyl ether is slightly soluble in water, and water is slightly soluble in diethyl ether. The amount of diethyl ether in a saturated water solution at 15 °C and at 25 °C is given as 8.43 and 6.05% w/w, respectively.[26] The amount of water in saturated diethyl ether at 20 °C is 1.264% w/w.[8]

17.2 Melting Point[25,26]

Weast[25] gives the freezing point of diethyl ether as

−116.2 °C.

The Merck Index[26] gives the following melting points for diethyl ether:

−116.3 °C (stable crystals)
−123.3 °C (metastable crystals).

17.3 Boiling Point[25,27]

Weast[25] records the boiling point of diethyl ether as

34.5 °C.

Diethyl ether will form azeotropes with a number of other compounds. Details are listed below of some binary ones[27]:

Wt% diethyl ether	Second component	Wt% second component	Boiling pt. azeotrope (°C)
23.5	Acetaldehyde	76.5	18.9
50.4	Propylene oxide	49.6	32.6
98.74	Water	1.26	34.15

17.4 Density/Specific Gravity[8,26]

The specific gravity of diethyl ether at several temperatures (referred to the density of water at 4 °C), is listed by Merck[26] as follows:

Temperature in °C	Specific gravity
0	0.7364
10	0.7249
20	0.7134
30	0.7019

The critical density of diethyl ether is given by Marsden and Mann[8] as 0.263 g/cm^3. From this the calculated critical volume is 3.802 cm^3/g.

17.5 Vapour Pressure[25,26]

Weast[25] lists a selection of vapour pressures of diethyl ether below atmospheric pressure:

Temperature in °C	Vapour pressure in mmHg
−74.3	1
−48.1	10
−27.7	40
−11.5	100
17.9	400

Weast[25] also provides the following list of temperatures corresponding with elevated vapour pressures:

Temperature in °C	Vapour pressure in atm
34.6	1
56.0	2
90.0	5
122.0	10
159.0	20

In addition, The Merck Index[26] lists the following saturated vapour pressures:

Temperature in °C	Vapour pressure in mmHg
0	184.9
10	290.8
20	439.8
50	1276 [*ca.* 1.7 atm]
70	2304 [*ca.* 3.0 atm]

The critical temperature and critical pressure of diethyl ether are recorded as 192.6 °C, and 35.6 atm [*ca.* 27,100 mmHg], respectively.[25]

17.6 Vapour Density[26]

2.55 (relative to air = 1).

17.7 Flash Point[23]

−45 °C, closed cup test.

17.8 Explosive Limits[23]

The limits of flammability of diethyl ether in air at normal atmospheric temperature and pressure are given by the National Fire Protection Association[23] as:

Lower limit: 1.9% v/v
Upper limit: 36.0% v/v.

The NFPA[23] give the ignition temperature of diethyl ether as 160 °C. However, they point out that different test conditions (and also different definitions of "ignition temperature") can result in widely varying values being quoted. They recommend, therefore, that any value should be treated as an approximation.

17.9 Viscosity[25]

The viscosity of liquid diethyl ether is recorded by Weast[25] as follows:

Temperature in °C	Viscosity in cp
−100	1.69
−80	0.958
−60	0.637
−40	0.461
−20	0.362
0	0.2842
17	0.240

Temperature in °C	Viscosity in cp
20	0.2332
25	0.222
40	0.197
60	0.166
80	0.140
100	0.118

Weast also lists the viscosity of diethyl ether vapour[25]:

Temperature in °C	Viscosity in μp
0	67.8
14.2	71.6
100	95.5
121.8	98.3
159.4	107.9
189.9	115.2
251.0	130.0
277.8	135.8

17.10 Concentration Conversion Factors

At 25 °C and 760 Torr (1 atm),

$1 \ mg/m^3 = 0.33 \ ppm$, and
$1 \ ppm = 3.03 \ mg/m^3$.

18. TOXICITY

18.1 General[5,6,14,19,28]

Diethyl ether has predominantly narcotic properties. It is moderately toxic taken orally, and of low toxicity when inhaled, though over-exposed individuals may experience drowsiness, vomiting, and unconsciousness with death resulting from severe over-exposure. It is mildly irritating to the eyes, nose and throat. It can be absorbed through the skin and is capable of producing a dry, scaly, fissured dermatitis if contact is prolonged or repeated. Chronic exposure may result in anorexia, exhaustion, headache, drowsiness, dizziness, excitation and psychic disturbances. It may also cause increased susceptibility to alcohol.[5,14]

The odour threshold is reported as ranging from $0.9900 \ mg/m^3$ [0.33 ppm] to $3.0000 \ mg/m^3$ [0.99 ppm].[28] The odour is described as sweet, pungent,[6] and the liquid as having a burning, sweetish taste.[19]

18.2 Toxicokinetics[2,10,18,29-33]

Keeley in ref. 10 states that diethyl ether vapour is absorbed almost instantly from the lungs and very promptly from the intestinal tract. Once in the blood stream it passes rapidly into the brain.[2] More than 90% of absorbed diethyl ether is expired unchanged. Metabolism is not extensive nor are the metabolites hepatotoxic. Small amounts of diethyl ether are excreted unchanged in urine, milk, sweat, and other body fluids.[18] Hathway in ref. 29 notes that as with other volatile anaesthetics, diethyl ether crosses the placenta rapidly and enters the foetal circulation.

The partition coefficients between diethyl ether and various tissues *etc.* are given as follows, ('oil' is described as a substance that resembles subcutaneous fat[30]): blood/gas 15.2; oil/gas 50.2; fat/gas 50.[30] Ref. 18 quotes: blood/gas 12; fat/blood 4.2; and approximately unity for brain/blood, heart/blood, liver/blood and muscle/blood. Kidney/blood is given as 0.8.[18]

Ref. 31 notes that the relatively high blood, fat and tissue solubilities of diethyl ether partly explain the relatively prolonged recovery period frequently observed after long diethyl ether anaesthesia. Sufficient diethyl ether may be stored in the body to maintain anaesthesia after administration of the ether has been discontinued.

As much as 5% of inhaled diethyl ether is metabolized, with the following metabolic pathway having been proposed[32]; *viz*: oxidative cleavage to acetaldehyde and ethanol which in turn are rapidly oxidized to acetate by well-characterized enzymes and thence to carbon dioxide. Work with rats pretreated with phenobarbital suggests that microsomal metabolism of diethyl ether is catalysed by a cytochrome P450-containing mono-oxygenase system in a reaction analogous to the *o*-dealkylation observed for other compounds such as ethoxycoumarin.[32] Autoradiography has demonstrated concentration of radioactively labelled diethyl ether in mouse liver and, with rats (apart from recovery as labelled carbon dioxide), an additional 2% appears as non-volatile radioactivity in the urine. Additional pathways are required to account for non-volatile metabolites.[33]

18.3 Human Acute Toxicity[2,14,17,19-21,34-42]

Ingestion

The fatal oral dose of ether is stated to be 30 ml[34]; 1–2 ounces [*ca.* 30–60 ml] is given as being possibly fatal[35]; while Sax[20] gives LDLo for a man as being 260 mg/kg. Gosselin *et al.*[35] state that symptoms produced are similar to ethanol intoxication except that the onset is more rapid and the duration is shorter. Also, because of diethyl ether's volatility, the stomach becomes promptly distended and this may embarrass breathing. Liquid diethyl ether has been administered in mixtures because of a reputed carminative action.[17]

Kirwin and Sandmeyer in ref. 2 note that humans rarely consume diethyl ether because of its irritating effect on mucous membranes. However, repeated consumption resulting in 'ether habit' and general debility is known.

Intravenous/intra-arterial injection

Diethyl ether may be injected intravenously as a measure of circulation time when evaluating a patient's cardiac status. Quantities of 0.5 ml mixed with an equal volume of saline are frequently employed.[35] In a case where intra-arterial injection was made in error there was immediate burning pain in arms and hands. Over the following 24 hours the extremity remained painful and oedema and cyanosis developed. There was no thrombosis of the large arteries but the soft tissue pressure occluded the vessels so that ischaemia and gangrene developed.[17,35]

Inhalation

The inhalational TCLo for humans is given as 200 ppm. This level is given by Nelson *et al.*[36] as that producing nasal irritation in a majority of subjects (about ten persons of mixed sexes exposed for 3–5 minutes). 300 ppm was objectionable as a working atmosphere and 100 ppm was the highest concentration which the majority estimated as being satisfactory for eight-hour exposure.

Over-exposure to diethyl ether may cause irritation to the eyes, nose and throat. It may also cause dizziness, drowsiness, unconsciousness and death.[21] Browning[37] writing in 1965 reported that acute toxic effects from its industrial use are very rare and that fatalities are practically unknown.

Browning[37] notes the case of a man employed in perfumery manufacture where diethyl ether was used as an extractant and who developed acute mania and uraemic convulsions. Browning further notes that the most frequent manifestation of less severe acute industrial poisoning has occurred in workers in the smokeless powder industry during World War I in the form of a so-called 'ether-jag'. This condition was observed more often in women than in men and consisted of hysterical singing and weeping, nausea, dizziness, mental confusion and sometimes unconsciousness. Men having attacks early in their employment apparently developed tolerance later. Similar effects have been noted in workers using ether as an extractant in conditions of poor ventilation who also complained of after-effects of the acute attacks, namely nausea, headache, irritability, mental confusion, lack of appetite, vomiting and excessive perspiration.

ACGIH[38] states that concentrations anaesthetic to humans range from 3.6–6.5 volumes percent in air; from 7–10 for respiratory arrest and that concentrations greater than ten are fatal. Thienes and Haley[34] note that for anaesthesia in man a blood concentration of 130 mg/100 ml is required.

Martindale[17] notes that diethyl ether has an irritant action on the mucous membrane of the respiratory tract, that it stimulates salivation and increases bronchial secretion and that laryngeal spasm may occur. It causes vasodilation which may lead to a severe fall in blood pressure, it reduces blood flow to the kidneys and increases capillary bleeding. The bleeding time is unchanged but the prothrombin time may be prolonged.

Leucocytosis occurs after ether anaesthesia and convulsions occasionally occur in children or young adults under deep ether anaesthesia. Recovery is slow from prolonged anaesthesia and postoperative vomiting commonly occurs. Acute overdosage of ether is characterized by respiratory failure followed by cardiac arrest.[17]

Patty in ref. 2 states that diethyl ether has been used extensively as a general anaesthetic with medical safety and that few reports of death [due to over-exposure] appear in the literature. Ref. 39 notes that unskilled persons with rudimentary equipment have been able to use it safely. However, Thienes and Haley[34] writing in 1972 state that far too many deaths occur during the induction stage of anaesthesia. Excessive respiratory movements deplete the blood of carbon dioxide and supply the blood with ether at a rapid rate. These combined effects account for the apnoea which follows the period of struggling. The circulatory apparatus is not depressed at this stage, usually the heart beat is rapid and full and the blood pressure elevated. In individuals with weakened hearts, acute decompensation may occur from the cardiac strain, and weakened cerebral blood vessels may rupture. Even in patients with normal heart and blood vessels the apnoea may be so prolonged as to result in asphyxial circulatory failure. Thienes and Haley continue that the irritation of the respiratory mucosa may cause reflex inhibition of respiration and, in rare instances, reflex inhibition of the heart. They also note that several authorities think that such a reflex has been responsible for ventricular fibrillation during induction.

Diethyl ether can induce the formation of hepatic enzymes that accelerate drug metabolism.[39] Martin[40] in a listing of some potentially lethal drug combinations, lists both neomycin and propanolol (Inderal) as interactants with diethyl ether.

Price and Dripps in ref. 39 note that it has been found that, in a significant number of cases, the function of a normal liver is mildly impaired after operations performed under diethyl ether but that where there was abnormal liver function prior to the operation, untoward hepatic effects were more frequent, more intense, longer in duration and were directly related to the degree of initial dysfunction. It was concluded that the postoperatively observed changes probably represented a part of the total response of the organism to stress rather than a direct toxic effect of diethyl ether[39] Dykes[41] states that the exact contribution that diethyl ether makes to the incidence of post-operative hepatic dysfunction in man remains unknown.

Light diethyl ether anaesthesia only slightly influences uterine activity and may be employed intermittently during parturition to coincide with periods of pain.[39]

Price and Dripps in ref. 39 note that although diethyl ether causes muscular relaxation by means of actions on the CNS, an additional effect due to neuromuscular blockade is observed at high concentrations. The blockade by diethyl ether is unlike that caused by d-tubocurarine but the effects of the latter upon the neuromuscular junction are increased in the presence of diethyl ether and the dose of the blocking agent should be reduced accordingly. Diethyl ether also appears to increase the neuromuscular blockade caused by antibiotics such as neomycin, streptomycin, polymyxin and kanamycin.

ECDIN[14] recommends that before working with the solvent, workers should be medically examined to ensure that they do not present any pre-existing condition that subsequent exposure might aggravate and that particular attention should be paid to CNS, skin and kidneys.

Skin contact

ACGIH[38] notes that diethyl ether has no deleterious effect on the skin provided contact is of short duration. However repeated [or prolonged] exposure causes drying and cracking due to extraction of oils. ECDIN[14] describes [prolonged or repeated] contact with the liquid as possibly producing a dry, scaly, fissured dermatitis.

Diethyl ether is used as a surface antiseptic and cleaning agent and is sometimes incorporated in rubefacient liniments.[19]

Eye contact

Grant[42] notes that diethyl ether causes a transitory smarting sensation if splashed in the eye or if the eye is subjected to a high vapour concentration, though momentary exposure does not generally cause injury. Prolonged exposure of the cornea to high concentration such as may be employed in general anaesthesia causes superficial epithelial injury from which recovery is usually prompt. It appears that corneal opacification from ether injury during general anaesthesia has occurred rarely but slight injury by ether vapour appears to render the corneal epithelium abnormally vulnerable to mechanical injury and to injury from substances such as alcohol.

18.4 Animal Acute Toxicity[2,34,37,42–47]

Browning[37] notes that the lethal concentration for dogs ranged from 670,000 to 800,000 ppm. For mice, in which species also has been found a high percentage of delayed deaths for which no pathological cause could be ascertained, lethal concentrations were 180,000 to 420,000 ppm, according to duration. Inefficient mixing of diethyl ether vapour and air may have caused dosage errors in some experiments.

A study has shown that, in the dog, anaesthesia occurs at a blood concentration of 110 mg diethyl ether per 100 ml blood and that respiration ceases at 185 mg/100 ml. As given in Section 18.3, anaesthesia in man requires a higher concentration of diethyl ether, *viz.* 130 mg/100 ml.[34]

Alexander in ref. 43 (on veterinary anaesthesia) notes that diethyl ether is a profound myocardial depressant. It does not predispose the heart to cardiac arrhythmias, so allowing moderate doses of epinephrine and norepinephrine to be administered during its inhalation without fear of precipitating ventricular arrhythmias. An atrioventricular nodal rhythm is not uncommon during more profound depths of anaesthesia.

The same ref. 43 notes that, during light diethyl ether anaesthesia, the metarterioles have an enhanced sensitivity to topically applied epinephrine and capillary vasomotion is increased. These changes apparently are not deleterious but vascular reactivity and vasomotion are progressively depressed during moderate and deep anaesthesia; arteriolar vasodilation occurs and the blood flow is slowed in the capillary bed. Recovery of the micro-circulation to its normal active state after cessation of diethyl ether administration is slow. Like other general anaesthetics, diethyl ether inhibits homeostatic peripheral vascular mechanisms and tends to decrease the resistance to haemorrhage or shock. Alexander further notes that diethyl ether anaesthesia produces alterations of pituitary and adrenal function, carbohydrate metabolism and acid–base balance and that these changes are most pronounced in the dog. Diethyl ether also depresses the hepatic and renal functions.

Alexander[43] observes that there is evidence that in the cat at least, as well as in man, normal diethyl ether anaesthesia produces muscular relaxation by a depression of the CNS rather than by a blockade of neuromuscular transmission. Diethyl ether also increases the intensity and duration of action of both d-tubocurarine and succinylcholine in the cat although intensification in the case of succinylcholine only occurred in 50% of humans studied. Alexander notes that, for this reason, the dosage of neuromuscular blocking agents should be determined with care during diethyl ether anaesthesia.

Clarke and Clarke[44] note that fatalities [in animal anaesthesia] are rare. Albuminuria, oliguria or even anuria may occur as a temporary abnormality resulting from the irritant action of the drug on the kidneys (the effect being most pronounced in dogs over four years old according to a 1920 report).

Browning[37] notes that after preliminary irritation, narcosis sets in rapidly. With lethal concentrations, death is due to respiratory paralysis. It has been stated that during anaesthesia there occurs acidosis with a compensating tendency to alkalosis during recovery.[37]

Kirwin and Sandmeyer in ref. 2 note reports of lethal concentrations for various animal species experiencing short-term exposures to diethyl ether by inhalation. For mice the lethal concentration was 133.4 mg/l [approx. 44,000 ppm] continuous for 97 minutes. Other results for the mouse indicate an LC_{50} of 127.4 mg/l [approx. 42,000 ppm] over three hours.[2,45] Data given in RTECS include[46]:

oral – rat	LD_{50}	1,215 mg/kg
inhalation – rat	LC_{50}	73,000 ppm for 2.5 hours
inhalation – mouse	LC_{50}	65,000 ppm for 1.5 hours
inhalation – dog	LCLo	76,000 ppm
inhalation – rabbit	LCLo	106,000 ppm

A study has shown that 32,000 ppm diethyl ether produced excitation and anaesthesia in mice. 64,000 ppm produced deep anaesthesia and 128,000 caused respiratory arrest. On removal from the anaesthetic atmosphere, spontaneous respiration occurred and normal breathing resumed.[2]

Another study evaluated the age-related susceptibility of adult and neonatal rats to diethyl ether. The median time of death was found to be 5 to 6.5 times greater for neonates than for adult rats and the mean concentration of diethyl ether in blood was 2.5–3 times greater in neonatal rats than in adult rats.[2] It is also stated that animals suffering from diethyl ether overdosage may be resuscitated with positive pressure respiratory assistance.[47] The oral LD_{50} in 14-day old, in young adult and in adult rats respectively was found to be 2.2, 2.4 and 1.7 ml per kg.

Diethyl ether in a standardised test on rabbits' eyes has caused slight reversible injury, graded two on a scale of one to ten.[42]

RTECS gives the following data:

skin   – rabbit: 360 mg – open Draize test – mild irritation found.

eye   – rabbit: 100 mg – moderate irritation.

skin   – guinea pig: 50 mg over 24 hours – severe irritation[46].

18.5 Human Sub-acute and Chronic Toxicity Including Reproductive Effects[4,6,36–38]

General

Ref. 6 notes that chronic exposure in some persons can lead to anorexia, exhaustion, headache, drowsiness, dizziness, excitation and mental disturbances. Kidney injury has been cited but this has been disputed. Repeated inhalation or swallowing may lead to "ether habit" with symptoms resembling chronic alcoholism. OSHA[4] adds jaundice, liver damage and increased susceptibility to ethanol; that psychic disturbances have been reported and that tolerance may be acquired. A slight increase in miscarriage rate has been reported in women exposed to various solvents, including diethyl ether.[6] Browning[37] notes that ether addiction through 'sniffing' had the result of either acute intoxication or, following prolonged abuse, a psychosis similar to chronic alcoholism. Albuminuria may result from diethyl ether and also polycythaemia in the blood, particularly in women.

It has been noted in Section 18.1 that a study[36] had elicited complaints of nasal irritation beginning at 200 ppm and that 300 ppm represented an objectionable working atmosphere. ACGIH[38] notes this together with other reports, as follows.

One report estimated that, at a concentration of 400 ppm, a man of average weight would absorb a maximum of 1.25 g and that the concentration in the blood would be about 0.018 g/l; this concentration being not associated with any signs of intoxication. The report continued that the inhalation of 2,000 ppm, if continued to equilibrium, would result in the absorption of approximately 6.25 g of diethyl ether and a blood concentration of 0.09 g/l, which would cause dizziness in some persons. Another report stated that unsatisfactory conditions will exist only if the concentration in air is greater than 500 ppm. A third report quoted by ACGIH stated that industrial exposures at 500–1,000 ppm or more [*sic*] did not result in demonstrable injury to health but a limit of 500 ppm seemed justifiable to avoid irritation and complaint.[38]

ACGIH[38] state that in view of the foregoing data and the fact that persons exposed experimentally did not have the opportunity to develop the tolerance observed in workers, a TWA-TLV of 400 ppm and a STEL of 500 ppm are suggested, and that regular exposure at these concentrations should cause no demonstrable injury to health nor produce irritation nor signs of narcosis among workers.[38]

OSHA[4] states that there are no data available on the effects of chronic eye contact in humans and that chronic poisoning by ingestion is very unlikely.

Carcinogenicity to Man

No reports on diethyl ether showing carcinogenicity to man were found.

Reproductive Effects in Man

A number of literature references on reproductive observations in anaesthetists *etc.* give no attributability to diethyl ether rather than other materials, neither is there evidence from other studies of adverse reproductive effects in man. The widespread use of ether anaesthesia during pregnancy has not shown a significant adverse effect, though the foetus is well known to be narcotized by maternal ether anaesthesia.

18.6 Animal Sub-acute and Chronic Toxicity Including Reproductive Effects[41,48]

General

Dykes in ref. 41 (published 1970) notes that experimental studies have shown that diethyl ether inhalation can produce fatty changes and extensive degeneration in the liver of the dog. Another study found that the maximum hepatic injury produced by oral administration to mice was minimal fatty degeneration. [Dykes quotes references but no data.] Dykes states that these animal studies place diethyl ether in a more favourable position than chloroform and halothane with regard to hepatotoxic potential.

Carcinogenicity to Animals

No reports of diethyl ether showing carcinogenicity in animals were found.

Reproductive Effects in Animals

Ref. 48 abstracts a presentation alleging embryotoxicity and foetal malformations of mice and, to a lesser extent, rats, due to maternal diethyl ether anaesthesia for one hour during early or late embryogenesis. [The information given is inadequate for proper assessment, but concludes that ether anaesthesia is not highly teratogenic to mice or rats.] No other published reports indicating adverse reproductive effects in animals have been found.

18.7 Mutagenicity[49]

From the report of an evaluation of a DNA polymerase-deficient mutant of *E. coli* for the rapid detection of carcinogens,[49] a positive result was obtained only for 'aged' diethyl ether (freshly distilled material having negative effect). It was considered that the positive result obtained was due to peroxide contamination. No other report has been found showing reliable evidence of mutagenicity in non-peroxidised diethyl ether.

18.8 Summary

[Diethyl ether is a volatile narcotic of moderate acute oral and inhalational toxicity, but low dermal toxicity. It is mildly irritating to skin by degreasing if contact is prolonged or repeated, moderately irritating to eyes as liquid or high vapour concentrations, and irritating to nose and throat at high vapour concentrations. Ether 'sniffing' may be addictive, but otherwise repeated or long-term over-exposure is unlikely to cause significant effects not found with acute over-exposure. No evidence of carcinogenicity, or of mutagenicity in the absence of peroxidation on storage, has been found. No authenticated evidence of adverse reproductive effects of over-exposure to diethyl ether alone have been found, though the foetus *in utero* is well known to be narcotized by maternal ether anaesthesia. The compound is largely excreted unchanged, though a small proportion is oxidized metabolically.]

19. MEDICAL / HEALTH SURVEILLANCE

A decision on the need for, and content of, medical surveillance should be based on an assessment of the possibility and extent of exposure in the work operation. In addition, medical examination may be directed to identifying any pre-existing or newly arising condition in the individual workers which might either be aggravated by subsequent exposure, or might confuse any subsequent medical assessment in the event of excessive exposure or an illness not related to exposure. A particular aspect to be considered is the identification of sensitive subjects not adequately protected by the control limit in operation. A professional medical judgement may be required on continuance of employment in the specified process. The following information is relevant in the case of this solvent.

Pre-employment Medical Examination

Employees should be reviewed by clinical history and appropriate physical examination for:

(*a*) skin disease, because diethyl ether contact may cause dermatitis or aggravate existing skin conditions.

(*b*) respiratory impairment, as diethyl ether is an irritant when inhaled.

Periodic Medical Examination

The same considerations apply as for pre-employment medical examination.
Diethyl ether has well known anaesthetic and irritative effects but there are no known delayed effects after a single exposure.

20. OCCUPATIONAL EXPOSURE LIMITS

[The Committee felt that a time-weighted average limit of 400 ppm and a short term limit of 500 ppm were appropriate for diethyl ether.]

REFERENCES

1. Council of Europe. 'Dangerous Chemical Substances and Proposals Concerning their Labelling'. ('The Yellow Book'.) 4th ed. Maisonneuvre, 1978.
2. Clayton, G.D. and F.E. Clayton, Eds. 'Patty's Industrial Hygiene and Toxicology'. 3rd ed. rev. Wiley Interscience, 1982.
3. Sax, N.I. 'Dangerous Properties of Industrial Materials'. 6th ed. Van Nostrand Reinhold, 1984.
4. Occupational Safety & Health Administration. Material safety data sheets from the Occupational Health Services database, OSHA, Washington D.C., 1986.
5. Sax, N.I., Ed.-in-Chief. Dangerous Properties of Industrial Materials. Report 1, No.6(1981):54. 'Ethyl ether'. Van Nostrand Reinhold.
6. The Royal Society of Chemistry. Laboratory Hazards Data Sheet No.40: 'Diethyl ether', 1985.
7. Graedel, T.E. 'Chemical Compounds in the Atmosphere'. Academic Press, 1978.
8. Marsden, C. and S. Mann, Eds. 'Solvents Guide'. 2nd ed. rev. Cleaver-Hume Press, 1963.
9. Lowenheim, F.A. and M.K. Moran. 'Faith, Keyes and Clark's Industrial Chemicals'. 4th ed. John Wiley & Sons, 1975.
10. Grayson, M., Exec. Ed. 'Kirk-Othmer Encyclopedia of Chemical Technology'. 3rd ed. John Wiley & Sons, 1979.
11. Grasselli, J.G. and W.M. Ritchey, Eds. 'Atlas of Spectral Data and Physical Constants for Organic Compounds'. 2nd ed. CRC Press, 1975.
12. National Institute for Occupational Safety & Health. 'NIOSH Manual of Analytical Methods'. 3rd ed. DHHS (NIOSH) Publication No.84-100. US Department of Health & Human Services, 1984.
13. Pellizzari, E.D. *Environmental Science & Technology*, **16**, No.11(1982):781.
14. Commission of the European Communities. 'Environmental Chemicals Data and Information Network (ECDIN)' Databank. Commission of the European Communities Joint Research Centre, ECDIN Group, I-21020 Ispra (Varese), Italy, 1986.
15. *Economista*, **101**, No.5011(1986):35. 'Green light for alcohol and ether as fuels in the EEC'. (Abstract).
16. Bretherick, L. 'Handbook of Reactive Chemical Hazards'. 3rd ed. Butterworths, 1985.
17. Reynolds, J.E.F. and A.B. Prasad, Eds. 'Martindale, The Extra Pharmacopoeia'. 28th ed. The Pharmaceutical Press, 1982.
18. Goodman, L.S. and A. Gilman, Eds. 'The Pharmacological Basis of Therapeutics'. 5th ed. MacMillan, 1975.
19. Osol, A., Ed. 'Remington's Pharmaceutical Sciences'. 16th ed. Mack Publishing Co., 1980.
20. Sax, N.I., Ed. 'Hazardous Chemicals Information Annual', No.1. Van Nostrand Reinhold Information Services, New York, 1986.
21. National Institute for Occupational Safety & Health/Occupational Safety & Health Administration. 'NIOSH/OSHA Occupational health guideline for ethyl ether'. DHHS (NIOSH) Publication. US Department of Health & Human Services/US Department of Labor, 1978.
22. Bretherick, L., Ed. 'Hazards in the Chemical Laboratory'. 4th ed. Royal Society of Chemistry, UK, 1986.
23. National Fire Protection Association. 'Fire Protection Guide on Hazardous Materials'. 9th ed. NFPA, Massachusetts, USA, 1986.
24. Warren, P.J., Ed. 'Dangerous Chemicals Emergency Spillage Guide'. 1st ed. Jensons (Scientific) Ltd., Leighton Buzzard, UK, 1985.
25. Weast, R.C., Ed.-in-Chief. 'CRC Handbook of Chemistry and Physics' ('The Rubber Handbook'). 67th ed. CRC Press, 1986-1987.
26. Windholz, M. *et al.*, Eds. 'The Merck Index'. 10th ed. Merck & Co., 1983.
27. Horsley, L.H. 'Azeotropic Data III'. Advances in Chemistry Series No.116, American Chemical Society, 1973.

28. Ruth, J.H. *American Industrial Hygiene Association Journal*, 47(1986):A-142. 'Odor thresholds and irritation levels of several chemical substances: a review'.
29. The Chemical Society. 'Foreign compound metabolism in mammals', Vol.3. A Specialist Periodical Report. London, 1975.
30. Astrand, I. *Scandinavian Journal of Work, Environment & Health*, 1(1975):199. 'Uptake of solvents in the blood and tissues of man: a review'.
31. DiPalma, J.R., Ed. 'Drill's Pharmacology in Medicine'. 4th ed. McGraw-Hill, 1971.
32. Chengelis, C.P. and R.A. Neal. *Biochemical Pharmacology*, 29(1980):247. Short communications: 'Microsomal metabolism of diethyl ether'.
33. Geddes, I.C. *British Journal of Anaesthesia*, **44,** No.9(1972):953. 'Metabolism of volatile anaesthetics'.
34. Thienes, C.H. and T.J. Haley. 'Clinical Toxicology'. 5th ed. Lea & Febiger, 1972.
35. Gosselin, R.E. *et al.* 'Clinical Toxicology of Commercial Products'. 4th ed. Williams and Wilkins, 1976.
36. Nelson, K.W. . Journal of Industrial Hygiene & Toxicology, 25(1943):282. 'Sensory response to certain industrial solvent vapors'.
37. Browning, E. 'Toxicity and Metabolism of Industrial Solvents'. Elsevier, 1965.
38. American Conference of Governmental Industrial Hygienists. 'Documentation of the threshold limit values for chemical substances in the workroom environment'. ACGIH, 1986.
39. Goodman, L.S. and A. Gilman, Eds. 'The Pharmacological Basis of Therapeutics'. 4th ed. The Macmillan Company, 1970.
40. Martin, E.W. 'Hazards of Medication'. 2nd ed. J.B. Lippincott, Philadelphia and Toronto, USA, 1978.
41. Dykes, M.H.M. *International Anesthesiology Clinics*, **8,** No.2(1970):241. 'Hepatotoxicity of anesthetic agents'.
42. Grant, W.M. 'Toxicology of the Eye'. 2nd ed. Charles C. Thomas, Illinois, USA, 1974.
43. Alexander, S.C. 'The Pharmacodynamics of Diethyl Ether'. In 'Textbook of Veterinary Anesthesia' (1971):62. Ed. L.R. Soma. Williams and Wilkins, Baltimore.
44. Clarke, E.G.C. and M.L. Clarke. 'Garner's Veterinary Toxicology'. 3rd ed. rev. Bailliere Tindall & Cassell, London, 1967.
45. Spector, W.S., Ed. 'Handbook of Toxicology', Vol.I. 'Acute toxicities of solids, liquids and gases to laboratory animals'. W.B. Saunders, Philadelphia, USA, 1956.
46. National Institute for Occupational Safety & Health. 'Registry of Toxic Effects of Chemical Substances (RTECS)'. DHHS (NIOSH) Publication No.84-101-6. US Department of Health & Human Services, April, 1986.
47. Sancilio, L.F. *et al. Journal of Pharmaceutical Sciences*, **57,** No.7(1968):1248. 'Simple method for resuscitating rats from ether overdosage'.
48. Schwetz, B.A. and B.A. Becker. *Toxicology & Applied Pharmacology*, 17(1970):275. 'Embryotoxicity and fetal malformations of rats and mice due to maternally administered ether'. (Abstract.)
49. Fluck, E.R. et al. Chemico-Biological Interactions, 15(1976):219. 'Evaluation of a DNA polymerase-deficient mutant of *E. coli* for the rapid detection of carcinogens'.

CONTENTS

1. CHEMICAL ABSTRACTS NAME

1,4-dioxane
(10th Collective Index)

2. SYNONYMS AND TRADE NAMES[1-7]

diethylene dioxide
diethylene-1,4-dioxide
1,4-diethylene dioxide
diethylene ether
diethylene oxide
di(ethylene oxide)
dioksan (Polish)
diossano-1,4 (Italian)
1,4-diossano (Italian)
dioxaan-1,4 (Dutch)
1,4-dioxaan (Dutch)
1,4-dioxacyclohexane
dioxan
dioxan-4 (German)
1,4-dioxan (German)
p-dioxan (Czech)
para-dioxan
dioxane
p-dioxane
para-dioxane
dioxanne (French)
1,4-dioxanne (French)
dioxyethylene ether
glycol ethylene ether
NCI-C03689
OHS26970
tetrahydro-1,4-dioxin
tetrahydro-*p*-dioxin
tetrahydro-*para*-dioxin
U108
UN 1165

3. CHEMICAL ABSTRACTS SERVICES REGISTRY NUMBER

123-91-1

4. NIOSH NUMBER

JG8225000

5. CHEMICAL FORMULA

$C_4H_8O_2$
(Molecular weight 88.12)

6. STRUCTURAL FORMULA

$$H_2C \overset{O}{\diagdown} CH_2$$
$$H_2C \diagdown_{O} CH_2$$

7. OCCURRENCE[1,8,9]

Graedel[8] notes that no data were available regarding concentrations of 1,4-dioxane in the atmosphere. The presence of any 1,4-dioxane would be due to its use as a solvent. In sampling of 14 water-treatment plants in Great Britain, 1,4-dioxane was detected as a micro-pollutant at one of them (the raw water here having been river water).[9] ECETOC[1] found no data on environmental concentrations in the available literature but points out that, in view of the relatively low tonnage produced and the rapid dilution promoted by its volatility and water solubility, very low environmental concentrations may be expected.

ECETOC,[1] commenting on biodegradability, stated that a definite conclusion could not then [1983] be drawn because of data variability and the limited number of results. The limited data indicated that less than 30% of 1,4-dioxane biodegrades in 20 days, thus indicating that the material is not easily biodegradable. The high water-solubility would seem to exclude the possibility of bioaccumulation.

Studies on the phototransformation in air of 1,4-dioxane indicate a half-life of 14 hours (diurnal time) at 25 °C.[1]

ECETOC[1] quote LC_{50}s (96 hr) of 10,000 ppm for freshwater fish and 6,700 ppm for saltwater fish and comment that these high values indicate a virtual absence of acute toxicity to fish at any likely environmental concentrations.

8. COMMERCIAL AND INDUSTRIAL 1,4-DIOXANE[1,2,10]

Hawley[10] notes the availability of reagent, technical, spectrophotometric and scintillation grades of 1,4-dioxane in the USA. IARC[2] states that the technical grade is 99.9% pure and gives the specification for a typical commercial product as:

Peroxides (as H$_2$O$_2$ 50 mg/kg max.
Acidity (as acetic acid)   0.01% by weight max.
Water   0.1% max.
2-Methyl-1,3-dioxolane   0.05% max.
Non-volatile matter   0.0025% max.
Substantially free from suspended matter.

Specifications quoted by IARC for Japanese material are:

Purity   99.99%
Boiling range   101–102 °C
Freezing point   11.7 °C
Density d$^{20}_4$ 1.0333
Refractive index n^{20} 1.4224
Water is present [unstated amount] as an impurity.

ECETOC[1] states that technical grade material has, typically, a purity of 99.5%, a water content of 0.1% and a peroxides content of 10 ppm.

9. SPECTROSCOPIC DATA[11]

Infrared, Raman, ultraviolet, both ^{1}H and ^{13}C NMR, and mass spectral data for 1,4-dioxane have been tabulated.[11]

10. MEASUREMENT TECHNIQUES[12–14]

The method recommended for the determination of 1,4-dioxane in workshop atmospheres is the NIOSH[12] activated charcoal trapping method for collection and concentration, followed by solvent extraction of the charcoal and a gas chromatographic (gc) analysis of the extract.

The collection tube is 7 cm long with a 4 mm internal diameter. It contains a total of 150 mg of activated charcoal, (20–40 mesh) divided into a front section of 100 mg and a rear section of 50 mg separated by a small plug of urethane foam. Air is sampled at a flow-rate of 0.01–0.2 l per minute for a total sample of 0.5–15 l by means of a small pump. The entire apparatus is portable so that it may be carried in a pocket with the sampling tube in the breathing zone (normally a coat lapel). The apparatus can also be static.

Carbon disulphide is used to extract the 1,4-dioxane, which is separated on a 6 m × 3 mm stainless steel column packed with 10% FFAP on Chromosorb W-HP (80–100 mesh) using flame ionization detection. For full details of the method see ref. 12.

The sampling method uses a small portable apparatus with no liquids and is therefore easy to handle and maintain. The analytical method is relatively specific with the additional advantage of flexibility of operating conditions.

Air samples are normally taken by means of suction using an electric pump capable of operating at low flow-rates (*i.e.*, 10–300 ml per minute). If higher flow-rates are used, breakthrough can occur (*i.e.*, the 1,4-dioxane is not completely adsorbed and some of it passes through the adsorbent). It is also convenient either to establish the trapping device in a static position, or to have it attached to the user in such a way that the breathing zone is sampled (this usually means attachment to the worker's lapel).

It should be noted that suitable personal samplers may be commercially available.

There appears to be little interest in the determination of 1,4-dioxane in air. It has been trapped on activated charcoal contained in personal badges and extracted with carbon disulphide and the infrared spectrum of the extract between 500 and 4,000 cm^{-1} has been used for quantitation.[13] This is a relatively non-specific method and was used to assess the recovery from the badges of 1,4-dioxane when present at 50 ppm in air.

There also appears to be little interest in the determination of 1,4-dioxane in body tissues. Blood or urine (4–5 mg) was heated in a sealed tube at 120 °C for two minutes and the volatiles, including 1,4-dioxane and ethyl acetate, were analysed by gc, the limit of detection being 0.01–0.02 μg.[14]

11. CONDITIONS UNDER WHICH 1,4-DIOXANE IS PUT ON THE MARKET

Production[1,2,15–18]

ECETOC[1] names the dehydration of diethylene glycol under acid conditions as the most important method for the manufacture of 1,4-dioxane; less frequently used is ethylene glycol.

$$HOC_2H_4OC_2H_4OH \quad \rightarrow \quad O{:}(CH_2CH_2)_2{:}O + H_2O$$

or

$$2HOC_2H_4OH \quad \rightarrow \quad O{:}(CH_2CH_2)_2{:}O + 2H_2O$$

The reaction is carried out at around 160 °C and at a pressure of 250–110 mbar. The most commonly used catalyst is sulphuric acid. Others used are phosphoric acid, toluene sulphonic acid, sodium hydrogen sulphate and strongly acidic ion-exchange resins.

Catalytic dimerization of ethylene oxide in the vapour phase is also used for the manufacture of 1,4-dioxane as is the dehydrochlorination of bis(2-chloroethyl) ether with strong, aqueous sodium hydroxide.[15]

$$2(CH_2CH_2){:}O \quad \rightarrow \quad O{:}(CH_2CH_2)_2{:}O$$

and

$$ClC_2H_4OC_2H_4Cl + 2NaOH \quad \rightarrow \quad O{:}(CH_2CH_2)_2{:}O + 2NaCl + H_2O$$

2-Chloroethyl-2′-hydroxyethyl ether has been used instead of bis-(2-chloroethyl) ether.[16]

$$ClC_2H_4OC_2H_4OH + NaOH \quad \rightarrow \quad O:(CH_2CH_2)_2:O + NaCl + H_2O$$

Commercial production of 1,4-dioxane in the USA was first reported in 1951 but semi-commercial quantities were made available in 1929.[17] US production in 1972 was 6.3 million kilograms, in 1973 it was 7.4 million kilograms and in 1982 was 6.8 million kilograms.[17]

In Japan, 600 thousand kilograms were produced in 1968 and 2.3 million kilograms in 1973. Japan exported 60–70 thousand kilograms in 1974 and about 100 thousand kilograms in 1975, chiefly to the UK and Australia.[2] Ref. 18 indicates Japanese demand being of the order of 10 million kilograms per year in 1984, growing by as much as 10% per annum.

ECETOC[1] states that Western European demand for 1,4-dioxane was [in 1983] estimated as totalling 1,500 metric tons [1.5 million kilograms] per annum [*sic*], with industrial consumption having been constant over the previous few years. ECDIN[15] gives EC production and consumption in 1984 both as being 70 kilotonnes [70 million kilograms] [*sic*].

Uses[1,2,4,10,15–17]

1,4-Dioxane is used as a solvent for cellulosics and a wide range of organic products, lacquers, paints, varnishes, paint and varnish removers, wetting and dispersing agents in textile processing, dye baths, stain and printing compositions, cleaning and detergent preparations, cements, cosmetics, deodorants, fumigants, emulsions and polishing compositions.[10] ECETOC[1] states that 1,4-dioxane is not an important solvent in the chemical industry itself but finds its use mainly outside. It is an excellent stabilizer for a number of chlorinated solvents and can protect active aluminium or aluminium-alloy surfaces.

Other uses include that of solvent in the purification of organic compounds; in molecular weight determinations and in the radio-immunoassay of glucogen after extraction of blood with 1,4-dioxane.[16] Ref. 17 notes that in the USA, about 90% of the then current [1985] production was used as a stabilizer for chlorinated solvents, particularly 1,1,1-trichloroethane, to neutralize small amounts of hydrogen chloride formed during its solvent use, and to deactivate metal surfaces and remove or complex trace amounts of metal chloride salts that may form.

IARC[2] stated that in Japan [in 1976] the major uses then were as a solvent, as a surface-treating agent for artificial leather and as a stabilizer for trichloroethylene.

When perfectly dry 1,4-dioxane can remain stable indefinitely. However, its hygroscopic nature and its ether linkages cause peroxide formation and other degradation products to form upon standing in the presence of moisture.[4]

1,4-Dioxane may be chlorinated to give 2,3-dichlorodioxane along with other chlorodioxanes.[15] It is used as an adjunct in the preparation of cyclophosphamide from the reaction of N,N-2-chloroethylphosphamide dichloride with propanol-amine in the presence of trimethylamine and dioxane.[15]

12. STORAGE, HANDLING AND USE PRECAUTIONS[19–23]

Dry 1,4-dioxane is non-corrosive, and iron, mild steel, copper or aluminium are suitable for plant and containers.[19] It should be stored away from heat and sunlight in tightly closed containers in dry, well ventilated areas. Containers should be bonded and earthed when the liquid is transferred and they should be periodically checked for leakage.[20] 1,4-Dioxane has a boiling point of 101.1 °C and is very flammable. Exposure to heat, sparks, open flames, moisture and strong oxidizers should be prevented. Peroxides, which may form during storage, especially under the influence of light, are explosive and must be destroyed chemically before distillation or evaporation in order to avoid the danger from their being concentrated in the residues.[19] As 1,4-dioxane is water-miscible, treatment with aqueous reductants such as ferrous sulphate or sodium sulphide, *etc.* may be impractical. However, the peroxides may be removed under anhydrous conditions by passage through a column of activated alumina which also removes water. The column must be freed from adsorbed peroxides by elution with methanol or water before the column-packing is discarded.[21]

[As with other flammable solvents the organization of storage facilities has to take into account two separate requirements. These are that the stored material be protected from a fire elsewhere on the premises and that the premises be protected from a fire involving the stored materials. On an industrial scale, sufficient physical separation and water spray curtains are often used effectively. On a laboratory scale, flammable solvents need to be stored in fire resistant, thermally insulated cupboards.

It is important to appreciate that a non-combustible barrier is not synonymous with a fire barrier. A fire barrier has to protect against both conducted and radiant heat. For example, a metal cupboard is useless on both these counts.] Stored 1,4-dioxane should be isolated from oxidizing materials and should not present a toxic hazard to personnel.

NIOSH[20] name the use of completely enclosed processes as the recommended method for control of 1,4-dioxane. Local exhaust ventilation used alone or in combination with process enclosure may also be effective [in keeping personnel exposures from being above the recommended limits] but should be designed to prevent the accumulation or recirculation of 1,4-dioxane in the workroom (and to prevent its re-entry *via* air intakes), to remove it from the breathing zones of users and to maintain vapour concentration below the recommended ceiling value.[20] Systems and materials should be adequately maintained and all relevant pollution regulations adhered to.

In addition, forced-draught ventilation systems should be equipped with remote manual controls. They should turn off automatically in the event of a fire in the work area.[20] Buildings in which dioxane is used and where it could form an explosive mixture with air should be explosion-proof. Explosion vents can be installed at appropriate positions, stair enclosures should be fire resistant and have self-closing fire doors.[20]

Valve protection covers should be in place when containers either are being moved or are disconnected and not in use. Containers should be moved only with proper equipment and in a secure manner. They should be handled and opened with care and with adequate ventilation and protective clothing in use.[20]

Process valves and pumps should be readily accessible. They should not be located in areas which are congested or of difficult access, nor in pits.[20] NIOSH[20] recommended that suitable protective clothing shall be worn by those coming into direct contact with liquid 1,4-dioxane. Clothing shall be impervious to penetration and resistant to degradation by the chemical. Neoprene-coated gloves, boots, overshoes and bib-type aprons that cover boot tops shall be provided where necessary. Persons entering confined space, e.g., pits and tanks, that are not known to be safe should wear impervious, supplied-air hoods or suits. Where heat-stress is likely to occur, air-supplied suits, preferably cooled, are recommended. (Such workers should be monitored from outside the hazard area by more than one person equipped for rescue). For eye protection, chemical-type goggles, safety glasses with splash shields or plastic face shields (eight inch minimum) made completely of 1,4-dioxane-resistant material should be used. Suitable eye wash apparatus and safety showers should be available.[20] [Note however that personnel should not be allowed to enter areas containing potentially hazardous concentrations of 1,4-dioxane unless all relevant legislation and official guidelines have been strictly adhered to. Suitable respiratory protective equipment should be used.]

Odour and irritation effects are regarded as not necessarily strong enough to constitute an adequate warning of a hazardous accumulation of vapour.[22] Ruth[23] gives the range of odour threshold levels as 20.16–972 mg/m³ [ca. 5.6–270 ppm], and the irritation concentration as 720 mg/m³ [ca. 200 ppm].

Note: Consideration should be given to informing unaware personnel that 1,4-dioxane is not to be confused with the highly toxic substance, 'dioxin' (2,3,7,8-tetrachlorodibenzo-p-dioxin; TCDD).

N.B. the caveat in Section 13 (final paragraph).

13. FIRE HAZARDS[6,21,24]

1,4-Dioxane is a flammable, water-miscible liquid of boiling point 101.1 °C. It constitutes a dangerous fire and a moderate explosion hazard when exposed to heat or flame. It has a flash point (closed cup) of 12 °C; an auto-ignition temperature of 180 °C and lower and upper explosive limits in air of 2.0% and 22% respectively (see Section 17). The vapour is about three times denser than air and, if in sufficient quantity, can flow a considerable distance to an ignition source and then 'flash back'. 1,4-Dioxane can form peroxides on exposure to air and these can explode if concentrated, e.g., by distillation.

Fire-exposed containers of 1,4-dioxane can explode when heated and, if feasible, should be cooled with flooding amounts of water applied from as far as possible until well after the fire is extinguished. Fires can be fought with alcohol-resistant foam, carbon dioxide, dry chemical or water spray. (Note that as the boiling point of water-dioxane azeotrope is only about 88 °C the addition of water may cause marked ebullition.) Immediate withdrawal is imperative in the event of rising sound from a venting safety device or any discolouration of a storage tank due to the fire.

Personnel should keep upwind of the incident and avoid breathing dioxane vapour and possibly toxic combustion products.

Ref. 24 notes that where a fire is fed by an uncontrolled flow of combustible liquid the decision on how or if to fight it will depend on the size and type of fire anticipated and must be carefully considered. This may call for special engineering judgement, particularly in large-scale applications.

[Notwithstanding the foregoing review of various authorities noted in Sections 12 and 13, it is essential that managers and others responsible for the planning, implementation and overseeing of personnel and plant safety should be familiar with the legal constraints and official guidelines applicable to them and that they liaise with their local emergency services in the planning of plant and storage facilities and in the preparation of contingency plans for dealing with fires and other emergencies. Managers should regularly monitor staff knowledge of, and ability to implement, emergency procedures and should ensure that equipment provided for use in emergencies is regularly inspected and maintained.]

14. HAZARDOUS REACTIONS[6,21]

1,4-Dioxane reacts almost explosively with Raney nickel catalyst above 210 °C.[21]

Decaborane (14) forms impact-sensitive mixtures with 1,4-dioxane.[21]

An explosion occurred after a final addition of perchloric acid was made, with continued heating, to a solution of boron trifluoride in aqueous 1,4-dioxane which had been evaporated with nitric acid (three portions) for analysis.

If refrigerated storage is not possible, the 1:1 addition complex of 1,4-dioxane with sulphur trioxide should only be prepared immediately before use as the complex sometimes decomposes violently on storing at ambient temperature.

The residue from sublimation of the complex of triethynylaluminium with 1,4-dioxane is explosive and should not be dried by heating.

1,4-Dioxane is susceptible to auto-oxidation when exposed to air. The peroxides so formed can explode if concentration (*e.g.*, by distillation) is effected.

An explosive compound may be formed on reaction of silver perchlorate with 1,4-dioxane.[6]

15. EMERGENCY MEASURES IN THE CASE OF ACCIDENTAL SPILLAGE[6,20,25]

All possible sources of ignition should be eliminated and, if possible, any leak should be stopped if this can be done without risk. The aim should be to isolate the material, *e.g.*, by closing doors to the spillage area, and by ventilation *e.g.*, by opening windows. Non-essential personnel should leave the area and those engaged in dealing with the hazard should wear appropriate apparel and breathing equipment (see Section 12).

Small spillages may be collected for reclamation or absorbed in vermiculite, dry sand, earth or similar unreactive material and placed in dioxane resistant, closable or coverable containers for later disposal. (Quantities of up to 1 l can be diluted greatly and disposed of to a sewer.) Larger spillages should be bunded far ahead of the spill for later disposal.

16. FIRST AID

General

The casualty should be removed from danger of further exposure into fresh air. Rescuers should ensure their own safety from inhalation and skin contamination where necessary. Conscious casualties should be asked for information about what has happened. Bear in mind that the casualty may lose consciousness at any time. Where necessary continued observation and care should be ensured. First aiders should take care not to become contaminated.

In view of the pronounced toxicity of this material, and its confirmed absorption through the skin as well as by inhalation and ingestion routes, medical attention should be obtained immediately. No specific antidote is available; treatment is symptomatic and supportive. Gastric lavage, treatment with activated charcoal, emesis and oxygen therapy may be administered by qualified medical personnel.

Inhalation

The conscious casualty should be kept at rest. The unconscious casualty should be placed in the recovery position and an open airway maintained. If breathing or heart beat stop, resuscitation should be commenced immediately. Medical aid should be obtained or the casualty removed to hospital immediately.

Skin Contamination

Remove contaminated clothing (which should be washed before re-use). Wash the skin with copious amounts of water, or soap and water, for at least 15 minutes. If necessary, seek medical aid. In the case of persistent skin irritation refer for medical advice.

Eye Contamination

Irrigate with copious amounts of water for at least ten minutes but avoid further damage to the eye from use of excessive force. Seek medical aid. Remember the possible presence of contact lenses, which may be affected by some solvents and may impede decontamination of the eye.

Ingestion

If the lips or mouth are contaminated rinse thoroughly with water. Do not induce vomiting. Approximately 250 ml of water should be given to dilute the chemical in the stomach. Give further supportive treatment as for inhalation. In the case of ingestion of significant amounts wash out mouth with water. Obtain immediate medical attention. Contact a hospital or poisons centre at once for advice. Note that symptoms may develop many hours after exposure so continued care and observation may be necessary.

In all cases note information on the nature of exposure, and give this to ambulance or medical personnel. Information for doctors is provided in some technical literature issued by manufacturers.

17. PHYSICO-CHEMICAL PROPERTIES

17.1 General[7,10,26,27]

1,4-Dioxane is a colourless, flammable liquid with a mild, ether-like odour.[7,26] Its odour is also described as pleasant.[26]

1,4-Dioxane is miscible with water, ethanol, diethyl ether and most organic solvents.[10,27]

17.2 Melting Point[26]

11.80 °C.

17.3 Boiling Point[26,28]

Merck[26] quotes the boiling point of 1,4-dioxane as

101.1 °C.

1,4-Dioxane forms azeotropes with a number of other compounds. Some binary ones (including two studies on water) quoted by ref. 28 are:

Wt% 1,4-dioxane	Second component	Wt% second component	Boiling pt azeotrope (°C)
44	Heptane	56	91.85
45	n-Propyl alcohol	55	95.3
82	Water	18	87.82
82.4	Water	17.6	87.8

17.4 Density/Specific Gravity[19,29]

Weast[29] quotes the specific gravity of 1,4-dioxane at 20 °C (referred to water at 4 °C) as 1.0337. Marsden and Mann[19] quote the specific gravity at 20 °C (referred to water at 20 °C) as 1.0356.

17.5 Vapour Pressure[29]

Weast[29] tabulates the following vapour pressures of 1,4-dioxane below atmospheric pressure:

Temperature in °C	Vapour pressure in mmHg
− 35.8 (solid)	1
− 1.2 (solid)	10
25.2	40
45.1	100
81.8	400

The critical temperature and critical pressure of 1,4-dioxane are recorded as 314.8 °C, and 51.4 atm [*ca.* 39100 mmHg], respectively.[29]

17.6 Vapour Density[24]

3.0 (relative to air = 1).

17.7 Flash Point[24]

12 °C, closed cup test.

17.8 Explosive Limits[24]

The limits of flammability of 1,4-dioxane in air at normal atmospheric temperature and pressure are given by the National Fire Protection Association[24] as:

Lower limit: 2.0% v/v
Upper limit: 22% v/v.

The NFPA[24] give the auto-ignition temperature of 1,4-dioxane as 180 °C. However, they point out that different test conditions (and also different definitions of "ignition temperature") can result in widely varying values being quoted. They therefore recommend that any value should be treated as an approximation.

17.9 Viscosity[26,30]

The viscosity of 1,4-dioxane is quoted as follows:

0.0131 p at 20 °C.[30]
0.0120 p at 25 °C.[26]

17.10 Concentration Conversion Factors

At 25 °C and 760 Torr (1 atm),

$1 \ mg/m^3 = 0.28 \ ppm$, and
$1 \ ppm = 3.60 \ mg/m^3$.

18. TOXICITY

18.1 General[2,4,7,16,31]

Liquid 1,4-dioxane is painful and irritating to the eyes and is irritating to the skin on prolonged or repeated contact. It can be absorbed through the skin in toxic amounts.[4] The vapour is a mucous membrane irritant. On prolonged over-exposure, 1,4-dioxane is toxic to the liver and particularly to the kidneys.[7,31] Liver necrosis and severe kidney damage were found in workers dying after two months of heavy exposure to 1,4-dioxane vapour. The onset of the poisoning was marked by drowsiness and headache, nausea, vomiting and irritation of the eyes and respiratory passages. In one fatal case there was also brain as well as kidney and liver damage.[7] The mild odour of the solvent is insufficient to provide forewarning of serious or fatal exposure.[4,7]

Inhalation studies in animals have, in addition to liver and kidney damage, produced lung damage, behavioural changes and increased blood counts. Large doses of 1,4-dioxane in drinking water administered to rats and guinea pigs have led to the development of tumours of, variously, nose, liver, lungs and gall-bladder.[2,16]. Chronic inhalation by rats of 1,4-dioxane vapour at 111 ppm failed to produce any signs of toxicity.[16]

18.2 Toxicokinetics[17,32-41]

Santodonato *et al.* in ref. 17 (published 1985) found no specific data in the published literature on the rate and efficiency of absorption of 1,4-dioxane, but considered that absorption by all routes could be inferred from various toxicological studies in laboratory rats and humans. Young *et al.*[32] in studies on humans and on rats report that human volunteers, exposed to 50 ppm 1,4-dioxane for six hours, eliminated it (in the urine), partially metabolized, with a half-life of less than one hour and with no evidence for saturation. Santodonato *et al.*[17] note this indicates that absorption across the respiratory tract occurs and also note (from Young *et al.* – ref. 17) that, based on calculated body burden, respiratory rates and volumes, an absorption efficiency of 100% was estimated for rats inhaling 1,4-dioxane at 50 ppm for six hours.

Work by Woo *et al.*,[33] using ^3H 1,4-dioxane with intraperitoneal injection, showed the solvent or its metabolites to distribute more or less uniformly among the liver, kidney, spleen, lung, colon and skeletal muscle of rats. The authors note that studies of the nature of the dioxane binding showed that the extent of 'covalent' binding (as measured by incorporation into lipid-free, acid-insoluble tissue residues) was significantly higher in the liver (the main carcinogenesis target tissue), spleen and colon than that in other tissues. Investigations of subcellular distribution in the liver indicated that most of the radioactivity was in the cytosol, followed by the microsomal, mitochondrial and nuclear fractions, the binding of dioxane to the macromolecules in the cytosol being mainly noncovalent. The percent covalent binding was highest in the nuclear fraction followed by mitochondrial and microsomal fractions and the whole homogenate. The authors continue that pretreatment of rats with inducers of microsomal mixed-function oxidases had no significant effect on the covalent binding of 1,4-dioxane to the various subcellular fractions of the liver. There was no microsome-catalysed *in vitro* binding of ^{3}H or ^{14}C dioxane to DNA under conditions which brought about substantial binding of ^{3}H benzo(*a*)pyrene.

Santodonato *et al.*[17] note that, based on the rapid rate of elimination of metabolites, accumulation of 1,4-dioxane following low-level exposure seemed unlikely. Work by Young *et al.*[32] highlights the very different situation with massive doses where they find that, regardless of the mechanism for the acute and chronic toxicity of 1,4-dioxane, it is clear that the pharmacokinetics are dramatically different for massive doses. The authors note that whereas a single dose of 10 mg/kg is rapidly metabolized and excreted in the urine of rats, a dose of 1,000 mg/kg saturates the metabolism of 1,4-dioxane to 2-hydroxyethoxyacetic acid [HEAA] [*v.i.*], causing prolonged retention of the dioxane *per se* in the body. In addition, daily administration of 1,4-dioxane at ·massive levels caused further alterations in the pharmacokinetics of the solvent and which may be explained in terms of induction of its metabolism to HEAA and possibly to other metabolites that are not formed at lower doses.

This study on rats by Young *et al.*[32] suggests that there is a metabolic plateau in the plasma concentration of 1,4-dioxane, starting at about 100 µg/ml plasma. A single oral dose of 100 mg/kg was probably above this plateau, with 10 mg/kg below it, while six hour inhalation of 50 ppm in air was also below. Doses giving plasma concentrations below the plateau level are metabolized and eliminated rapidly (in the urine) without toxicological manifestations, while doses of 1,4-dioxane that result in plasma levels above 100 µg/ml are removed from the body progressively more slowly because of saturation of metabolism and, when given chronically, result in pathological alterations. The authors further observe that rats that have been given high dosages and retain 1,4-dioxane in the body because detoxification is saturated, excrete a larger fraction of the dose in the breath, and, when the dose is given repeatedly, evoke metabolizing enzyme induction to survive. Such rats are biochemically changed causing pathological alterations of the liver that alone may be precursors of cancer. The authors continue that the altered fate of 1,4-dioxane given repeatedly at high dose levels creates conditions in the rat that may be related to its toxicity but that do not exist after lower doses. They therefore consider that toxicological studies in rats at dose levels that swamp the detoxification mechanism for dioxane have little relevance to assessing the hazard of small doses encountered in the work environment which are eliminated rapidly. The authors also note that the 100 µg/ml region in the above study may constitute a 'threshold' for the detoxification of the solvent, below which rapid metabolism and elimination will not lead to untoward consequences, and above which the metabolism is saturated. They observe though that even if there is no 'absolute' threshold, the data show that there will be a change in the dose–response curve that makes linear extrapolation inappropriate. (See also Sections 18.5 and 18.6.)

Woo *et al.*[34] identified 1,4-dioxane-2-one (or the ring-opened equivalent,

2-hydroxyethoxyacetic acid (HEAA)) as a single, major urinary metabolite in the urine of rats to which 1,4-dioxane had been administered intraperitoneally. The amount excreted was dose-dependent and time-dependent, reaching a maximum between 20 and 28 hours after the 1,4-dioxane administration. 40–60% of the radioactivity was recovered from the urine within 48 hours when 50 µCi of [14]C 1,4-dioxane together with 300 mg unlabelled 1,4-dioxane/100 g body-weight was administered to rats. Of the ^{14}C labelled compounds present in the urine, 73–86% could be recovered as 1,4-dioxane-2-one (or HEAA) and the rest mainly as 1,4-dioxane. Diethylene glycol also gave the same metabolite when administered to rats, but it was excreted almost entirely within the first 16 hours, suggesting that diethylene glycol may be an intermediate stage in the metabolism of 1,4-dioxane. Other workers are noted in ref. 34 as having identified HEAA as the major metabolite but it is pointed out that such compounds usually exist either as salts or, if in the acid form, as lactones.

Ref. 35 observes that the biotransformation processes for 1,4-dioxane are not well understood and that, (quoting a 1982 publication) no intermediate products have been found. Other studies by Woo *et al.* are cited showing that biotransformation is increased (induced) by phenobarbital and is therefore very probably dependent on cytochrome P450.

Woo *et al.*[34] note that the biological significance of 1,4-dioxane metabolism remains to be elucidated. Preliminary experiments on the acute toxicity of 1,4-dioxane-2-one indicate that it is considerably more toxic than 1,4-dioxane and the possibility that this lactone may be a proximate carcinogen was (1977) under study. Fishbein[36] notes that a reactive free radical or a carbonium ion may arise in the metabolism of 1,4-dioxane and that this may represent a proximate carcinogen.

Fishbein[36] also notes that the metabolism of 1,4-dioxane may be closely related to its toxicity and/or carcinogenicity since acute toxicity studies show that a number of agents that modify the LD_{50} of the chemical also modify its metabolism in the same manner.

Stott *et al.*[37] observe that it is noteworthy that dioxane-induced hepatic carcinogenesis occurs only at those high doses which also induce mixed function oxidase enzyme induction and hepatotoxicity, while at lower doses, neither of these toxicologic/pharmacologic effects nor carcinogenicity occur.

Young *et al.* in ref. 32 state that their data (on rats) could be described by a one-compartment open system model with parallel first-order (urinary and pulmonary excretion) and Michaelis–Menten (metabolism) elimination kinetics. At saturation, the maximum velocity of metabolism of 1,4-dioxane to 2-hydroxyethoxyacetic acid [or 1,4-dioxane-2-one] was about 18 mg/kg/hour. Multiple daily oral doses of 1,000 mg/kg but not of 10 mg/kg were excreted more rapidly than equivalent single doses, indicating that, at high daily doses, dioxane induced its own metabolism.

Young *et al.*[38] observe that the use of the rat as a toxicological model for 1,4-dioxane in humans is supported by the finding that the urine of workers exposed to the chemical contains HEAA which also is the major metabolite of 1,4-dioxane in rats. In ref. 32 Young *et al.* further note that the adequacy of using the rat as a model depends on the pharmacokinetic and metabolic similarities between the rat and humans as well as their relative sensitivity to the toxic effects of 1,4-dioxane. Both species, when exposed to 50 ppm for six hours (see refs. 32 and 39), eliminate dioxane with a half-life of about one hour with no evidence of saturation (see also Section 18.6) and rats which have repeatedly been exposed to such concentrations [and to 111 ppm[40]] have not manifested toxic effects.

Ref. 41, in citing the above mentioned (ref. 39) study on humans by Young *et al*, notes that the authors determined the pharmacokinetics and metabolism of 1,4-dioxane in human volunteers following a single six-hour exposure to 50 ppm 1,4-dioxane vapour. Samples of blood and urine collected during and after the exposure were analysed for 1,4-dioxane and HEAA. The plasma concentration–time curve for 1,4-dioxane and HEAA for humans exposed to 50 ppm 1,4-dioxane vapour for six hours indicates that the dynamics of 1,4-dioxane uptake and elimination in humans could be described by a one-compartment open system with zero-order uptake and first-order elimination. A simulation of repeated daily exposures to 50 ppm 1,4-dioxane for 8 hours/day indicated that 1,4-dioxane would never accumulate in concentrations above those attained after a single eight-hour exposure as long as the exposure concentration of 1,4-dioxane was 50 ppm or less. It is noted that combined pharmacokinetic and chronic toxicity data for 1,4-dioxane in rats indicated that animal toxicity was observed only in situations in which the metabolic elimination of 1,4-dioxane was saturated by the repeated administration of high dose levels. That is, toxicity was associated with non-linear pharmacokinetics.

Because there was no evidence of non-linear pharmacokinetics in the disposition of 1,4-dioxane in humans exposed to 50 ppm 1,4-dioxane vapour, and this exposure does not exceed the metabolic threshold, the authors concluded that exposure of humans to 1,4-dioxane at a concentration of 50 ppm would not cause adverse effects, even if exposure was on a repeated basis.

18.3 Human Acute Toxicity[1,4,39,42,43]

Ingestion

No data on human ingestion of 1,4-dioxane were found.

Inhalation

Yant *et al.*[42] report a study in which five persons were exposed to air containing 1,600 ppm of 1,4-dioxane vapour for a period of ten minutes. They noted slight irritation of the nose and throat. The alcohol-like odour of 1,4-dioxane was easily noticeable at first, but decreased in intensity with continued exposure. Nasal irritation persisted throughout the test. No vertigo was noted. One person complained of "upset stomach" after exposure. The atmosphere was not intolerable but easily noticeable. An exposure to 5,500 ppm for one minute produced a burning sensation in the nose and throat in the same five persons, three of whom noticed a slight vertigo which disappeared quickly on cessation of exposure.

A study on 12 volunteers exposed for 15 minutes to various concentrations of

1,4-dioxane found that 200 ppm was the highest acceptable one. At 300 ppm, irritation of eyes, nose and throat was observed.[1,4]

When volunteers were exposed to 50 ppm 1,4-dioxane in air for 6 hours, a total of 5.4 ± 1.1 mg/kgbw was absorbed, the maximum amount in the body at any one time being 1.2 ± 0.2 mg/kgbw. The only toxic effect reported was eye irritation.[39]

Skin contact

The liquid is irritating to the skin with prolonged or repeated contact.[4]

A study, *in vitro*, using excised human skin, showed that up to 3.2% of applied liquid solvent may penetrate the skin under occlusive conditions but only up to 0.3% when allowed to evaporate. The velocity of skin penetration is independent of the presence of other polar or non-polar components.[1]

Eye contact

The liquid is painful and irritating to the eyes.[4]

In the Yant *et al.*[42] study referred to above (*Inhalation*), an immediate slight burning of the eyes accompanied by lacrimation were noted and which persisted throughout the test at 1,600 ppm for ten minutes. When the same five persons were exposed for one minute to 5,500 ppm 1,4-dioxane vapour, there was irritation to the eyes resulting in blinking, squinting and lacrimation.

Ref. 43 states that 1,4-dioxane vapour has caused no serious disturbances by external contact with the eye and that irritation is noted only at concentrations in air of greater than 200 ppm (*v.s.*). Exposure to 50 ppm in air caused eye irritation[39] (*v.s. Inhalation*).

18.4 Animal Acute Toxicity[4,42,44-48]

The following relevant data on acute toxicity are reported[4,44]:

oral	– rat LD_{50} :	4,200–7,350 mg/kg
ihl	– rat LC_{50} :	46 g/m³ [*ca.* 12,750 ppm] over two hours
ihl	– rat LC_{50} :	14,260 ppm over 4 hours
oral	– mouse LD_{50} :	5,700 mg/kg
ihl	– mouse LC_{50} :	65 g/m³ [*ca.* 18,000 ppm] over two hours
oral	– cat LD_{50} :	2,000 mg/kg
oral	– rabbit LD_{50} :	2,000–2,100 mg/kg
oral	– guinea pig LD_{50} :	1,270–3,150 mg/kg.

Early studies quoted by Browning[45] give the lethal inhalation dose for guinea pigs exposed for three hours or more [sic] as 1,000–3,000 ppm; for mice [exposure period unstated] as 8,000 ppm, and for mice, rats, guinea pigs and rabbits as 4,000–11,000 ppm [exposure period unstated]. In the guinea pig study, 2,000 ppm was tolerable without serious symptoms but higher concentrations produced eye, nose and lung irritation.[46] Rowe and Wolf[4] observe that apart from marked irritation of the mucous membranes at high concentrations, deaths occurring during exposure or shortly afterward were usually due to respiratory failure because of lung oedema. Deceased animals also exhibited congestion of the brain and delayed deaths were usually due to pneumonia. Microscopic examination of animals dying days after exposure as well as of those sacrificed several days after apparently recovering almost always showed liver and kidney injuries.

Toxic effects found in cats exposed to concentrations of 12,000 ppm for 7 hours, 18,000 ppm for 4.3 hours, 24,000 ppm for 4 hours and 31,000 ppm for 3 hours were loss of equilibrium, lacrymation and increased salivation. Narcotic effects also were observed, the onset of which depended on dioxane concentration. After exposure, activity of all the cats gradually decreased before eventual death. Autopsy revealed fatty livers and inflamed respiratory organs. Browning[45] citing the same study notes that the lethal dose by inhalation by cats was 10,900 ppm with deaths sometimes delayed for several days. Similar narcotic effects have been found in the rabbit.[4]

Rowe and Wolf in ref. 4 observe that symptoms resulting from single-dose ingestion progress from weakness, depression, incoordination and coma to death. Postmortem examinations reveal haemorrhagic areas in the pyloric region of the stomach, bladders distended with urine, enlarged kidneys, slight proteinuria but no haematuria. Microscopic changes in the liver and kidneys were generally of a type seen in diethylene glycol poisoning.

RTECS[44] gives the dermal LD_{50} for rabbits as 7,600 mg/kg. RTECS also reports mild irritation (open Draize test) by 515 mg on rabbit skin.

Liquid 1,4-dioxane caused irritation and transient corneal injury when tested in rabbits' eyes.[4,44,47] Ten microlitres in the eyes of guinea pigs produced irritancy.[44,48]

Intraperitoneal injection of 1,4-dioxane into female rats resulted in an estimated approximate average lethal dose of 1,500 mg/kg. Toxic effects, where displayed, of 500 mg/kg were narcosis, weakness, diarrhoea, liver damage, kidney damage and peritoneal adhesions.[48]

Rowe and Wolf 4 note work in which intravenous injection of 1,4-dioxane in guinea pigs, rabbits and cats caused a selective action on the convoluted tubules of the kidney, characterized by acute hydropic degeneration and liver cell degeneration. Deaths were due to uraemia caused by intrarenal obstruction and anuria.[4] The intravenous LD_{50} value for the rabbit is 1,500 mg/kg.[44]

LC_{50} values for acute inhalational toxicity (for two hours) to rat and mouse are noted in RTECS[44] (v.s.).

Yant et al.[42] subjected guinea pigs to concentrations of 1,4-dioxane vapour of 1,000, 2,000, 3,000, and 10,000 ppm for periods of one minute to 480 minutes and to 30,000 ppm for up to 540 minutes. The symptoms variously exhibited were irritation of the eyes and nose, retching movements, changes in the respiration and apparent narcosis. No symptoms were exhibited by control animals.

Signs of nasal irritation were immediately evident at all concentrations. Eye irritation (squinting and lacrymation) was manifest at 2,000 ppm after about five minutes. Both nasal and eye irritation increased in intensity with increasing concentration and both decreased in apparent intensity as the time of exposure to a particular concentration was prolonged.

Retching movements occurred in some animals only. The symptom was irregular in its time of occurrence, was manifest only at 10,000 ppm and above, and ceased when the animals were in an apparent unconscious condition.

Changes in respiration were noted only in the guinea pigs exposed to 30,000 ppm for 45 minutes or more. As near as could be ascertained, the respiration was first laboured, then became shallow and rapid during which time occasional gasping was noted. As a terminal condition the respiration became shallow and slow until death ensued.

Narcosis or apparent unconsciousness was produced within 87 to 141 minutes' exposure to 30,000 ppm dioxane vapour.

18.5 Human Sub-acute and Chronic Toxicity Including Reproductive Effects[4,16,17,37,39,45,46,49-51]

General

Six deaths have been attributed to intoxication following (probably sub-acute) over-exposure to 1,4-dioxane vapour, with possible additional percutaneous absorption in the sixth.

The first five deaths referred to above were of male employees in a British silk working plant in 1933.[4,49,50] The indications were that the deceased were exposed to high (but unknown) concentrations over a relatively short period of time. Two months before the acute episodes, the dioxane-using process was changed in a way that required the men to put their heads into a vat containing the solvent. Four of the five men reportedly worked unusually long shifts prior to the onset of symptoms, which were irritation of the upper respiratory passages, coughing, irritation of the eyes, drowsiness, vertigo, headache, anorexia, stomach pains, nausea, vomiting, uraemia with diminished and eventual absence of urine output, coma and death within two weeks. There were other, non-fatal cases in which there was anorexia, nausea and vomiting. No jaundice was present but in one non-fatal case the liver was palpable, with a positive, indirect Van den Bergh reaction and a trace of albumin and a few red corpuscles in the urine. Other workers also showed a trace of albuminuria and some leucocytosis.[4,16,45,49]

Browning[45] also notes that autopsy of these deaths revealed macroscopic enlargement of the kidneys with areas of haemorrhage on the surface, and an enlarged and pale liver not stained with bile. Microscopically the kidneys showed necrosis of the outer part of the cortex, vascular in distribution and with haemorrhage at the edge of the necrotic zone. The liver showed complete necrosis of the inner half or two thirds of each lobule, but no fatty degeneration. In one case the lungs were oedematous. Three of the men showed a leucocytosis of 24,000, 21,400 and 38,000 respectively, with neutrophilia.

Barber[50] states that the workers were believed sometimes to have licked their fingers at work but, insofar as this may be true, it had gone on from the start of the process and was not increased by the changes implemented two months prior to the accidents and which undoubtedly had increased the amount of 1,4-dioxane vapour inhaled.

The ultimate cause of death was attributed to haemorrhagic nephritis. Barber[50] notes that there was such acute damage to the kidneys as to suggest one large dose of the poison absorbed from the stomach (though this did not happen). Autopsy also revealed oedema of the brain. Observers have noted from this incident that 1,4-dioxane does not have warning properties adequate to prevent short-term exposures dangerous to life.[4,45,50]

The sixth fatal case reported was of a worker who was exposed for one week to about 470 ppm of 1,4-dioxane in the air. This worker also used the solvent to wash glue from his hands from an open bucket between his knees, which probably resulted in some solvent absorption through the skin. Other workers with similar exposure were unaffected. Autopsy revealed damage to kidneys, liver and brain.[4,46]

As an effective fat solvent, 1,4-dioxane can cause eczema on prolonged and repeated contact with the skin.[4] It was not noticed as being specifically irritant however. Rowe and Wolf[4] note a report of a woman who had experienced a severe skin burn from exposure to isoprene and who later developed contact eczema after possible exposure to dioxane. The cause of this response was not clear.

Rowe and Wolf[4] note three epidemiological studies in 1,4-dioxane manufacturing plants. In one of these, 74 workers who had worked in the factory for 3 to 41 years were studied. Analysis of the workroom air in 1974 (some two years before the report was published) showed an exposure level of 14.24 ppm of dioxane. Of the 74 workers, 24 were still working at the plant, 26 were working elsewhere, 15 had retired and 12 were dead. The studies included extensive medical and physical examinations, chromosome analysis in six of the actively working employees and a careful analysis of the cause of death of the 12 deceased. No evidence was found of exposure to 1,4-dioxane having caused adverse effects on the workers.

Another study (in 1975) was made of 165 employees working in a 1,4-dioxane manufacturing plant between April 1954 and June 1959 when exposure levels ranged up to 32 ppm. No adverse effects due to dioxane exposure were found.

In addition to the physical and medical examinations, plus cause of death evaluation where appropriate, the urine of five of the workers was analysed for 1,4-dioxane and for 2-hydroxyethoxyacetic acid (HEAA). Virtually all the dioxane excreted was metabolized to HEAA (the ratio of dioxane to HEAA being 1 to 118) suggesting that under the exposure conditions experienced, the metabolic pathway in humans was not saturated. This led the investigators to believe that exposure up to 32 ppm posed no significant health hazard.

A similar study involving 80 workers at a dioxane manufacturing plant again found no evidence of adverse effects from their dioxane exposure. Complete physical examinations, chest X-rays, electrocardiograms, a series of liver profile tests and, where appropriate, a careful review of cause of death were conducted. Exposure levels ranged from 0.05 to 51 ppm.

See Section 18.2 regarding the study in ref. 39 on human exposure to 50 ppm for six hours with no toxic effects.

Reproductive Effects in Man

No data on reproductive effects were noted.

Carcinogenicity to Man

Santodonato *et al.*,[17] in commenting on the foregoing three epidemiological studies, note that all three lacked sufficient cohort size and number of cases to enable identification of low-level cancer risk, the low number of cancer deaths observed being not significantly different from what would be expected normally.

In connection with possible carcinogenicity to humans, Dietz *et al.*[51] note that Stott *et al.* in ref. 37 have provided new data on the mechanism associated with 1,4-dioxane tumorigenesis. In studies of the effects of repeated 1,4-dioxane administration to rats at a tumorigenic dose level (1,000 mg/kg/day), Stott *et al.* observed an increase in hepatic DNA synthesis, cellular swelling, and histopathology, indicative of hepatocellular hypertrophy and some regenerative activity. No treatment-related effects on DNA repair or DNA alkylation were noted in animals given 1,000 mg 1,4-dioxane/kg. 1,4-Dioxane was also negative in *in vitro* bacterial mutagenicity tests. These data indicate a lack of direct genotoxicity by 1,4-dioxane. Collectively, these results have indicated the likelihood that 1,4-dioxane is tumorigenic in rats *via* an epigenetic mechanism of tumorigenesis. Based upon these data, several possible epigenetic mechanisms of tumorigenicity appear to be possible for 1,4-dioxane. Recurrent cytotoxicity may result in an enhanced level of spontaneously occurring mutations and the incidence of carcinomas. Alternatively, an induction of the mixed-function oxidase enzyme system in conjunction with a prolonged hepatocellular hypertrophic response in rats suggests a possible mechanism similar to that observed with other mixed-function-oxidase-inducing agents. In any event, only high dose levels of 1,4-dioxane, associated with a non-linear pharmacokinetic behaviour (see Section 18.2) and toxicity in animals over a prolonged period of time, are expected to be capable of causing an increase in tumorigenicity. When exposure to these high, saturating dose levels is avoided, 1,4-dioxane is not expected to pose a significant carcinogenic risk to humans.

18.6 Animal Sub-acute and Chronic Toxicity Including Reproductive Effects[1,2,4,32,40,46,52]

General

(i) Inhalation

Torkelson *et al.*[40] report studies

"in which groups of 24 male and 24 female rats, 3 male and 3 female rabbits, and 2 female dogs received 130–136 7-hour exposures in 180–195 days to 50 ppm dioxane vapour in air. In addition, 7 male and 8 female guinea pigs received 82 exposures in

118 days to 50 ppm dioxane vapour in air, and groups of 12 rats and 2 rabbits of each sex received 133–136 7-hour exposures to 100 ppm. There were no adverse effects when the exposed groups were compared to control groups on the basis of appearance, demeanour, growth, mortality, hematological and clinical chemical studies, organ weights, or gross and microscopic pathological examination."

ACGIH[46] reports work in which the repeated inhalation by rabbits of 800 ppm 1,4-dioxane vapour for 30 days resulted in fatal kidney injury to some.
(See also *Carcinogenicity to Animals.*)

(ii) Skin Contact

Rowe and Wolf in ref. 4 cite work in which kidney and liver injury in rabbits and guinea pigs resulted from repeated topical application of 1,4-dioxane which, in sufficient amounts, has been shown to cause unsteadiness and inco-ordination.
(See also *Carcinogenicity to Animals.*)

(iii) Ingestion

Various studies on carcinogenicity caused by ingestion of 1,4-dioxane have shown non-neoplastic lesions at some dosages. These are noted below in *Carcinogenicity to Animals (q.v.).*

Carcinogenicity to Animals

(i) Dermal application

IARC[2] notes a study in which groups of 30 male and 30 female Swiss–Webster mice received thrice weekly paintings for 60 weeks of an unspecified concentration of 1,4-dioxane in acetone applied on the clipped dorsal skin. One skin sarcoma and one malignant lymphoma were observed. In similar groups of mice, skin paintings were preceded one week earlier by the application of 50 μg 7,12-dimethylbenz(*a*)anthracene (DMBA). Four males and five females survived the 59 weeks of treatment. Among 15 mice examined, eight skin tumours were observed. These were two papillomas, two squamous-cell carcinomas and four sarcomas. In addition, 24 other tumours (mainly malignant lymphomas and lung tumours) occurred. Eight skin papillomas and one malignant lymphoma occurred in 55 animals receiving 50 μg DMBA followed by thrice weekly paintings with acetone alone. The increase in the incidence of skin tumours was significant (p < 0.01) in the two-stage skin carcinogenesis experiment, showing that 1,4-dioxane was active as a promoter.[1,2]
(See also *General: (ii) Dermal toxicity.*)

(ii) Oral administration

Various studies have been undertaken with a view to ascertaining the carcinogenicity of 1,4-dioxane. Routes of administration have included inhalation and cutaneous but most studies were of administration of the dioxane in drinking water.[1]

In an NCI study, B6C3F1 mice were exposed for 90 weeks to 1,4-dioxane in drinking water at average dose levels of 830 or 720 mg/kg/day for males and at 860 or 380 mg/kg/day for females. Groups consisted initially of 50 animals of each sex/dose level. The NCI report on this work[52] notes that in each sex, the Fischer exact test shows that the incidence of either hepatocellular adenomas or carcinomas in any of the dosed groups are significantly higher (p < ±0.014) than that in the matched controls, the statistical conclusion being that the incidence of this tumour in male and female mice is associated with the administration of the 1,4-dioxane. In male mice the incidence of hepatocellular carcinoma was 24 out of 47 test animals (51%) high dose; 18/50 (36%) low dose; 2/49 (4%) matched control. With female mice the respective figures were 29/37 (78%), 12/48 (25%) and 0/50 (0%). For hepatocellular adenoma the results for male mice were 4/47 (9%) high dose; 1/50 (2%) low dose and 6/49 (12%) matched control. For females the respective figures were 6/37 (16%), 9/48 (19%) and 0/50 (0%).

The NCI bioassays[52] found that nasal cavity carcinoma was significantly increased in male and female Osborne-Mendel rats ingesting 1,4-dioxane at average daily doses of 530 and 240 mg/kg (males) and 640 and 350 mg/kg (females) over 110 weeks, there being initially 35 animals in each sex-dose group. The incidence of this carcinoma for males was 16/34 (47%) high dose; 12/33 (36%) low dose; 0/33 (0%) matched control. For female rats the incidence was 8/35 (23%) high dose; 10/35 (29%) low dose and 0/34 (0%) matched control. The Fischer exact test showed that the incidences in both the dosed groups were significantly higher (p < ±0.003) than that in the matched controls and the statistical conclusion was that in both sexes of rats, this tumour was associated with the administration of the 1,4-dioxane.

The incidence of hepatocellular adenomas was significantly higher (p < ±0.001, Fischer exact test) for both the low and high dose groups of female rats than in the matched controls. Results for this neoplasm were: female, high dose 11/32 (34%); low dose 10/33 (30%); matched control 0/31 (0%). (For male rats the respective figures were: high dose 1/33 (3%); low dose 1/32 (3%) and matched control 2/31 (6%).) These results for male rats were not statistically significant. Those for female rats gave the statistical conclusion that the incidence of this tumour was associated with the administration of the 1,4-dioxane.

The NCI carcinogenicity study also elicited non-neoplastic responses at some levels of ingestion. In rats such responses were observed in the kidneys (tubular degeneration) of males at both dose levels (27/33 high dose, 20/31 low dose, 0/31 matched control); in the livers (hyperplasia) of both sexes, especially at high dosage (males: high dose 11/33, low dose 3/33, matched control 5/31; females: high dose 17/32, low dose 11/33, matched control 7/31), and in the stomachs (ulceration) of males (high dose 5/30, low dose 5/28, matched control 0/31). Dosed rats had higher incidences of pneumonia than the controls (males: high dose 14/33, low dose 15/31, matched control 8/30; females: high dose 25/32, low dose 5/34, matched control 6/30). The development of nasal carcinomas is noted as possibly having been a contributing factor to the incidences of pneumonia. In the NCI study, mice also showed increased incidence of pneumonia (males: high dose 17/47, low dose 9/50, matched control 1/49; females: high dose 32/36, low dose 33/47, matched control

2/50). Rhinitis was observed in 2% of males (both dosages) and in 21% of high dose and 14% of low dose females, with matched controls 0/49 for males and 0/50 for females. The NCI notes that hepatic cytomegaly was commonly observed in dosed mice.

IARC[2] cites other studies on carcinogenicity due to 1,4-dioxane. A group of 26 male Wistar rats was given 1% 1,4-dioxane in their drinking water for 63 weeks, the total dosage being 130 g per animal. Six of the rats developed liver tumours ranging from small neoplastic nodules to multifocal hepatocellular carcinomas, with miscellaneous tumours in two other rats, while one lymphosarcoma occurred in nine controls.

Four groups of 28–32 Sprague–Dawley rats were given 0.75, 1.0, 1.4 or 1.8% 1,4-dioxane in their drinking water for 13 months and the animals were killed after 16 months. The total doses were 104–256 g per animal. One rat receiving 0.75%, one receiving 1.0%, two receiving 1.4% and two receiving 1.8% 1,4-dioxane developed tumours of the nasal cavity. These were mainly squamous-cell carcinomas with areas containing adenocarcinomas in two cases. Hepatomas and hepatocellular carcinomas developed in three rats receiving 1.4% and in twelve rats receiving 1.8%. All treated groups developed microscopic lesions described as 'incipient hepatomas'. A subcutaneous fibroma occurred 'on the back of the nose' in one of thirty controls.

IARC[2] also notes a study on which four groups of 60 male and 60 female Sherman rats were given 0, 0.01, 0.1 or 1% 1,4-dioxane in their drinking water for up to 716 days. The daily doses were: for males, 0, 8–12, 59–113, 914–1,229 mg/kgbw; for females, 0, 18–20, 130–160, 1,416–2,149 mg/kgbw. The 50% survival time at the highest dose was 16 months compared with 22 months in other groups. IARC notes that at the highest level, ten hepatocellular carcinomas, two cholangiomas and three squamous-cell carcinomas of the nasal cavity were observed, compared with one hepatocellular carcinoma in a rat receiving 0.1% dosage. No statistically significant increase in the incidence of tumours was seen in rats given the two lower dose levels. ACGIH,[46] citing the same study, notes that both sexes of rats receiving 0.1% 1,4-dioxane exhibited inconsistent degrees of renal tubular sloughing and hepatocellular degenerative changes, but no significant increases in neoplasms.

In another oral administration study noted by IARC,[2] 22 male guinea pigs received drinking water containing 0.5–2% 1,4-dioxane such that normal growth was maintained over 23 months. All animals were killed within 28 months. The total dose was 588–623 g per animal. Two animals had carcinomas of the gall-bladder and three had hepatomas. The 10 untreated controls had no liver tumours reported.

A study of noticeably short duration (40–43 weeks) was conducted wherein no tumours occurred in groups of 50 male and 50 female B6C3F1 mice receiving 0.5 or 1% 1,4-dioxane in their drinking water.[2]

(iii) Inhalational administration

A "lifetime" (two year) study by Torkelson et al.[40] of chronic inhalational toxicity to rats was instituted with a view to ascertaining the carcinogenic propensity of 1,4-dioxane under conditions more amenable to extrapolation to long-term inhalational exposure in man than was possible with the results from earlier, high-dosage oral studies that had been performed. Torkelson *et al's* paper 40 was published in 1974, before the NCI study referred to above but after the other studies on oral administration to rats mentioned above as being noted by IARC in ref. 2.

Torkelson *et al.*'s summary of their work states that

> "repeated 7 hour daily exposures to an average concentration of 0.4 mg/l (111 ppm) 1,4-dioxane vapours were given 5 days/week for two years to groups of 288 male and 288 female Wistar strain rats. Groups of 192 male and 192 female rats, similarly exposed to filtered room air served as controls. There were no observable compound-related effects with respect to demeanour, growth or mortality rate. Hematological studies during the 16th and 23rd months of exposure revealed no deviation from controls. Clinical chemical values determined at the termination of the experiment were within normal limits. Gross microscopic examination of the major organs and tissues revealed no treatment-related lesions. In this study, no hepatic or nasal carcinomas were observed.... . The incidence of all types of tumour in the control and in the exposed groups was not statistically different."

The vapour concentration used in this study was selected by virtue of the then TLV of 100 ppm v/v as set by the ACGIH. The authors state that the selection of this concentration was also supported by previous studies in their laboratory [referred to above in 18.6, *General, (i) Inhalation*].

Torkelson *et al.*[40] observed that although several assumptions were required, it was of value to calculate and compare the possible daily dose of 1,4-dioxane (as mg/kg/day) received in their study with those ingested *via* the drinking water in the study on Sherman rats [*vide supra*]. Given the assumptions made, it was calculated that the dose of 1,4-dioxane received by male rats exposed to atmospheres containing 0.4 mg/litre for seven hours, was very similar to that received *via* ingestion of drinking water containing 0.1% dioxane, 94 mg/kg/day. In the oral study on Sherman rats this level did produce pathological change in the liver though this did not occur in Torkelson *et al.*'s inhalational study.

Factors which might explain this discrepancy were noted by Torkelson *et al*, *viz*: the inhalation study was over five days per week; the ingestion one was over seven days per week, with no opportunity for recovery during the 2-day weekends. The assumption of 100% absorption is made in comparing the studies but in fact it may have been less in the respiratory tract. Absorption from the gastrointestinal tract may have exposed the liver to a higher concentration of 1,4-dioxane than absorption from the lungs. There may have been a strain difference between the Wistar rats of the inhalation study and the Sherman rats of the ingestional one (though Torkelson *et al.* give reason for considering this unlikely).

Torkelson *et al.* observe that both studies support the hypothesis that liver injury precedes tumour development in the liver.

[The work of Young *et al.*[32] noted in Section 18.2 (Toxicokinetics), wherein there appears to be either an actual or at least an effectual threshold exposure value below which toxic effects were not observed, may be of relevance.]

ACGIH[46] observes that in 1976 1,4-dioxane had been classed by the TLV Committee as an animal tumorigen of such low potential as to be of no practical significance as an occupational carcinogen at or around the former TLV of 50 ppm. [The present ACGIH level, set in 1981, is 25 ppm and in 1984 its STEL of 100 ppm was recommended for deletion pending additional toxicological data and industrial hygiene experience becoming available.]

ACGIH however also observes that "NIOSH took an opposite viewpoint in its criteria document for dioxane, published in 1976, recommending a 1 ppm standard based on '... belief that dioxane can cause tumours in exposed workers and on the belief that information allowing the derivation of a safe exposure limit is not now available".[46]

Reproductive Effects in Animals

Rowe and Wolf[4] note that teratogenic studies were conducted on rats and mice given, by inhalation, 1,1,1-trichloroethane containing 3.5% 1,4-dioxane. The airborne dioxane concentration was calculated to be 32 ppm. There was no evidence of maternal, embryonal or foetal toxicity nor any signs of teratogenicity at this level. The studies were inconclusive because of the nature of the product tested. ECETOC[1] notes a multigeneration dominant lethal and teratology study in which mice were dosed with 1,1,1-trichloroethane, stabilized with 3% 1,4-dioxane, given in drinking water. The maximum dose rate of 1,4-dioxane was 30 mg/kgbw/day and none of the treated mice showed abnormalities in their reproduction or development. Again, this study was inconclusive because of the nature of the test material and the fact that the top dose was not shown to be approaching a maternally toxic dose of 1,4-dioxane.

18.7 Mutagenicity[4,36,37,53]

Fishbein[36] noted that the mechanism of the carcinogenic action of 1,4-dioxane had not been unequivocally elaborated and that the compound was found not to bind to DNA.

Stott *et al.*[37] found that lack of genotoxic activity by 1,4-dioxane and its cytotoxicity at tumorigenic dose levels suggests a nongenetic mechanism of liver tumour formation in rats. They report that treatment of rats with tumorigenic dose levels of 1,4-dioxane (1 g/kg/day) in drinking water for 11 weeks resulted in a 1.5 times increase in hepatic DNA synthesis. Cytotoxicity was not detected in rats dosed orally with non-tumorigenic levels (10 mg/kg/day). Alkylation of hepatic DNA and DNA repair were not detected in rats dosed with 1 g [14]C 1,4-dioxane by gavage nor did 1,4-dioxane elicit a positive response in the Ames *Salmonella typhimurium* mutagenesis assay. No increase in background reversion rate was observed with or without a metabolic activating system over a broad range (0 to 103 mg 1,4-dioxane per plate) using strains TA1535, 1537, 1538, 98 and 100. The top dosage used was the maximum amount of pure 1,4-dioxane possible per assay

(0.1 ml/plate) and was bacteriostatic in assays without metabolic activation. Primary rat UDS assays *in vitro* (Williams method) were also negative.

Rowe and Wolf[4] note briefly other work using *S. typhimurium* (strains TA1535, 1537 and 1538) and *Saccharomyces cereyisiae*. Negative results were obtained in all cases both with and without metabolic activation. They also refer briefly to a study wherein no chromosomal changes were found in six workers exposed to 1,4-dioxane in a manufacturing plant.

In a study by Kurl *et al.*[53] the effect of 1,4-dioxane on transcription was studied. Both hepatic RNA polymerase A and B activities were measured after single dose intravenous injection of eight week old male Sprague–Dawley rats with either 10 mg/rat or 100 mg/rat of 1,4-dioxane. Both the polymerase A and B activities were speedily depressed (*ca.* 10 minutes after injection) with both dosages (to about 75–80% of control in the lower dose case and to about 50–55% of control in that of the higher dose). With the lower dose, RNA polymerase A activity recovered and increased to 40% above control at about one hour after injection. The polymerase B activity returned to control value at one hour post-injection but then underwent a secondary decrease to about 65% control value at four hours post-injection. It thereafter recovered to be slightly below control at 24 hours whereas the polymerase A activity (which had dropped to control values at about two hours post-injection) was about 35% above control values at this time.

With the higher (100 mg/rat) dosage, recovery after the initial (and greater) inhibition followed a similar pattern. The polymerase A activity rose to 30% above control values at one hour post-injection; decreased to about 60% of control at four hours and then rose to control values at 24 hours.

18.8 Summary

[1,4-Dioxane shows moderate acute toxicity by inhalation, skin absorption or ingestion, and is also cumulative in action, prolonged over-exposure causing kidney and liver damage. The liquid is irritating to the eyes, and may cause skin irritation on prolonged or repeated contact, while vapour concentrations above 200 ppm irritate the respiratory tract and eyes. Repeated high doses cause cancer in animals. At low doses, the compound is rapidly metabolized to 2-hydroxyethoxyacetic acid, which is then excreted in the urine, but high doses overload this metabolic system, leading to accumulation of unchanged 1,4-dioxane in the body. There is thus a metabolic threshold dose above which harmful cumulative effects, including cancer, may occur; there is evidence that single or daily exposure to 50 ppm in air or to about 10 mg/kg orally is below this threshold and unlikely to cause cumulative toxic or carcinogenic effects, or acute effects other than reported eye irritation at 50 ppm in air.]

19. MEDICAL / HEALTH SURVEILLANCE

A decision on the need for, and content of, medical surveillance should be based on an assessment of the possibility and extent of exposure in the work operation. In addition, medical examination may be directed to identifying any pre-existing or newly arising condition in the individual workers which might either be aggravated by subsequent exposure, or might confuse any subsequent medical assessment in the event of excessive exposure or an illness not related to exposure. A particular aspect to be considered is the identification of sensitive subjects not adequately protected by the control limit in operation. A professional medical judgement is required on continuance of employment in the specified process. The following information is relevant in the case of this solvent.

Pre-employment medical examination

A comprehensive occupational and medical history should be taken and a physical examination carried out; in both cases particular attention should be paid to the respiratory, hepatic and renal systems. If the work is to involve the use of respiratory protective equipment, fitness for this should be confirmed.

Specific clinical tests which have been recommended include urinalysis and liver and kidney function tests.

As 1,4-dioxane can cause dermatitis, examination of the skin is advisable.

Periodic medical examinations

Annual reviews should include update of medical and occupational history and a physical examination. Clinical tests as noted above may be carried out when appropriate. No specific biological monitoring tests are available, although there is a limited ability to metabolise the solvent to 2-hydroxyethoxyacetic acid (HEAA), the major urinary metabolite. Advice of an occupational physician should be sought on the maintenance of medical records in view of the recognition of 1,4-dioxane as an animal carcinogen.

20. OCCUPATIONAL EXPOSURE LIMITS

[The Committee felt that a time-weighted average limit of 25 ppm and a short term limit of 50 ppm were appropriate for 1,4-dioxane. This substance may produce serious toxic effects by absorption through the skin (see Introduction).]

REFERENCES

1. European Chemical Industry Ecology & Toxicology Centre. 'Joint assessment of commodity chemicals' No.2: 1,4-Dioxane. ECETOC, 1983.
2. International Agency for Research on Cancer. IARC Monographs on the Evaluation of Carcinogenic Risk of Chemicals to Man 11(1976). 'Cadmium, nickel, some epoxides, miscellaneous industrial chemicals & general considerations on volatile anaesthetics'.
3. Council of Europe. 'Dangerous Chemical Substances and Proposals Concerning their Labelling'. ('The Yellow Book'.) 4th ed. Maisonneuvre, 1978.
4. Clayton, G.D. and F.E. Clayton, Eds. 'Patty's Industrial Hygiene and Toxicology'. 3rd ed. rev. Wiley Interscience, 1982.

5. Sax, N.I. 'Dangerous Properties of Industrial Materials'. 6th ed. Van Nostrand Reinhold, 1984.
6. Occupational Safety & Health Administration. Material safety data sheets from the Occupational Health Services database, OSHA, Washington D.C., 1986.
7. National Institute for Occupational Safety & Health/Occupational Safety & Health Administration. 'NIOSH/OSHA Occupational health guideline for dioxane'. DHHS (NIOSH) Publication. US Department of Health & Human Services/US Department of Labor, 1978.
8. Graedel, T.E. 'Chemical Compounds in the Atmosphere'. Academic Press, 1978.
9. Fielding, M. *et al.* 'Organic Micropollutants in Drinking Water'. Technical Report TR 159. WRC Environmental Protection, Water Research Centre, UK, 1981.
10. Hawley, G.G., Ed. 'The Condensed Chemical Dictionary'. 10th ed. Van Nostrand Reinhold, 1981.
11. Grasselli, J.G. and W.M. Ritchey, Eds. 'Atlas of Spectral Data and Physical Constants for Organic Compounds'. 2nd ed. CRC Press, 1975.
12. National Institute for Occupational Safety & Health. 'NIOSH Manual of Analytical Methods'. 3rd ed. DHHS (NIOSH) Publication No.84-100. US Department of Health & Human Services, 1984.
13. Baker, B.B., jr. *American Industrial Hygiene Association Journal*, **43**, No.2(1982):98.
14. Sopikov, N.F. and A.I. Gorshunova. *Gigiena i Sanitariya*, **44**, No.5(1979):59.
15. Commission of the European Communities. 'Environmental Chemicals Data and Information Network (ECDIN)' Databank. Commission of the European Communities Joint Research Centre, ECDIN Group, I-21020 Ispra (Varese), Italy, 1986.
16. International Labour Office. 'Encyclopaedia of Occupational Health and Safety'. 3rd rev. ed. ILO, 1983.
17. Santodonato, J. *et al.* 'Monograph on Human Exposure to Chemicals in the Workplace: 1,4-Dioxane'. Report No.SRC-TR-84-659. National Cancer Institute, USA, 1985.
18. *Japan Chemical Week*, **26**, No.1292(1985):3. 'Toho Chemical expands dioxane capacity to 6,000 ton/y'. (Abstract.)
19. Marsden, C. and S. Mann, Eds. 'Solvents Guide'. 2nd ed. rev. Cleaver-Hume Press, 1963.
20. National Institute for Occupational Safety & Health. 'NIOSH Recommended standard for occupational exposure to dioxane'. US Department of Health, Education & Welfare, 1977.
21. Bretherick, L. 'Handbook of Reactive Chemical Hazards'. 3rd ed. Butterworths, 1985.
22. The Royal Society of Chemistry. Laboratory Hazards Data Sheet No.29: 'Dioxane', 1984.
23. Ruth, J.H. *American Industrial Hygiene Association Journal*, 47(1986):A-142. 'Odor thresholds and irritation levels of several chemical substances: a review'.
24. National Fire Protection Association. 'Fire Protection Guide on Hazardous Materials'. 9th ed. NFPA, Massachusetts, USA, 1986.
25. Warren, P.J., Ed. 'Dangerous Chemicals Emergency Spillage Guide'. 1st ed. Jensons (Scientific) Ltd., Leighton Buzzard, UK, 1985.
26. Windholz, M. *et al.*, Eds. 'The Merck Index'. 10th ed. Merck & Co., 1983.
27. Reynolds, J.E.F. and A.B. Prasad, Eds. 'Martindale, The Extra Pharmacopoeia'. 28th ed. The Pharmaceutical Press, 1982.
28. Horsley, L.H. 'Azeotropic Data III'. Advances in Chemistry Series No.116, American Chemical Society, 1973.
29. Weast, R.C., Ed.-in-Chief. 'CRC Handbook of Chemistry and Physics' ('The Rubber Handbook'). 67th ed. CRC Press, 1986-1987.
30. Grayson, M., Exec. Ed. 'Kirk-Othmer Encyclopedia of Chemical Technology'. 3rd ed. John Wiley & Sons, 1979.
31. Last, J.M., Ed. 'Maxcy-Rosenau Public Health and Preventive Medicine'. 11th ed. Appleton-Century-Crofts, New York, 1980.
32. Young, J.D. *et al. Journal of Toxicology & Environmental Health*, 4(1978):709. 'Dose-dependent fate of 1,4-dioxane in rats'.

33. Woo, Y.-T. *et al. Life Sciences*, **21**, No.10(1977):1447. 'Tissue and subcellular distribution of ³H-dioxane in the rat and apparent lack of microsome-catalyzed covalent binding in the target tissue'.

34. Woo, Y.-T. *et al. Archives of Pharmacology*, **299**, No.3(1977):283. 'Structural identification of *p*-dioxane-2-one as the major urinary metabolite of *p*-dioxane'.

35. Swedish Criteria Group for Occupational Standards. *Arbete och Halsa*, 1983:36. 'Scientific basis for Swedish occupational standards IV'.

36. Fishbein, L. *The Science of the Total Environment*, **17**, No.2(1981):97. 'Carcinogenicity and mutagenicity of solvents - I: Glycidyl ethers, dioxane, nitroalkanes, dimethylformamide and allyl derivatives'.

37. Stott, W.T. *et al. Toxicology & Applied Pharmacology*, **60**, No.2(1981):287. 'Differentation of the mechanisms of oncogenicity of 1,4-dioxane and 1,3-hexachlorobutadiene in the rat'.

38. Young, J.D. *et al. Toxicology & Applied Pharmacology*, **38**, No.3(1976):643. '1,4-Dioxane and β-hydroxyethoxyacetic acid excretion in urine of humans exposed to dioxane vapors'.

39. Young, J.D. *et al. Journal of Toxicology & Environmental Health*, 3(1977):507. 'Pharmacokinetics of 1,4-dioxane in humans'.

40. Torkelson, T.R. *et al. Toxicology & Applied Pharmacology*, **30**, No.2(1974):287. '1,4-Dioxane – II. Results of a 2-year inhalation study in rats'.

41. Braun, W.H. and J.M. Waechter, jr. *Journal of Animal Science*, **56**, No.1(1983):235. 'Sources of uncertainty in pharmacokinetic prediction'.

42. Yant, W.P. *et al. Public Health Reports*, 45(1930):2023. (Washington D.C.) 'Acute response of guinea pigs to vapors of some new commercial organic compounds: VI. Dioxan'.

43. Grant, W.M. 'Toxicology of the Eye'. 2nd ed. Charles C. Thomas, Illinois, USA, 1974.

44. National Institute for Occupational Safety & Health. 'Registry of Toxic Effects of Chemical Substances (RTECS)'. DHHS (NIOSH) Publication No.84-101-6. US Department of Health & Human Services, April, 1986.

45. Browning, E. 'Toxicity and Metabolism of Industrial Solvents'. Elsevier, 1965.

46. American Conference of Governmental Industrial Hygienists. 'Documentation of the threshold limit values for chemical substances in the workroom environment'. ACGIH, 1986.

47. Carpenter, C.P. and H.F. Smyth, jr. *American Journal of Ophthalmology*, 29(1946):1363. 'Chemical burns of the rabbit cornea'.

48. Sanderson, D.M. *Journal of Pharmacy & Pharmacology*, 11(1959):150. 'A note on glycerol formal as a solvent in toxicity testing'.

49. Buffler, P.A. *et al. Journal of Occupational Medicine*, **20**, No.4(1978):255. 'Mortality follow-up of workers exposed to 1,4-dioxane'.

50. Barber, H. *Guy's Hospital Report*, 84(1934):267. 'Haemorrhagic nephritis and necrosis of the liver from dioxan poisoning'.

51. Dietz, F.K. *et al. Drug Metabolism Reviews*, **13**, No.6(1982):963. 'Nonlinear pharmacokinetics and their impact on toxicology: illustrated with dioxane'.

52. National Cancer Institute. Carcinogenesis Technical Report Series No.80: 'Bioassay of 1,4-dioxane for possible carcinogenicity'. National Institutes of Health, US Department of Health, Education & Welfare, 1978.

53. Kurl, R.N. *et al. Archives of Toxicology*, **49**, No.1(1981):29. 'Effects of dioxane on RNA synthesis in the rat liver'.

CONTENTS

1. CHEMICAL ABSTRACTS NAME

acetic acid ethyl ester
(10th Collective Index)

2. SYNONYMS AND TRADE NAMES[1 – 5]

acetato di etile (Italian)
acetic ester
acetic ether
acetidin
acetoxyethane
essigester (German)
ethyl (acétate d') (French)
ethyl acetate (IUPAC)
ethyl acetic ester
ethyl ethanoate
ethylacetat (German)
etile (acetato di) (Italian)
etyl acetaat (Dutch)
octan etylu (Polish)
vinegar naphtha

3. CHEMICAL ABSTRACTS SERVICES REGISTRY NUMBER

141-78-6

4. NIOSH NUMBER

AH5425000

5. CHEMICAL FORMULA

$C_4H_8O_2$
(Molecular weight 88.1)

6. STRUCTURAL FORMULA

$$CH_3 - COO - C_2H_5$$

7. OCCURRENCE[6-8]

Ethyl acetate is reported to be present in some natural fruit aromas, in the petals of magnolia fuscata, and in rum and rum ether distillates.[6]

The following concentrations (ppm) of ethyl acetate have been reported in food products[6] [further details of the specific food products tested are not given]:

Non-alcoholic beverages	67
Alcoholic beverages	50 – 65
Ice cream, ices, *etc.*	99
Confectionery	170
Baked goods	170
Gelatins and puddings	200
Chewing gum	1400

The following sources of emission of ethyl acetate to the atmosphere have been identified[7]:

> Animal waste treatment
> Glue manufacture and use
> Micro-organisms
> Petroleum manufacture
> Petroleum storage of products containing ethyl acetate
> Plant volatile [*sic*]
> Tobacco smoke
> Whisky manufacture.

Ethyl acetate is photolytically degraded in the atmosphere.[8]

8. COMMERCIAL AND INDUSTRIAL ETHYL ACETATE

Commercial ethyl acetate is available in various grades, for example:

85 – 88% pure
95 – 98% pure
99% pure anhydrous.

Commercial ethyl acetate may be denatured for example with:

calel ethatate	0.125%
methyl isobutyl ketone	1%
or	
wood alcohol	5%

Technical grade ethyl acetate of 99 – 100% purity is widely used. It has a boiling range of 76 – 78 °C and a maximum acidity content of 0.1%.

See also Section 11.

9. SPECTROSCOPIC DATA[9]

Infrared, Raman, ultraviolet, nuclear magnetic resonance (^{1}H) and mass spectral data for ethyl acetate have been tabulated.[9]

10. MEASUREMENT TECHNIQUES[10-17]

The method recommended for the determination of ethyl acetate in workshop atmospheres is the NIOSH[10] activated charcoal trapping method for collection and concentration, followed by solvent extraction of the charcoal and a gas chromatographic (gc) analysis of the extract.

The collection tube is 7 cm long with a 4 mm internal diameter. It contains a total of 150 mg of activated charcoal (20 – 40 mesh) divided into a front section of 100 mg and a rear section of 50 mg separated by a small plug of urethane foam. Air is sampled at a flow-rate of 200 ml per minute for 10 minutes by means of a small pump. The entire apparatus is portable so that it may be carried in a pocket with the sampling tube in the breathing zone (normally a coat lapel). The apparatus can also be static.

Carbon disulphide is used to extract the ethyl acetate, which is separated on a 3.05 m (10 ft) × 3.18 mm (0.125 in) stainless steel column packed with 5% FFAP on Supelcoport using flame ionization detection. For full details of the method see ref. 10.

The sampling method uses a small portable apparatus with no liquids and is therefore easy to handle and maintain. The analytical method is relatively specific with the additional advantage of flexibility of operating conditions.

Air samples are normally taken by means of suction using an electric pump capable of operating at low flow-rates (*i.e.*, 30 – 300 ml per minute). If higher flow-rates are used breakthrough can occur (*i.e.*, the ethyl acetate is not completely adsorbed and some of it passes through the adsorbent). It is also convenient either to establish the trapping device in a static position, or to have it attached to the user in such a way that the breathing zone is sampled (this usually means attachment to the worker's lapel).

It should be noted that suitable personal samplers may be commercially available.

The most popular accurate measurement technique, usually following carbon disulphide extraction of the ethyl acetate, is gc, although acetone has been claimed to be a better solvent,[11]

A similar trapping method to the recommended one used a low flow-rate of sampled air of 11 ml per minute and the ethyl acetate was extracted with carbon disulphide.[12] The carbon disulphide solution was then subjected to infrared (ir) analysis in a 6 mm cell with CsBr windows. The method was used to determine ethyl acetate in admixture with *p*-xylene.

After sampling from air, ethyl acetate was separated from acetone by gc on a 3 m × 3 mm column packed with 5% SE-30 on Chromaton N-AW at 70 °C using nitrogen as carrier gas and flame ionization detection[13] The method was capable of detecting 4 ng of ethyl acetate.

In another gc method, ethyl acetate was separated from acetone, toluene and xylene in a sample of workshop air after trapping on a 20 cm × 4 mm column of activated charcoal.[14] Gc was carried out on a column packed with 10% dioctyl phthalate on Chromosorb W (80 – 100 mesh) operated at 86 °C.

In a measurement of ethyl acetate and toluene vapour concentrations in workplace air during the production of cellulose lacquer film, the vapour was trapped in ice-cold *o*-xylene.[15] Gc was used to separate the solvents on a 3 m × 3 mm column packed with 10% polyoxyethylene glycol 1500 on Chromaton N AW-HMDS at 80 °C and using flame ionization detection.

A piezoelectric detector has been used in the separation of ethyl acetate, toluene and nitrobenzene in air samples.[16] The indicator was a silica crystal resonator which was coated with a special liquid-crystalline compound.

Very little attention has been paid to the determination of ethyl acetate in body tissue. Tissue, blood or urine (4 – 5 mg) was heated for two minutes at 120 °C in a sealed tube and the volatiles, including ethyl acetate and 1,4-dioxane, were subjected to gc determination.[17] The limit of detection was 0.01 – 0.02 μg.

11. CONDITIONS UNDER WHICH ETHYL ACETATE IS PUT ON THE MARKET[18-24]

Production

Ethyl acetate is manufactured by the action of acetic acid on ethanol in the presence of a catalyst (*e.g.*, sulphuric acid). The reaction reaches equilibrium at about 67% conversion and higher yields are obtained by forcing the reaction to completion. This is achieved by removing the water formed during the reaction and by using an excess of one reactant, generally ethanol.

There are many variations on the basic manufacturing process, the main variable being the concentration of acetic acid employed. Acetic acid concentrations between 8 and 80% have been used. The sulphuric acid is usually between 63 and 96%, and the ethanol is usually 95%. The process may be either continuous or batch, though the former is more common. The ratio of reactants varies with the type of process and the equipment used.

Continuous processes generally give a yield of 99 to 100% based on the acetic acid used. Yields from batch processes are of the order of 95%. It is necessary to remove the copper salts formed if copper columns are employed. Other impurities such as higher-boiling esters may also be formed depending on the purity of the raw materials.

ILO[20] record that ethyl acetate can also be produced from anhydrous acetaldehyde in the presence of aluminium ethoxide. The ethyl acetate is then purified by distillation.

Individual plants manufacturing ethyl acetate generally produce from about 5 to 45 million kilograms per year.

According to ref. 23 the total production of ethyl acetate in the EC in 1980 was 140 kilotonnes, while in the USA in 1983 it was about 95 kilotonnes.

Production of ethyl acetate was once considered a good indicator of general industrial activity, mainly because of its use in protective coating. However according to ref. 24 methyl ethyl ketone has tended to replace ethyl acetate as a low-boiling constituent of lacquers. This is expected to limit increases in production of the ester. [Attention is drawn to the fact that ref. 24 was published in 1975.]

Uses

The principal use of ethyl acetate is as a solvent. To a lesser extent it is used as an additive to 'lift' food flavours. In particular it is used as a solvent for resistant coatings and linings, lacquers and varnishes, plastics, fats, nitrocellulose and celluloid, dyes and acrylic inks, and aeroplane dopes.

A much smaller proportion of the ethyl acetate produced is employed in chemical syntheses. It is utilised in the production of perfumes, pharmaceuticals, smokeless powder (nitrocellulose containing about 13.1% nitrogen) and other explosives, photographic materials, synthetic silk and leather.

According to ref. 24 [published 1975], the combined uses of ethyl acetate as a solvent account for about 97% of its total use, with use in chemical synthesis accounting for the remaining 3% (use as a solvent for coatings is by far the major single use, accounting for some 70% of the total).

Other minor applications of ethyl acetate include its use as an artificial fruit essence and as a textile cleaner. It is also used as a stabilizer for carbon tetrachloride, as a laboratory solvent and as an agent for splitting viruses for the preparation of vaccines.

12. STORAGE, HANDLING AND USE PRECAUTIONS[4,5,25]

Although ethyl acetate is stable at normal temperatures and pressures it can form explosive mixtures with air at ambient temperatures. It may be ignited by heat, sparks or flames. Containers of ethyl acetate may explode when heated and ethyl acetate vapour may constitute a flammable and/or explosive hazard in confined spaces (*e.g.*, in sewers). Large escapes may present similar hazards outdoors. (See also Sections 13, 14 and 15.)

Care must be taken to ensure with storage, process and drying plant that hazard-free conditions exist regarding fire and explosion risks and exposure of personnel to ethyl acetate. [As with other flammable solvents the organization of storage facilities has to take into account two separate requirements. These are that the stored material be protected from a fire elsewhere on the premises and that the premises be protected from a fire involving the stored materials. On an industrial scale, sufficient physical separation and water spray curtains are often used effectively. On a laboratory scale, flammable solvents need to be stored in fire resistant, thermally insulated cupboards.

It is important to appreciate that a non-combustible barrier is not synonymous with a fire barrier. A fire barrier has to protect against both conducted and radiant heat. For example, a metal cupboard is useless on both these counts.]

Process enclosure or local exhaust ventilation should be provided when the material is in use in order to keep personnel exposures from being above the recommended limits. Although, in general, use of ethyl acetate does not require protective clothing, repeated or prolonged exposure should be avoided and suitable protective gloves should be worn when using it. [If there is any possibility of the liquid entering the eyes suitable means should be provided for irrigating the eyes.] It is recommended that ethyl acetate should be kept in tightly sealed containers away from heat and protected from fire and direct sunlight.

OSHA[25] states that persons with a history of chronic respiratory disease, skin disease or anaemia may be at an increased risk from exposure to ethyl acetate.

[Personnel should not be allowed to enter areas containing potentially hazardous concentrations of ethyl acetate unless all relevant legislation and official guidelines have been strictly adhered to. Suitable respiratory protective equipment should be used.] Such exposed personnel should be able to communicate with each other and with those (*i.e.*, more than one person) monitoring the work from outside the hazardous area. Monitors should be suitably equipped to effect rescue (possibly with lifelines to exposed personnel) and should themselves ensure, so far as is possible, that one of their number remains outside the hazard area in the event of an emergency.

N.B. the caveat in Section 13 (final paragraph).

13. FIRE HAZARDS[4,5,25 – 28]

Ethyl acetate is a readily flammable liquid and presents a dangerous fire and explosion hazard. It has a boiling point 77 °C, the flash point is − 4 °C (closed cup) and the lower and upper explosive limits in air are 2.0 and 11.5% v/v. The auto-ignition temperature is about 426 °C. It has limited solubility in water (about 8% at 20 °C) and a specific gravity of 0.9 (see Section 17). The vapour density is about three times that of air and the vapour, if in sufficient quantity, may therefore flow to a remote ignition source and then 'flash back'.

Containers of ethyl acetate can explode when heated and, if feasible, should be cooled with flooding quantities of water from as far as possible until well after the fire is extinguished.

Small fires may be fought with alcohol-resistant foam, dry powder, carbon dioxide, vaporizing liquid or water spray. Note though that, as the boiling point of water–ethyl acetate azeotrope is only about 70.4 °C, the addition of water may cause marked ebullition. Jets of water may scatter burning liquid.

Large fires may be fought with dry chemical, carbon dioxide or alcohol-resistant foam. Ref. 4 states that the use of dry chemical, where it can get into a tank of ethyl acetate, is not recommended. Immediate withdrawal is imperative in the event of rising sound from a venting safety device or any discolouration of a storage tank due to the fire. Personnel should keep upwind of the incident and avoid breathing the vapours of ethyl acetate and its possibly toxic combustion products which may include carbon monoxide.

Ref. 26 notes that where a fire is fed by an uncontrolled flow of combustible liquid the decision on how or if to fight it will depend on the size and type of fire anticipated and must be carefully considered. This may call for special engineering judgement, particularly in large-scale applications.

[Notwithstanding the foregoing review of various authorities noted in Sections 12 and 13, it is essential that managers and others responsible for the planning, implementation and overseeing of personnel and plant safety should be familiar with the legal constraints and official guidelines applicable to them and that they liaise with their local emergency services in the planning of plant and storage facilities and in the preparation of contingency plans for dealing with fires and other emergencies. Managers should regularly monitor staff knowledge of, and ability to implement, emergency procedures and should ensure that equipment provided for use in emergencies is regularly inspected and maintained.]

14. HAZARDOUS REACTIONS[3,5,25,27,29]

Ethyl acetate is stable at normal temperatures and pressures. It is not known to polymerize.[25]

OSHA[5] states that ethyl acetate is incompatible with nitrates, strong oxidizing agents, strong alkalis and strong acids. Contact with these materials may lead to fire and explosion. Bretherick[27,29] states that ethyl acetate may ignite or explode in contact with lithium tetrahydroaluminate or potassium t-butoxide. It may also react vigorously with chlorosulphonic acid, oleum, and with mixtures of lithium tetrahydroaluminate and 2-chloromethylfuran.

OSHA[5] notes that toxic gases and vapours (*e.g.*, carbon monoxide) may be released in fires involving ethyl acetate (see also Section 13). Ethyl acetate may attack some forms of plastics, rubber and coatings.

15. EMERGENCY MEASURES IN THE CASE OF ACCIDENTAL SPILLAGE[4,5,25,30]

See also Sections 12, 13 and 14.

If ethyl acetate is spilled all ignition sources should be eliminated immediately and personnel instructed to keep well away from the spillage area. The general aim should be to ventilate the area of the spill while at the same time isolating the material itself. It will be necessary to decide whether to control the spillage on site or to evacuate the building and summon the fire service. If possible windows should be opened and doors closed as the area is evacuated. Every effort should be made to prevent the ethyl acetate entering a confined space, such as a drain or sewer, because of the possibility of explosion. In the event of such a hazard appearing, the water authority must be informed.

OSHA[5] recommends that clothing contaminated with ethyl acetate should be removed immediately and not re-used until the contamination is removed. Clothing wet with ethyl acetate should be stored in sealed containers until discarded or suitably cleaned.

If the spillage is less than about 0.5 l the ethyl acetate should be absorbed on paper or inert absorbent. The solvent should then be allowed to evaporate from the paper in an appropriate facility (*e.g.*, a fume cupboard). Alternatively absorb the ethyl acetate on sand or other absorbent material and place in containers for later disposal.

For spillages up to about 25 l absorb the ethyl acetate onto sand, vermiculite or other absorbent (*e.g.*, by pouring sand over the spill). Transfer the contaminated absorbent to buckets. Remove the material to a safe open area and evaporate to the atmosphere. The site of the spillage should be thoroughly washed with water and detergent, and the washings discharged to a sewer (not to a 'soakaway' nor a land drain).

Large spillages should be bunded ahead of the spill. Possible ignition sources should be eradicated. The hazard area should be identified and isolated. Non-essential personnel should be banned from it.

In cases where the spillage might have environmental effects the appropriate authorities including the water authority must be informed immediately. Attempts should be made to contain the ethyl acetate and to isolate it from possible ignition sources and oxidizing materials. It should be approached from upwind.

In water, ethyl acetate may form colourless slicks that float and dissolve at a moderate rate. The dissolved ester is degradable in the presence of competent organisms, *e.g.*, sewage, and undergoes slow hydrolysis to form ethanol and acetic acid both of which are also degradable. Slicks can be treated with oil skimming equipment and sorbent foams. Carbon or peat may be used on dissolved material. Ethanol formed by hydrolysis rapidly biodegrades, while the much more slowly biodegradable acetic acid can be neutralized with sodium bicarbonate. For the purposes of beach and shore restoration ethyl acetate may be burnt off.

16. FIRST AID

General

The casualty should be removed from danger of further exposure into fresh air. Rescuers should ensure their own safety from inhalation and skin contamination where necessary. Conscious casualties should be asked for information about what has happened. Bear in mind that the casualty may lose consciousness at any time. Where necessary continued observation and care should be ensured. First aiders should take care not to become contaminated. Note that ethyl acetate is flammable.

Inhalation

The conscious casualty should be kept at rest. The unconscious casualty should be placed in the recovery position and an open airway maintained. If breathing or heart beat stops, resuscitation should be commenced immediately. Medical aid should be obtained or the casualty removed to hospital immediately.

Skin contamination

Remove contaminated clothing. Wash with copious amounts of water, or soap and water, for at least 15 minutes. If necessary, seek medical aid. In the case of persistent skin irritation refer for medical advice.

Eye contamination

Irrigate with copious amounts of water for at least ten minutes but avoid further damage to the eye from use of excessive force. Seek medical aid. Remember the possible presence of contact lenses, which may be affected by some solvents and may impede decontamination of the eye.

Ingestion

If the lips or mouth are contaminated rinse thoroughly with water. Do not induce vomiting. Give further supportive treatment as for inhalation. In the case of ingestion of significant amounts wash out mouth with water. Obtain immediate medical attention. Contact a hospital or poisons centre at once for advice. Remember that symptoms may develop many hours after exposure so continued care and observation may be necessary.

In all cases note information on the nature of exposure, and give this to ambulance or medical personnel. Information for doctors is provided in some technical literature issued by manufacturers.

17. PHYSICO-CHEMICAL PROPERTIES

17.1 General[18,31]

Ethyl acetate is a colourless, flammable, volatile liquid. It has a fruity odour, and is said to have a pleasant taste when diluted.[18,31]

The solubility of ethyl acetate in water at 0 °C is 10.08% w/w and at 20 °C is 7.94% w/w. The solubility of water in ethyl acetate at 0 °C and at 20 °C is 2.28% w/w and 3.01% w/w respectively.[31] Ethyl acetate is miscible with ethanol, diethyl ether, acetone and chloroform.[18]

17.2 Melting Point[32]

The melting point of ethyl acetate is given by Weast[32] as

-83.6 °C.

17.3 Boiling Point[32,33]

Weast[32] records the boiling point of ethyl acetate as

77.06 °C.

Ethyl acetate forms azeotropes with a number of other compounds. Binary ones formed with water and also with some common organic solvents are as follows[33]:

Wt% ethyl acetate	Second component	Wt% second component	Boiling pt azeotrope (°C)
54	Cyclohexane	46	72.8
56	Cyclohexane	44	71.6
74.2	Ethanol	25.8	72.18
69.02	Ethanol	30.98	71.81
91.53	Water	8.47	70.38

17.4 Density/Specific Gravity[31,32]

The specific gravity of ethyl acetate at 20 °C (relative to water at 4 °C) is 0.9003.[32]

The following values for the specific gravity of ethyl acetate relative to water at 20 °C are given by Marsden and Mann[31]:

Temperature in °C	Specific gravity
0	0.9260
10	0.9140
20	0.9020
30	0.8905
40	0.8790
50	0.8675
60	0.8550
70	0.8430

The critical density of ethyl acetate is given by Marsden and Mann[31] as 0.308 g/cm^3. From this the calculated critical volume is 3.247 cm^3/g.

17.5 Vapour Pressure[32]

A selection of vapour pressures of ethyl acetate below one atmosphere is given by Weast[32]:

Temperature in °C	Vapour pressure in mmHg
−43.4	1
−13.5	10
9.1	40
27.0	100
59.3	400

Weast[32] also gives the following values, for pressures at and above atmospheric pressure:

Temperature in °C	Pressure in atm
77.1	1
100.6	2
136.6	5
169.7	10
209.5	20

The critical temperature and critical pressure of ethyl acetate are recorded as 250.4 °C, and 37.8 atm [*ca.* 28700 mmHg] respectively.[32]

17.6 Vapour Density[18]

3.04 (relative to air = 1).

17.7 Flash Point[18,26]

−4 °C, closed cup test.[26]
7.2 °C, open cup.[18]

17.8 Explosive Limits[26]

The limits of flammability of ethyl acetate in air at normal atmospheric temperature and pressure are given by the National Fire Protection Association[26] as:

Lower limit: 2.0% v/v
Upper limit: 11.5% v/v.

The NFPA[26] give the ignition temperature of ethyl acetate as 426 °C. However, they point out that different test conditions (and also different definitions of "ignition temperature") can result in widely varying values being quoted. They therefore recommend that any value should be treated as an approximation.

17.9 Viscosity[32]

The viscosity of liquid ethyl acetate is recorded by Weast[32] as follows:

Temperature in °C	Viscosity in cp
0	0.582
8.96	0.516
10	0.512
15	0.473
20	0.455

Temperature in °C	Viscosity in cp
25	0.441
30	0.400
50	0.345
75	0.283

Weast also lists the viscosity of ethyl acetate vapour[32]:

Temperature in °C	Viscosity in μp
0	68.4
100	94.3
128.1	101.8
158.6	109.8
192.9	119.5
212.5	126

17.10 Concentration Conversion Factors

At 25 °C and 760 Torr (1 atm),

$1 \text{ mg/m}^3 = 0.28 \text{ ppm, and}$
$1 \text{ ppm} = 3.60 \text{ mg/m}^3.$

18. TOXICITY

18.1 General[4,5,8,25,34]

[Some reports relevant to the toxicology of ethyl acetate are difficult to evaluate because they appear to concern exposure to solvent mixtures of unknown composition.]

Ethyl acetate has the reputation of being one of the least toxic of volatile organic solvents (*e.g.*, refs. 8, 34). However, attention is drawn to the possibility that it may be more toxic than commonly supposed (*e.g.*, ref. 34). Patty[8] reports no record of any saturated aliphatic acetate having given rise to any health problems during their manufacture, handling or end use.

Ethyl acetate can affect the body if it is inhaled or swallowed, or if it comes into contact with the eyes or skin. It has a pleasant, fruity smell and, when diluted, a pleasant taste. The odour recognition concentration in air is about 50 ppm [4,5]. Ethyl acetate can be tasted in water and milk at 6 and 4.7 ppm, respectively.[8]

Ethyl acetate is a CNS depressant and concentrations above about 2,000 ppm are anaesthetic.[4] Severe over-exposure may lead to weakness, drowsiness, unconsciousness and death. The IDLH concentration for man is about 10,000 ppm. Ethyl acetate is moderately irritating to the eyes, mucous membranes, and respiratory tract. In quantities greater than about 0.5 ml it is said[4] to cause small burns [presumably to mucous membranes].

Sax[4] states that ethyl acetate has no potential for accumulation. It is slightly toxic to man on chronic exposure by any route. Prolonged inhalation can lead to renal or hepatic damage. Prolonged contact may cause skin irritation, dermatitis, conjunctival irritation and corneal clouding.

Persons with a history of chronic respiratory disease, skin disease or anaemia may be at increased risk from the toxic effects of ethyl acetate.

18.2 Toxicokinetics[8,35]

Ethyl acetate is relatively soluble in plasma and therefore passes readily through the alveoli. It is probably hydrolysed by simple chemical hydrolysis or by esterases in the liver and plasma, as well as being partially excreted in the urine and exhaled. Hathway and Bedford[35] report that ethyl acetate is biotransformed to ethanol in rat blood with $t_{0.5}$ *in vitro* and *in vivo* transformation times of 65, and 5 to 10 minutes, respectively. The transformation was assumed to take place esterically. Patty[8] reports studies which showed that for exposures to 217 to 770 ppm in man there was a linear relationship between inhaled ethyl acetate and ethanol in the alveolar air; respiratory retention of ethyl acetate is high and respiratory elimination practically negligible. In the rat, at inhaled concentrations above 2,000 ppm, ethyl acetate hydrolysis apparently exceeded ethanol oxidation, leading to ethanol accumulation in blood. *In vivo*, hydrolysis to ethanol occurs rapidly.

18.3 Human Acute Toxicity[3,5,8,25,28,36-40]

See also Section 18.1.

Sax[3] states that ethyl acetate is irritating to mucous surfaces (in particular the eyes and respiratory passages). It is also a CNS depressant and mildly narcotic. Ingestion of small quantities may cause sore throat, abdominal pain and diarrhoea.[25] Severe over-exposure may cause weakness, drowsiness, headache and unconsciousness.[5]

Patty[8] concludes that in general aliphatic esters have lower anaesthetic potency than most chlorinated solvents or diethyl ether. However they are more potent than ethanol, acetone and pentane, causing narcosis with slow recovery. Inhalation of 400 – 600 ppm for two to three hours caused no narcosis.[36] The IDLH for man is about 10,000 ppm.[40]

ACGIH[28] report that subjects not accustomed to ethyl acetate found the odour objectionable at 200 ppm. At 400 ppm there was mild eye, nose and throat irritation. Ruth[37] reports lowest and highest reported odour thresholds for ethyl acetate of 0.0196 and 665 mg/m³, respectively. However it is stated that ethyl acetate is a mild narcotic even at concentrations of 46,000 mg/m³ [*sic*] [equivalent to about 13,000 ppm]. Patty[36] reports that ethyl acetate vapour is irritant to eyes and respiratory tract at 400 – 800 ppm. The liquid may cause defatting and cracking of the skin but skin sensitization is negligible. Lehmann and Flury[38] report that in man ethyl acetate is the least irritant in the series of acetate esters – benzyl > amyl > butyl > propyl > ethyl.

ACGIH < reports that workers exposed to between 375 and 500 ppm for several months showed no unusual signs or symptoms.

Ethyl acetate is absorbed through the lungs and GIT. The extent to which it is absorbed through the intact skin is not known.[39] Prolonged skin contact may cause dryness and irritation. Persons previously exposed may develop sensitization dermatitis possibly due to impurities.[25,36] The main pathological findings are marked hyperaemia of the upper respiratory tract, petechial haemorrhages in the epicardium and pleura, hyperaemia of the spleen and kidney, and haemorrhagic gastritis. The affected organs had a definite odour of ethyl acetate.[39]

18.4 Animal Acute Toxicity[3,5,8,25,28,36,39]

ACGIH[28] states that the acute toxicity of ethyl acetate is low by all routes of administration.

OSHA[5] states that ethyl acetate has a narcotic effect in animals at concentrations greater than 5,000 ppm. [Comparison with ref. 39 suggests that the animals concerned were white mice.] Ref. 39 reports that for cats the minimum narcotic concentration is 12,000 ppm for five hours exposure. However, deeply anaesthetised cats frequently died after temporary recovery. Intravenous injection of 117 mg/kg caused narcosis in rabbits. Large doses [*sic*] caused a fall in blood pressure in rabbits whereas intravenous injections of small doses (4.5 ml of 5% solution) were found to have no effect.

Rabbits injected with about 50 mg/kg of ethyl acetate exhibited marked stimulation of the respiration which lasted some 15 to 30 minutes. A similar effect was found in morphinised rabbits. Doses of 100 mg/kg led to stimulation followed by depression and decrease of respiratory volume.[39]

According to ref. 39 the main toxic effects of ethyl acetate are CNS depression, moderate dyspnoea and primary irritation. The most characteristic pathological findings in animals killed by exposure to ethyl acetate are pulmonary oedema, haemorrhage and hyperaemia of the respiratory tract. In addition to these effects guinea pigs exposed to about 21,500 ppm for 10 hours exhibited congestion of liver and kidneys.

Splashes to the eye may cause painful conjunctival irritation and when applied direct to rabbit's eyes the liquid caused temporary corneal injury.

The following summary of animal acute toxicity data for ethyl acetate may be helpful:

orl	– rat	LD_{50}	=	11 g/kg (ref. 3)
			=	5.6 to > 11 g/kg (ref. 36)
ihl	– rat	LC_{50}	=	1,600 ppm/8 hrs (ref 3)
			=	200 mg/l (refs. 36, 39)
orl	– mus	LD_{50}	=	440 mg/kg (ref. 36)
ihl	– mus	LCLo	=	31,000 mg/m³/103 mins (ref. 3)
		LC_{50}	=	10,000 ppm/45 mins (ref. 36)

ihl	– cat	LCLo	=	61,000 mg/m^3 (ref. 3)
			=	> 20,000 ppm (ref. 36)
orl	– rbt	LD$_{50}$	=	4,935 mg/kg (refs. 3, 36)
ihl	– gpg	LCLo	=	77 g/m^3/60 mins (ref. 39)
			=	> 2,000 ppm (ref. 36)

Patty[8] reports that when injected intraperitoneally (1 ml/kg) toluene and ethyl acetate were less toxic in combination than in isolation. The toxicity of the ester was increased in combination with propylene oxide, propylene glycol, formalin, morpholine, ethylene glycol or ethanol.

18.5 Human Sub-acute and Chronic Toxicity Including Reproductive Effects[5.25.36.38.39.41–44]

General

Hamilton and Hardy[41] conclude that as a group ester lacquer solvents (primarily ethyl and butyl acetates) have relatively insignificant toxic properties. However Lehmann and Flury[38] suggested in 1943 that ethyl acetate is not as harmless as is sometimes assumed. Grant[42] reports that prolonged inhalation of ethyl acetate may damage the lungs, liver, kidney and heart. Spencer *et al.*[43] state that the possible role of agents such as ethyl acetate (*e.g.*, as potentiators) in hexacarbon neurotoxicity has only begun [1980] to be investigated.

Patty[36] states that no cumulative effects are apparent in man other than irritation.

OSHA[5.25] states that it has received no reports of chronic systemic effects in man and that no data are available concerning the effects of chronic inhalation of ethyl acetate. Browning[44] states that there is no evidence that ethyl acetate has a cumulative effect, although it has been reported to cause pulmonary oedema and to have a tendency to cause habituation. Prolonged exposure may lead to some loss of smell.

Von Oettingen[39] reports that in rare instances ethyl acetate may cause sensitization leading to inflammation of the mucous membranes and eczematous eruptions.

Lehmann and Flury[38] report the occurrence of inflamed gums and skin diseases in persons washing their hands frequently with a mixture of ethanol and ethyl acetate. (The authors also report that impure ethyl acetate vapour caused a strong burning sensation in the eyes but that this decreased when pure ethyl acetate was used.) Grant[42] reports that at 400 ppm ethyl acetate vapour irritates the human eye. The vapour produced vacuoles in the corneal epithelium of cats but these were not observed in thirty workers chronically exposed to 15 to 50 mg of ethyl acetate plus 20 to 80 mg of amyl acetate, per litre of air. The workers exhibited hyperaemia of the bulbar conjunctiva but no abnormalities of the cornea. The author reports that systemic absorption is not known to affect the eyes.

Hamilton and Hardy[41] state that there have been rare reports of visceral injury, and effects on the bone marrow and nervous system associated with exposure to aliphatic ester vapours. However the authors conclude that careful examination of the exposure and diagnostic data in such cases almost always suggest that the ester(s) was not necessarily responsible for the observed effect. Browning[44] notes that a number of ill-effects were attributed to ethyl acetate at times when it was in intensive use and when due care may not have been exercised. Furthermore impurities or other solvents may have been present.

Carcinogenicity to Man

No data were found concerning any carcinogenic effects of ethyl acetate in man.

Reproductive Effects in Man

No data were found concerning any reproductive effects of ethyl acetate in man.

18.6 Animal Sub-acute and Chronic Toxicity Including Reproductive Effects[3,5,36,38,39,42,44–46]

See also Sections 18.1, 18.2, 18.4.

General

Patty[36] notes that aliphatic esters in general do not show marked cumulative effects in animals or man. (The exception is their irritant effects on eyes and respiratory tract.) Animal pathology after chronic or acute exposure to aliphatic esters is limited to vascular congestion, respiratory irritation, and loss of weight at high daily exposures. Rats given 1 ml/kg intraperitoneally for eight days showed increased blood pyruvate, lactate and glycolytic enzymes. Guinea pigs showed no notable effects following 65 inhalational exposures to 2,000 ppm.

Sax[3] reports that chronic poisoning with ethyl acetate leads to secondary anaemia, leucocytosis and cloudy swelling, and fatty degeneration of the viscera.

Lehmann and Flury[38] report that [in mice ?] inhalation of 15 to 16 mg/l for six hours per day for seven days led to irritation of the mucous membranes followed by habituation. However there were no definite narcotic symptoms. Browning[44] reports that in guinea pigs 2,000 ppm also failed to produce narcotic symptoms. After-effects found were decreased appetite, loss of weight and fatigue. Lehmann and Flury[38] state that repeated exposure to ethyl acetate leads to blood changes *i.e.*, an increase in the number of erythrocytes, and increase in neutrophils at the expense of lymphocytes. In general there was no increase in haemoglobin and no general change in the white blood corpuscles. There was no apparent liver or kidney injury (as indicated by changes in the urine).

Rabbits exposed to about 4,450 ppm for one hour per day for 40 days suffered secondary oedema, leucocytosis, hyperaemia, cloudy swelling and variable fatty degeneration of various organs. The animals showed liver and kidney damage, a reduction in haemoglobin and a decreased red blood cell count. Neither the white blood cell count nor the level of phosphorus in the blood were affected. However there was reduction in blood chloride and normal calcium levels with hypopotassaemia and hypermagnesaemia. Marked hyperplasia of the reticulo-endothelial system was found together with fatty degeneration and a trend towards hypertrophy of the splenic follicles as a consequence of increased blood destruction.[39]

Cats exposed to ethyl acetate vapour for six to eight hours per day for a week exhibited an increase in red blood cells but not haemoglobin. There was no leucocytosis but a relative increase in segmented neutrophiles compared to lymphocytes.[39]

OSHA[5] reports that rabbits repeatedly exposed to 4,450 ppm of ethyl acetate for one hour per day for 40 days developed anaemia with leucocytosis, as well as liver and kidney damage.

Grant[42] reports that rabbits exposed to concentrations of ethyl acetate vapour that would scarcely be tolerated by human beings [sic] exhibited no corneal damage. Exposure was for eight hours per day for five days per week for up to seven weeks. However there were signs of conjunctival irritation.

Carcinogenicity to Animals

Few studies appear to have been undertaken to investigate the possible carcinogenicity of ethyl acetate to animals.

RTECS[45] refers to one study [ref. 46] in mice. An intraperitoneal dose of 3,600 mg/kg delivered three times per week for eight weeks was found to be the lowest published toxic dose. [However the report of the original study gave no indication of carcinogenic effect.]

Reproductive Effects in Animals

No data were found concerning any reproductive effects of ethyl acetate in animals.

18.7 Mutagenicity[45]

Few studies appear to have been made into the possible mutagenic effects of ethyl acetate.

RTECS[45] reports effects in *Saccharomyces cerevisiae* at 24,400 ppm and in hamster fibroblasts at 9 g/l.

18.8 Summary

[Ethyl acetate is of low acute toxicity, the main systemic effect of high doses by ingestion or inhalation being narcosis, perhaps with headache. Non-narcotic vapour concentrations of 400 – 800 ppm may irritate the respiratory tract and the eyes. The liquid may irritate eyes and mucous membranes, while prolonged or repeated skin contamination may cause degreasing and chapping; skin sensitization is unlikely unless associated with impurities. We found no evidence of systemic cumulative effects in man and no observations indicating carcinogenicity or adverse reproductive effects. The compound is readily metabolized to ethanol.]

19. MEDICAL / HEALTH SURVEILLANCE

A decision on the need for, and content of, medical surveillance should be based on an assessment of the possibility and extent of exposure in the work operation. In addition, medical examination may be directed to identifying any pre-existing or newly arising condition in the individual workers which might either be aggravated by subsequent exposure, or might confuse any subsequent medical assessment in the event of excessive exposure or an illness not related to exposure. A particular aspect to be considered is the identification of sensitive subjects not adequately protected by the control limit in operation. A professional medical judgement may be required on continuance of employment in the specified process. The following information is relevant in the case of this solvent.

Pre-employment and Periodic Medical Examination

At both initial and periodic medical reviews, particular attention will be directed to the state of the skin, respiratory function, and information on hepatic and renal systems. Where exposure has exceeded limit values, or in the event of severe skin contamination, a further medical review may be necessary. Concentrations in excess of 400 ppm cause irritation of the eyes and upper respiratory tract. There are no reported delayed effects. Biological monitoring by measurement of ethanol and ethyl acetate in exhaled breath has been noted as a possible indication of exposure but confounding factors may be present.

20. OCCUPATIONAL EXPOSURE LIMITS

[The Committee felt that a time-weighted average limit of 400 ppm and a short term limit also of 400 ppm were appropriate for ethyl acetate. At this level there may be some initial irritant effects but there is no evidence of systemic effects.]

REFERENCES

1. Council of Europe. 'Dangerous Chemical Substances and Proposals Concerning their Labelling'. ('The Yellow Book'.) 4th ed. Maisonneuvre, 1978.
2. International Union of Pure and Applied Chemistry. 'Nomenclature of Organic Chemistry'. 1979 ed. Pergamon Press, 1979.
3. Sax, N.I. 'Dangerous Properties of Industrial Materials'. 6th ed. Van Nostrand Reinhold, 1984.
4. Sax, N.I., Ed. 'Hazardous Chemicals Information Annual', No.1. Van Nostrand Reinhold Information Services, New York, 1986.
5. National Institute for Occupational Safety & Health/Occupational Safety & Health Administration. 'NIOSH/OSHA Occupational health guideline for ethyl acetate'. DHHS (NIOSH) Publication. US Department of Health & Human Services/US Department of Labor, 1978.

6. Furia, T.E. and N. Bellanca, Eds. 'Fenaroli's Handbook of Flavor Ingredients'. 2nd ed. CRC Press, 1975.
7. Graedel, T.E. 'Chemical Compounds in the Atmosphere'. Academic Press, 1978.
8. Clayton, G.D. and F.E. Clayton, Eds. 'Patty's Industrial Hygiene and Toxicology'. 3rd ed. rev. Wiley Interscience, 1982.
9. Grasselli, J.G. and W.M. Ritchey, Eds. 'Atlas of Spectral Data and Physical Constants for Organic Compounds'. 2nd ed. CRC Press, 1975.
10. National Institute for Occupational Safety & Health. 'NIOSH Manual of Analytical Methods'. 2nd ed. DHEW (NIOSH) Publication No.77-157-A. US Department of Health, Education & Welfare, 1977.
11. Mariotti, M. *Bollettino dei Chimici dei Laboratori Provinciali*, **28**, No.7(1977)(III):193.
12. Diaz-Rueda, J. *et al. Applied Spectroscopy*, **31**, No.4(1977):298.
13. Mal'tseva, G.A. and Y.N. Talakin. *Gigiena i Sanitariya*, **50**, No.9(1985):55.
14. Mukhtarova, M. *Khigiena i Zdraveopazvane*, **23**, No.2(1980):173.
15. Firsova, O.V. *et al. Khimicheskie Volokna*, 3(1980):56.
16. Mierzwinski, A. and Z. Witkiewicz. *Chemia Analityczna (Warsaw)*, **30**, No.3(1985):429.
17. Sopikov, N.F. and A.I. Gorshunova. *Gigiena i Sanitariya*, **44**, No.5(1979):59.
18. Windholz, M. *et al.*, Eds. 'The Merck Index'. 10th ed. Merck & Co., 1983.
19. Hawley, G.G., Ed. 'The Condensed Chemical Dictionary'. 10th ed. Van Nostrand Reinhold, 1981.
20. International Labour Office. 'Encyclopaedia of Occupational Health and Safety'. 3rd rev. ed. ILO, 1983.
21. Grayson, M., Exec. Ed. 'Kirk-Othmer Encyclopedia of Chemical Technology'. 3rd ed. John Wiley & Sons, 1979.
22. Furia, T.E., Ed. 'Handbook of Food Additives'. 2nd ed. CRC Press, 1980.
23. Commission of the European Communities. 'Environmental Chemicals Data and Information Network (ECDIN)' Databank. Commission of the European Communities Joint Research Centre, ECDIN Group, I-21020 Ispra (Varese), Italy, 1986.
24. Lowenheim, F.A. and M.K. Moran. 'Faith, Keyes and Clark's Industrial Chemicals'. 4th ed. John Wiley & Sons, 1975.
25. Occupational Safety & Health Administration. Material safety data sheets from the Occupational Health Services database, OSHA, Washington D.C., 1986.
26. National Fire Protection Association. 'Fire Protection Guide on Hazardous Materials'. 9th ed. NFPA, Massachusetts, USA, 1986.
27. Bretherick, L., Ed. 'Hazards in the Chemical Laboratory'. 4th ed. Royal Society of Chemistry, UK, 1986.
28. American Conference of Governmental Industrial Hygienists. 'Documentation of the threshold limit values for chemical substances in the workroom environment'. ACGIH, 1986.
29. Bretherick, L. 'Handbook of Reactive Chemical Hazards'. 3rd ed. Butterworths, 1985.
30. Warren, P.J., Ed. 'Dangerous Chemicals Emergency Spillage Guide'. 1st ed. Jensons (Scientific) Ltd., Leighton Buzzard, UK, 1985.
31. Marsden, C. and S. Mann, Eds. 'Solvents Guide'. 2nd ed. rev. Cleaver-Hume Press, 1963.
32. Weast, R.C., Ed.-in-Chief. 'CRC Handbook of Chemistry and Physics' ('The Rubber Handbook'). 67th ed. CRC Press, 1986-1987.
33. Horsley, L.H. 'Azeotropic Data III'. Advances in Chemistry Series No.116, American Chemical Society, 1973.
34. American Conference of Governmental Industrial Hygienists. 'Documentation of the threshold limit values for chemical substances in the workroom environment'. 4th ed. ACGIH, 1982.
35. The Chemical Society. 'Foreign compound metabolism in mammals', Vol.4. A Specialist Periodical Report. London, 1977.
36. Fassett, D.W. and D.D. Irish, Eds. 'Patty's Industrial Hygiene and Toxicology', Vol.II: 'Toxicology'. 2nd rev. ed. Wiley Interscience, 1962.
37. Ruth, J.H. *American Industrial Hygiene Association Journal*, 47(1986):A-142. 'Odor thresholds and irritation levels of several chemical substances: a review'.

38. Lehmann, K.B. and F. Flury, Eds. 'Toxicology and Hygiene of Industrial Solvents'. The Williams & Wilkins Company, Baltimore, USA, 1943.
39. Von Oettingen, W.F. *American Medical Association Archives of Industrial Health*, **21**, No.1(1960):46/34. 'The aliphatic acids and their esters: toxicity and potential dangers'.
40. National Institute for Occupational Safety & Health. 'Pocket Guide to Chemical Hazards'. 5th printing. DHEW (NIOSH) Publication No.78-210. US Department of Health & Human Services, 1985.
41. Hamilton, A. and H.L. Hardy. 'Industrial Toxicology'. 3rd ed. Publishing Sciences Group, Massachusetts, USA, 1974.
42. Grant, W.M. 'Toxicology of the Eye'. 2nd ed. Charles C. Thomas, Illinois, USA, 1974.
43. Spencer, P.S. *et al. CRC Critical Reviews in Toxicology*, **7**, No.4(1980):279. (Ed. L. Golberg.) 'The enlarging view of hexacarbon neurotoxicity'. CRC Press.
44. Browning, E. 'Toxicity and Metabolism of Industrial Solvents'. Elsevier, 1965.
45. National Institute for Occupational Safety & Health. 'Registry of Toxic Effects of Chemical Substances (RTECS)'. DHHS (NIOSH) Publication No.84-101-6. US Department of Health & Human Services, April, 1986.
46. Stoner, G.D. *et al. Cancer Research*, 33(1973):3069. 'Test for carcinogenicity of food additives and chemotherapeutic agents by the pulmonary tumor response in strain A mice'.

CONTENTS

1. CHEMICAL ABSTRACTS NAME

methanol
(10th Collective Index)

2. SYNONYMS AND TRADE NAMES[1-5]

alcool metilico (Italian)
alcool méthylique (French)
carbinol
colonial spirit
columbian spirit
columbian spirits
denaturing grade wood alcohol
hydroxymethane
manhattan spirits
metanolo (Italian)
méthanol (French)
methyl alcohol
methylalcohol (Dutch)
methylalkohol (German)
methyl hydroxide
methylol
metylowy alkohol (Polish)
monohydroxymethane
OHS14280
pyroxylic spirit
U154
UN 1230
wood alcohol
wood naphtha
wood spirit
wood spirits

3. CHEMICAL ABSTRACTS SERVICES REGISTRY NUMBER

67-56-1

4. NIOSH NUMBER

PC1400000

5. CHEMICAL FORMULA

CH_4O
(Molecular weight 32.05)

6. STRUCTURAL FORMULA

$$CH_3\!\!-\!\!OH$$

7. OCCURRENCE[6-8]

Methanol occurs naturally, being a metabolite in man, animals and plants. It is readily biodegradable both under aerobic and anaerobic conditions.[6] Man-made sources of methanol in the atmosphere include cigarette smoke (in which it can have a level of 700 ppm) and the exhaust from gasoline engines (in which levels of 0.1–0.6 ppm are known).[7] Some other man-made and natural sources include wood[7] and wood uses, such as pulping, combustion and charcoal manufacture[8]; animal waste; refuse and plastics combustion; volcanoes; usage and storage of methanol.[8] Ref. 6 quotes data giving the methanol content of various wines and spirits as ranging up to 325 mg/l.

Graedel[8] notes that methanol is also a product of atmospheric alkane chemistry. Concentrations as high as 100 ppb have been found and, at that level, methanol is classifiable as a major atmospheric trace species.

It is believed that the initial atmospheric reaction is hydrogen abstraction by the hydroxyl radical. For methanol, a plausible sequence is[8]

$$CH_3OH \xrightarrow{HO\cdot} CH_2OH\cdot \xrightarrow{O_2} O_2CH_2OH\cdot \xrightarrow{NO} OCH_2OH\cdot \xrightarrow{O_2} H\text{-}COOH\cdot$$

8. COMMERCIAL AND INDUSTRIAL METHANOL[9-12]

Methanol is a clear, colourless, volatile, flammable liquid with a mild odour.[9,10] Ref. 11 lists available grades as, "technical; chemically pure (99.85%); electronic (for cleaning and drying components); fuel".

Wade *et al.* in ref. 12 state that, in the US, sales-grade methanol must normally meet US Federal Grade Specifications A or AA. The specification for Grade A is

methanol content, wt%, min.	99.85
acetone and aldehydes, ppm, max.	30
acid (as acetic acid), ppm, max.	30
water content, ppm, max.	1,500
specific gravity, (d_{20}^{20})	0.7928
permanganate time, min.	30
odour	characteristic

distillation range at 101 kPa (760 mmHg)	1 °C, must include 64.6 °C
colour, platinum-cobalt scale, max.	5
appearance	clear – colourless
residual on evaporation, g/100 ml	0.001
carbonizable impurities; colour platinum-cobalt scale, max.	30

Grade AA differs in specifying an acetone maximum (20 ppm), a minimum for ethanol (10 ppm), and in having a more stringent water content specification (1,000 ppm, maximum).

9. SPECTROSCOPIC DATA[13]

Infrared, Raman, ultraviolet, both ^{1}H and ^{13}C NMR, and mass spectral data for methanol have been tabulated.[13]

10. MEASUREMENT TECHNIQUES[14-24]

The method recommended for the determination of methanol in workshop atmospheres is the NIOSH[14] silica gel trapping method for collection and concentration, followed by aqueous extraction of the silica gel and a gas chromatographic (gc) analysis of the extract.

The collection tube is 7 cm long with a 4 mm internal diameter. It contains a total of 150 mg of silica gel (20 – 40 mesh) divided into a front section of 100 mg and a rear section of 50 mg separated by a small plug of urethane foam. Air is sampled at a flow-rate of 0.2 l per minute for 25 minutes by means of a small pump. The entire apparatus is portable so that it may be carried in a pocket with the sampling tube in the breathing zone (normally a coat lapel). The apparatus can also be static.

Water is used to extract the methanol, which is separated on a 10 ft × 0.125 in stainless steel column packed with 10% FFAP on Chromosorb W-AW (80 – 100 mesh) using flame ionization detection. For full details of the method see ref. 14.

The sampling method uses a small portable apparatus with no liquids and is therefore easy to handle and maintain. The analytical method is relatively specific with the additional advantage of flexibility of operating conditions.

Air samples are normally taken by means of suction using an electric pump capable of operating at low flow-rates (*i.e.*, 30 – 300 ml per minute). If higher flow-rates are used, breakthrough can occur (*i.e.*, the methanol is not completely adsorbed and some of it passes through the adsorbent). It is also convenient either to establish the trapping device in a static position, or to have it attached to the user in such a way that the breathing zone is sampled (this usually means attachment to the worker's lapel).

It should be noted that suitable personal samplers may be commercially available.

Most methods for the collection of methanol use trapping on activated charcoal or simultaneous collection and oxidation of methanol to formaldehyde. Air containing methanol, toluene and *o*-xylene was passed at a flow-rate of no greater than 500 ml per minute (maximum 3 l) through a 7 cm × 4 mm tube containing 50 – 100 mg activated charcoal.[15] The organic compounds were extracted with dimethylformamide and the extract was analysed by gc on a 3 m × 4 mm column packed with 10% Carbowax 20M on Chromosorb W (80 – 100 mesh). The column temperature was 120 °C and detection was by flame ionization.

Methanol, together with toluene was also trapped on activated charcoal from air samples taken in a driver's cab of a vehicle, the engine of which was fuelled by gasoline–methanol.[16] The methanol and toluene were extracted with chlorobenzene and the extract was subjected to gc on a 3 m × 3 mm column packed with 8% 1,2,3-tris-(2-cyanoethoxy)propane on Chromaton N and detection was by flame ionization. The limit of detection for methanol was 1 µg per cubic metre of air.

Methanol, together with acetone, toluene and ethyl acetate was recovered from 10 l of air at a flow-rate of 11 ml per minute by passage through a tube containing 150 mg of activated charcoal.[17] Carbon disulphide was used to extract the organic compounds which were determined by infrared spectrometry using caesium bromide windows.

Methanol from air was absorbed by acidified potassium permanganate solution whereby it was simultaneously oxidized to formaldehyde.[18] The yellow dye produced with 4-nitroaniline was spectrophotometrically determined at 395 nm.

Aqueous potassium permanganate acidified with phosphoric acid was used to absorb methanol from air with the simultaneous oxidation to formaldehyde.[19] After the addition of *p*-aminoazobenzene and sulphur dioxide, the resulting pink dye was spectrophotometrically determined at 505 nm. The limit of detection was 5 µg per ml air.

Air was drawn at 0.5 l per minute through a water trap and an aliquot of this was treated with 75% sulphuric acid and 0.5% potassium permanganate.[20] The formaldehyde formed by the oxidation of the methanol was reacted with 10% acidic chromotropic acid and sodium sulphite and the resulting dye was spectrophotometrically determined at 570 nm. The limit of detection was 0.15 mg per ml air.

Infrared lasers have been used to selectively detect trace organic gases including methanol.[21] The air sample at 8 Torr was introduced to the 20 l capacity sample cell which comprised two highly polished and optically flat stainless steel electrodes spaced 1 mm apart. A square wave was applied after the optimum maximum and minimum voltages had been established, and laser radiation was detected synchronously by a (mercury, cadmium) Te detector. The detection limit was in the region of 0.03 – 1.1 ppm.

Blood serum was deproteinised, propanol was added as internal standard and acetone and aliphatic alcohols were separated by gc using a pre-column of 3% OV-1 on Gas Chrom Q and an analytical 30M capillary column.[22] The capillary column was packed with SPB-1 and was operated at 35 °C using flame ionization detection. Methanol and other aliphatic alcohols plus acetone were separated in less than three minutes.

Nuclear magnetic resonance has been used to determine methanol and acetone in blood serum at a level of less than 1 mM.[23]

Methanol and its metabolite, formaldehyde were determined in blood by a gc method.[24] After the addition of propanol as internal standard, chromatography was performed on a 2 m × 3 mm column packed with 25% PEG 1000 plus 2% potassium hydroxide on Chromosorb W (60 – 80 mesh) at 80 °C and using flame ionization detection.

11. CONDITIONS UNDER WHICH METHANOL IS PUT ON THE MARKET

Production[10,12,25]

Modern industrial production of methanol is based on synthesis from pressurized mixtures of hydrogen, carbon monoxide and carbon dioxide gases in the presence of metallic heterogeneous catalysts. The pressure used depends on the catalyst. Two hundred and fifty to three hundred and fifty atmospheres are regarded as high, 100 – 250 atmospheres as medium and 50 – 100 atmospheres as low pressures. High pressure processes commenced in Germany in 1923, the catalyst system used being a zinc oxide–chromium oxide one. In the late 1960s medium and low pressure plants became feasible with the successful development of highly active, durable, copper–zinc oxide catalysts.[12] Wade *et al.* in ref. 12 (published 1979) estimated that by 1981, 80% of US capacity would be using low pressure technology. A then incipient process innovation was a three-phase process wherein an inert liquid fluidises the catalyst and removes heat of reaction, giving higher single-pass conversions.

Methanol formation from hydrogen and mixtures of carbon oxides is given by

$$CO + 2H_2 \rightarrow CH_3OH \qquad \Delta H_{298\,K} = -90.77 \text{ kJ/mol}$$

$$CO_2 + 3H_2 \rightarrow CH_3OH + H_2O \qquad \Delta H_{298\,K} = -49.52 \text{ kJ/mol}$$

The product is purified by distillation. Typically, US Federal Grade A (see Section 8) specification requires two columns and Grade AA requires three columns though an individual manufacturer's procedure may entail the use of fewer or more columns than usual.

Natural gas and petroleum residues are common feedstocks for producing the synthesis gas for the above reactions. Other suitable feedstocks are naphtha and coal. Natural gas, petroleum residues and naphtha accounted for 90% of worldwide methanol capacity with (in 1980) natural gas alone accounting for 70%, petroleum residues 15% and naphtha 5%.[12]

With natural gas (principally methane) as feedstock, this is first desulphurised, usually to less than 0.25 ppm (by volume) of hydrogen sulphide. It is then mixed with steam, preheated to 425 – 500 °C and fed into a reformer where it is passed through externally fired tubes containing a nickel-impregnated ceramic catalyst.[12] Product from the reformer, a mixture of carbon monoxide, carbon dioxide, hydrogen, methane and steam, exits at, typically, 840 – 880 °C and 7 – 17 atmospheres. Significant amounts of fuel are necessitated by the highly endothermic overall reaction. The hydrogen is in excess of that stoicheiometrically necessary but this has the advantage of improving the effectiveness of the catalyst. The excess hydrogen is vented in the subsequent methanol synthesis and used as a fuel in the reforming step. The reactions involved in the reforming are:

Steam reforming of

$$CH_4 + H_2O \rightleftharpoons CO + 3H_2 \qquad \Delta H_{298\,K} = 206.2 \text{ kJ/mol}$$

Hydrocracking of heavy hydrocarbons

$$C_nH_{(2n + 2)} + (n - 1)H_2 \rightarrow nCH_4 \qquad \Delta H_{298K}^{ethane} = -65.07 \text{ kJ/mol}$$

Water gas shift

$$CO + H_2O \rightarrow CO_2 + H_2 \qquad \Delta H_{298\,K} = -41.25 \text{ kJ/mol}$$

With petroleum residue as feedstock, this is fed into a partial oxidation reactor where it reacts adiabatically with compressed oxygen and without catalyst to form a mixture of carbon oxides and hydrogen. The reaction is exothermic and sufficient heat is released to cause reforming of the remaining feedstock as described by the above equations. To provide a more appropriate hydrogen to carbon ratio, some of the carbon monoxide is shifted to carbon dioxide, which latter is then removed.[12] Impurities (soot, hydrogen sulphide, carbonyl sulphide and heavy metals) in the product gas need to be removed.[12]

The processing of naphtha feedstocks is nearly identical to that of the reforming of natural gas except that naphtha requires vaporization before being desulphurised.[12] The desulphurisation may be more complex and a different type of reforming catalyst is required. Copper catalysts' sensitivity to poisons requires careful purification of feed streams.

Wade *et al.* in ref. 12 observe that methanol production from coal accounted for under 2% of worldwide capacity in 1980. However the method was attracting serious attention as being independent of oil and natural gas supplies. Ref. 10 gives a reaction scheme in which coal and water produce first coke and then blue-water gas (containing about equal proportions of hydrogen and carbon monoxide) which is then purified and fortified with addition hydrogen from coke-oven gas prior to conversion to methanol.

World production of methanol in 1981 was eight million tonnes.[25] In 1980, EC production was 2.50 million tonnes. In 1983 it was 2.53 million tonnes and usage was 3.87 million tonnes.[25]

Methanol is transported in road and rail tankers, barges, drums, barrels, cans and bottles.[10]

Uses[5,9,10,12]

Important industrial uses of methanol are dehydrogenation and oxidative dehydrogenation to formaldehyde over silver or molybdenum–iron oxide catalysts;

carbonylation to acetic acid catalysed by cobalt or rhodium; acid catalysed elimination of water to give dimethyl ether, and acid catalysed reaction with isobutylene to form methyl t-butyl ether (MTBE) which is used increasingly as a gasoline octane improver. Methanol is used to form the methyl esters of organic and inorganic acids and it gives the various methylamines on direct reaction with ammonia. It has considerable use as a solvent[12] and as a denaturant for ethanol. Methanol itself is a clean-burning fuel and ref. 9 notes that its fuel use, either directly or as MTBE, consumed about 10% of 1983 production. Acetic acid manufacture accounted for about 8%.

Ref. 10 (published 1975) gave the following [USA] use pattern for methanol:

Formaldehyde	45%
Dimethyl terephthalate	10%
Solvents	10%
Methyl methacrylate	8%
Methylamines	4%
Methyl halides	4%
Acetic acid	4%
Miscellaneous and exports	15%

Rowe and McCollister in ref. 5 observe that seventy two occupations that offer exposure to methanol have been reported by the US Department of Labour.

Potential uses cited by Wade *et al.* in ref. 12 included gasoline manufacture in the MTG (methanol to gasoline) process and in single cell protein (SCP) production, SCPs being used as animal feed additives.

12. STORAGE, HANDLING AND USE PRECAUTIONS[12,26-29]

Methanol is volatile and flammable. It is miscible with water in all proportions. Ref. 26 observes that it is slightly corrosive to metals, especially to brass and bronze, this action being greatly accelerated by air and water. Plant and containers of steel are generally satisfactory for a reasonable period but aluminium or lead-lined vessels are more resistant.

Care must be taken with storage, process and drying plant to ensure that hazard-free conditions exist regarding fire and explosion risks and exposure of personnel to methanol. [As with other flammable solvents the organization of storage facilities has to take into account two separate requirements. These are that the stored material be protected from a fire elsewhere on the premises and that the premises be protected from a fire involving the stored materials. On an industrial scale, sufficient physical separation and water spray curtains are often`used effectively. On a laboratory scale, flammable solvents need to be stored in fire resistant, thermally insulated cupboards.

It is important to appreciate that a non-combustible barrier is not synonymous with a fire barrier. A fire barrier has to protect against both conducted and radiant heat. For example, a metal cupboard is useless on both these counts.]

Wade *et al.* in ref. 12 state that methanol may be stored and handled in clean carbon-steel equipment; that storage tanks should be constructed having an internal floating roof with an inert gas pad to minimize vapour emissions and that, because of methanol's flammability, tanks are usually bunded and protected by a foam-type fire extinguishing system (either carbon dioxide or dry chemical). Inert gas and not air may be used when pressure loading or unloading, but pumping is preferable.[12] ILO adds that ramped sills should be constructed at doorways of rooms within which methanol might escape from storage vessels.[27] Flameproof exhaust ventilation should be used and care should be taken that ventilation discharge is not close to air intakes.

Adequate ventilation should be provided to keep personnel exposures from being above the recommended limits. Personnel handling methanol without laboratory containment should be provided with and required to use adequate protective equipment to prevent repeated or prolonged skin contact with the chemical. Such equipment includes impervious clothing, gloves, face shields (eight inch minimum), splash-proof safety goggles and, in areas where vapour concentrations cannot be brought down to safe levels, suitable respiratory protective equipment should be provided. In such high vapour concentration areas all equipment should be spark and explosion proof. (For example, a static discharge in a plastics-lined metal tank ignited a 30:70 methanol – water mixture.[28]) [Note however that personnel should not be allowed to enter areas containing potentially hazardous concentrations of methanol unless all relevant legislation and official guidelines have been strictly adhered to.] Such exposed personnel should be able to communicate with each other and with those (*i.e.*, more than one person) monitoring the work from outside the hazardous area. Monitors should be suitably equipped to effect rescue (possibly with lifelines to exposed personnel) and should themselves ensure, so far as is possible, that one of their number remains outside the hazard area in the event of an emergency.

N.B. the caveat in Section 13 (final paragraph).

13. FIRE HAZARDS[1,30,31]

Methanol is a flammable, water-miscible liquid of boiling point 64.7 °C and with a flash point of 11 °C (closed cup). The lower and upper explosive limits in air are 6.0 and 36% v/v. The auto-ignition temperature is about 385 °C. The vapour density is about 11% greater than that of air (see Section 17).

Containers of methanol can explode when heated and, if feasible, should be cooled with flooding quantities of water applied from as far as possible until well after the fire is extinguished.

Small fires may be fought with dry chemical, carbon dioxide, water spray, alcohol-resistant foam or by dilution with plenty of water. Large fires can be attacked with water spray, fog or alcohol-resistant foam. Jets of water may scatter burning liquid notwithstanding its water miscibility. Water used in fire-fighting should be bunded for later disposal. Immediate withdrawal from the area is imperative in the case of rising sound from a safety venting device or in the case of discolouration of a storage tank due to the fire. Personnel should keep upwind of the fire and avoid breathing methanol vapour and possibly toxic combustion products.

Ref. 31 notes that where a fire is fed by an uncontrolled flow of combustible liquid the decision on how or if to fight the fire will depend on the size and type of fire anticipated and must be carefully considered. This may call for special engineering judgement, particularly in large-scale applications.

[Notwithstanding the foregoing review of various authorities noted in Sections 12 and 13, it is essential that managers and others responsible for the planning, implementation and overseeing of personnel and plant safety should be familiar with the legal constraints and official guidelines applicable to them and that they liaise with their local emergency services in the planning of plant and storage facilities and in the preparation of contingency plans for dealing with fires and other emergencies. Managers should regularly monitor staff knowledge of, and ability to implement, emergency procedures and should ensure that equipment provided for use in emergencies is regularly inspected and maintained.]

14. HAZARDOUS REACTIONS[3,28]

Bretherick[28] includes the following hazards.

Methanol interacts violently with acetyl bromide, hydrogen bromide being evolved.

A violent reaction ensued when methanol was used to clean a syringe that had been used for a dilute alkylaluminium solution.

Methanol reacts violently with ether-containing beryllium hydride even at $-196\,°C$.

The rapid autocatalytic dissolution of aluminium, magnesium or zinc in 9:1 methanol – carbon tetrachloride mixture may be vigorous enough to be hazardous. An induction period of two hours may occur which, in the case of zinc powder, can be eliminated by traces of copper(II) chloride, mercury(II) chloride or chromium(III) bromide.

Cyanuric chloride has reacted violently and uncontrollably with methanol, this being attributed to the absence of an acid acceptor to prevent the initially acid-catalysed (and later auto-catalysed), exothermic reaction of all three chlorine atoms simultaneously.

Reaction of diethyl zinc with methanol is explosively violent and ignition ensues.

When disposing of oxidized potassium metal stocks by the addition of t-butanol to small portions of the potassium in xylene (under a hood), it is dangerous to replace the t-butanol with methanol.

The (vigorous) reaction of magnesium with methanol is often subject to a lengthy induction period. Sufficient methanol must be present to absorb the sometimes violent release of energy.

Mixtures of methanol with powdered magnesium or aluminium are capable of extremely powerful detonation.

Distillation with barium perchlorate gives the highly explosive methyl perchlorate.

The reaction of bromine with methanol is vigorously exothermic and is even more so with ethanol containing 5% methanol (industrial methylated spirits).

A mild explosion and ignition resulted from the passage of chlorine through cold, recovered (but not fresh) methanol, the formation of methyl hypochlorite apparently being catalysed by an impurity.

Several explosions involving methanol and sodium hypochlorite were attributed to methyl hypochlorite formation, especially in the presence of acids or other esterification catalyst.

A saturated solution of lead perchlorate in dry methanol exploded violently on being disturbed. Methyl perchlorate may have been involved.

Methanol ignited when used to clean a pestle and mortar that had been used to grind coarse chromium trioxide.

Methanol has been used as a propellant fuel with nitric acid with which it also readily forms the explosive ester, methyl nitrate.

Phosphorus(III) oxide (P_4O_6) reacts very violently with methanol.

Contact of 1.5 g portions of potassium t-butoxide with drops of methanol caused ignition after two minutes.

OSHA[3] notes violent reaction with calcium carbide and possible ignition of methanol in the presence of catalytic amounts of nickel.

Mixtures of methanol and chloroform, when brought into contact with sodium, sodium hydroxide or sodium methoxide can give rise to vigorous and sometimes explosive reaction.

15. EMERGENCY MEASURES IN THE CASE OF ACCIDENTAL SPILLAGE[3,32,33]

All ignition sources should be eliminated and any leak stopped if this can be done without risk. Water spray may be used to reduce vapours.[3] Very small spills may be mopped up with plenty of water and run to waste (a sewer) diluting greatly with running water,[32] or else such spillage may be absorbed on paper, sand or vermiculite and allowed to evaporate in a fume cupboard with a flameproof extractor until the whole of the ducting is clear of vapour.

For spillages up to *ca.* 25 l, the liquid may be absorbed onto sand or vermiculite followed by transfer in suitable containers [*e.g.*, covered buckets] to a safe, open area where atmospheric evaporation can take place. The site of the spillage should be washed thoroughly with water and biodegradable detergent.[33] The washings should be discharged to a sewer and not to a 'soakaway' nor a land drain. Larger spillages should be bunded far ahead of the spill for later disposal.[3] Protective wear, appropriate to the degree of spillage, should be worn and only necessary personnel allowed to enter the hazard area.

Warren[33] notes that "*ideally* all hydrocarbons and related flammable organic chemicals should be burned in an incinerator with an afterburner".

16. FIRST AID

General

Methanol is appreciably toxic. The casualty should be removed from danger of further exposure into fresh air. Rescuers should ensure their own safety from inhalation and skin contamination where necessary. Conscious casualties should be asked for information about what has happened. Bear in mind that the casualty may lose consciousness at any time. Where necessary, continued observation and care should be ensured. First aiders should take care not to become contaminated. Note that methanol is flammable.

Inhalation

The conscious casualty should be kept at rest. The unconscious casualty should be placed in the recovery position and an open airway maintained. If breathing or heart beat stops, resuscitation should be commenced immediately. Medical aid should be obtained or the casualty removed to hospital immediately.

Skin contamination

Remove contaminated clothing. Wash with copious amounts of water, or soap and water, for at least 15 minutes. If necessary, seek medical aid. In case of persistent skin irritation refer for medical advice.

Eye contamination

Irrigate with copious amounts of water for at least ten minutes but avoid further damage to the eye from use of excessive force. Seek medical aid. Remember the possible presence of contact lenses, which may be affected by some solvents and may impede decontamination of the eye.

Ingestion[34,35]

If the lips or mouth are contaminated rinse thoroughly with water. Do not induce vomiting. In the case of ingestion of significant amounts wash out mouth with water, and give copious quantities of water to drink. Give further supportive treatment as for inhalation. Obtain immediate medical attention. Contact a hospital or poisons centre at once for advice. Remember that symptoms may develop many hours after exposure so continued care and observation may be necessary. The usual fatal dose is between 30 and 250 ml, but 15 ml of a 40% solution has been known to prove fatal.[34,35]

In all cases note information on the nature of exposure, and give this to ambulance or medical personnel. Information for doctors is provided in some technical literature issued by manufacturers. Antidotes may be available for retention at the place of work for use by persons competent to treat casualties.

17. PHYSICO-CHEMICAL PROPERTIES

17.1 General[36.37]

Methanol is a colourless, mobile, flammable liquid.[36] When pure, it has a slightly alcoholic odour, though the odour of crude methanol is described as pungent and repulsive.[36] Methanol is described as having a burning taste.[37]

Methanol is miscible with water and most organic solvents, including ethanol, benzene, diethyl ether and ketones.[36] Methanol is generally a better solvent than ethanol, particularly for dissolving many inorganic salts.[36]

17.2 Melting Point[36]

$-97.8\,°C$.

17.3 Boiling Point[36.38]

Merck[36] gives the boiling point of methanol as

$64.7\,°C$.

Methanol forms azeotropes with many compounds. Some binary ones formed with common organic solvents are as follows[38]:

Wt % methanol	Second component	Wt % second component	Boiling pt azeotrope (°C)
38 – 39.1	Benzene	60.9 – 62	57.50 – 58
29	Carbon disulphide	71	39.8
20.56	Carbon tetrachloride	79.44	55.7
14	Cyclopentane	86	38.8
38	Trichloroethylene	62	59.3

17.4 Density/Specific Gravity[26.36]

Merck[36] lists the specific gravity of methanol at several temperatures (referred to water at 4 °C) as:

Temperature in °C	Specific gravity
0	0.8100
15	0.7960
20	0.7915
25	0.7866

The critical density of methanol is given by Marsden and Mann[26] as $0.272\ g/cm^3$. From this the calculated critical volume is $3.677\ cm^3/g$.

17.5 Vapour Pressure[39]

A list of vapour pressures of methanol below atmospheric pressure is given by Weast[39]:

Temperature in °C	Pressure in mmHg
−44.0	1
−16.2	10
5.0	40
21.2	100
49.9	400

Methanol vapour pressures at and above atmospheric pressure are also listed[39]:

Temperature in °C	Pressure in atm
64.7	1
84.0	2
112.5	5
138.0	10
167.8	20
203.5	40
224.0	60

Weast[39] also gives the critical temperature and critical pressure of methanol as 240 °C and 78.5 atm [59700 mmHg], respectively.

17.6 Vapour Density[36]

1.11 (relative to air = 1).

17.7 Flash Point[31]

11 °C, closed cup test.

17.8 Explosive Limits[31]

The limits of flammability of methanol in air at normal atmospheric temperature and pressure are given by the National Fire Protection Association[31] as:

Lower limit: 6.0% v/v
Upper limit: 36% v/v.

The NFPA[31] give the ignition temperature of methanol as 385 °C. However, they point out that different test conditions (and also different definitions of "ignition temperature") can result in widely varying values being quoted. They therefore recommend that any value should be treated as an approximation.

17.9 Viscosity[39]

The viscosity of liquid methanol is given by Weast[39] as follows:

Temperature in °C	*Viscosity in cp*
−98.30	13.9
−84.23	6.8
−72.55	4.36
−44.53	1.98
−22.29	1.22
0	0.82
15	0.623
20	0.597
25	0.547
30	0.510
40	0.456
50	0.403

Weast[39] also lists the viscosity of methanol vapour:

Temperature in °C	*Viscosity in μp*
66.8	135.0
111.3	125.9
217.5	162.0
311.5	192.1

17.10 Concentration Conversion Factors

At 25 °C and 760 Torr (1 atm),

$1 \text{ mg/m}^3 = 0.76$ ppm, and
$1 \text{ ppm} = 1.31 \text{ mg/m}^3$.

18. TOXICITY

18.1 General[5,40−44]

Methanol is a moderate irritant. Its acute lethal oral toxicity is low[5] though variation in this is high in humans, ranging from death due to 15 ml of 40% methanol to survival from ingestion of a quantity in the range 500 – 600 ml.[40] Sub-lethal doses though can cause severe effects on the central nervous system, the liver, and particularly to the visual system.[5] Most non-occupational cases of human poisoning occur through ingestion. Occasional cases are caused by skin contact or inhalation, this last being the most likely route in industrial operations.[5,40] Rowe and McCollister in ref. 5 state that it is the dose and resulting blood concentration, rather than the route of exposure that counts as regards toxicity. This manifests itself as CNS effects plus a marked and possibly lethal acidosis together with toxic optic retrobulbar neuritis, a specific effect of methanol poisoning which may leave the victim permanently blind or temporarily blind to a partial or total degree.[41,42] This metabolic acidosis and ocular toxicity is observed in humans and monkeys but not to a great extent in the non-primates for which CNS depression is the main manifestation.[42]

Information on chronic toxicity is less clear. Andrews and Snyder in ref. 43 observe that animal studies are sparse and those of non-primates may not be meaningful to humans in some cases. Some human studies cited are indicated as being uncertain because of extraneous complicating factors or because of insufficient information.

Rowe and McCollister in ref. 5 state that methanol does not have suitable warning odour or irritating properties except at high concentrations.[5] Ruth[44] gives the observed range for odour threshold as approximately 13 mg/m^3 [*ca.* 10 ppm] to approximately 26,800 mg/m^3 [*ca.* 20,500 ppm], with irritancy being manifest at approximately 22,900 mg/m^3 [*ca.* 17,500 ppm].

18.2 Toxicokinetics[5,6,35,40,42,43,45]

Andrews and Snyder in ref. 43 cite studies indicating that methanol is rapidly and well absorbed *via* inhalational, oral and topical routes following which it is rapidly distributed to organs according to the distribution of body water. Ref. 45 notes that in inhalational studies, when volunteers were exposed to methanol concentrations of 103 – 284 mg/m^3 [approx. 79 – 217 ppm] for eight hours, lung retention was calculated to be about 55%, this value being independent of time, exposure level or the increase of lung ventilation with work. Other studies have shown a skin uptake, on the lower arm, of 0.19 mg/cm^2 per minute, and numerous cases of poisoning indicate that methanol is easily absorbed from the digestive tract.[45] Rowe and McCollister in ref. 5 note human studies where, following oral doses of 71 – 84 mg/kg, urine concentrations peaked rapidly in about an hour and then declined exponentially, reaching the blank or control value in 13 – 16 hours. It is further observed that with these small dosages, an average of only 0.7% of the amount administered was accounted for in the urine, and that, with amounts in a few expired air studies only slightly larger, metabolism is indicated as accounting for most of the administered methanol. Ref. 40 states that peak absorption *via* ingestion occurs within 30 – 60 minutes, depending on whether or not there is food in the stomach.

Methanol's distribution throughout tissues being in relation to their water content results in the aqueous and especially the vitreous humours of the eye acquiring very high concentrations of the chemical, and quite elevated concentrations occur in gastric juices and cerebrospinal fluid. It has also been found that the concentration of methanol in gastric juice was 5 – 12 times greater than in blood even ten days after ingestion.[35] Ref. 6 notes that methanol passes through membranes more easily than does water. In two cases of lethal methanol ingestion, the methanol levels in blood and cerebrospinal fluid were found to be identical. In another case, of a deceased alcoholic, the blood and the tissue levels of methanol appeared virtually identical at 4.2 and 3.9 – 4.2 mg/g respectively.[6] Methanol is capable of passing the blood–brain barrier in adult rhesus monkeys. It is extracted by the brain from the blood to the same extent as water.[6] In sheep and guinea pigs, the placental transfer of ^{14}C-labelled methanol exceeded the transfer of tritiated water.[6]

Blood normally has a methanol content, derived from endogenous production and dietary sources. Blood concentrations from these sources are approximately 1.5 mg/l.[40]

Elimination of methanol from the blood appears to be very slow in all species, especially when compared with ethanol.[42] Tephly and McMartin[42] found that with monkeys given 1 g/kg methanol, the rate of elimination from the blood was linear with time but first-order decay curves were obtained with doses of 3 g/kg and peak blood methanol levels over 300 mg/dl. The higher dose was associated with a half-life of about 24 hours but this could be increased to about 49 hours by the administration of inhibitors of methanol oxidation.[42] Ethanol inhibits methanol oxidation being itself metabolized five to seven times more quickly.[35] The ethanol has a higher affinity for the alcohol dehydrogenase (ADH) which enzyme is primarily responsible for methanol oxidation in the liver. The result is that methanol is eliminated primarily by extra-hepatic routes when ethanol is present.[40].

Normally, only a small amount of methanol is excreted *via* breath, sweat and urine, the primary means of elimination in humans being by oxidation to formaldehyde, thence to formic acid and carbon dioxide.[40]

It has been noted in Section 18.1 that metabolic acidosis and ocular toxicity are observable in human beings and monkeys but not to a great extent in the non-primates. The possibility of formaldehyde being the toxic agent in methanol poisoning has been examined. However, this substance is highly reactive with proteins and other endogenous compounds containing active hydrogen atoms and can form adducts with cellular constituents, leading to the formation of stable intermediates.[42] In addition it is likely that the formaldehyde-oxidizing capabilities of the liver are extremely high. The conversion can apparently proceed as readily in humans as in rats which latter species does not exhibit the methanol poisoning uniquely present in humans and monkeys.[42] Other studies have failed to demonstrate the presence of formaldehyde in blood, urine or tissues obtained from methanol-intoxicated animals or from methanol-poisoned humans.[42] Tephly and McMartin[42] add that, although it is possible that formaldehyde may be responsible for certain of the toxic findings in methanol poisoning, it is unlikely that it can be generated in the liver and delivered intact to the optic nerve, and that it is formate that appears to be the major factor in the metabolic acidosis seen in monkeys and humans poisoned with methanol. Other possibilities involving formaldehyde *per se* can be postulated and more studies are required on its possible role in methanol poisoning in man, but it seems unlikely to be the responsible agent for ocular toxicity. Formate itself can produce ocular toxicity in the monkey.[42]

Studies of methanol poisoning in non-primate animals are difficult to relate to toxicity in humans. Doses in non-primates that would be expected to engender human toxicity symptoms cause only intoxication similar to that caused by ethanol. It has been found though that rhesus and pig tail monkeys are able to provide models for humans intoxication by methanol.[40] Monkeys, treated with formate alone, developed toxic effects in the optic nerve similar to those found in humans, probably by inhibiting cytochrome oxidase in the optic nerve, disturbing the flow of the axoplasm and so giving rise to the pathological condition of the eye.[40] Rats metabolise formate at a high rate and thus do not accumulate it sufficiently to manifest human methanol toxicity. The oxidation of formate to carbon dioxide is likely to be caused by a folate-dependent system in the liver of rats, monkeys and probably in humans, the level of which appears to be critical for the formate metabolism.[40] Rats rendered folate-deficient oxidize formate at a markedly slowed rate, and administration of methanol to such rats leads to high formate levels and severe metabolic acidosis.[42]

Various studies have shown that the catalase–hydrogen peroxide system can play a role in the oxidation of formate.[42] However, it was found that aminotriazole (a potent catalase inhibitor) had no effect on methanol oxidation to carbon dioxide in the monkey nor in the rat, although it resulted in some inhibition of the rate of formate oxidation in folate-deficient rats, suggesting that in this case the catalase-hydrogen peroxide system may be serving as an alternative pathway.[42]

Studies have shown that aminotriazole altered neither the rate of methanol oxidation to carbon dioxide nor of formate oxidation nor of the half-life of formate in the blood of monkeys. However, in folate-deficient monkeys, the rate of formate metabolism was approximately 50% lower than in the controls and when 0.5 g/kg of methanol was given to folate-deficient monkeys, the level of blood formate was more than twice that observed in the controls. Formate oxidation in monkeys was stimulated though by the administration of either folic acid or 5-formyltetra-hydrofolate (5-THF).

These and other studies with 5-THF on monkeys are cited by Tephly and McMartin as demonstrating that the severity of methanol toxicity in monkeys is correlated with the accumulation of formate in the blood and that this can be significantly modified by procedures which provide the monkey with more folate. The results suggest that there is a reciprocal relationship between the formate oxidation rate and the hepatic folate level of the animal and suggest the possible use of folates for the treatment of human methanol toxicity[42] In connection with treatment of methanol poisoning, Schneck[35] notes that 4-methylpyrazole has been found to inhibit alcohol dehydrogenase specifically and may therefore prove to be of use in therapy.

18.3 Human Acute Toxicity[3,5,34,35,40−43,46−53]

Eye irritation was experienced by a worker exposed for 25 minutes to 950 – 1,100 ppm (averaging 1,025 ppm) methanol vapour while operating a duplicating machine.[46] Headache is reported to have resulted from exposure to 800 – 1,000 ppm for four hours associated with the use of duplicating machine fluid.[47] OSHA[3] notes that liquid methanol may cause irritation to the eye, and cause superficial corneal lesions. Grant[48] notes that corneal opacities are alleged following external contact of liquid methanol with the eye but states that this must be far from the rule, tests on rabbit eyes indicating that the danger is slight.

OSHA[3] also notes that contact with liquid methanol can produce defatting of the skin and a mild dermatitis and that it is readily absorbed through intact skin, causing optic neuritis and acidosis. Rowe and McCollister in ref. 5 and Browning[49] note reports of ocular disturbances and blindness caused by repeated rubbing of the skin with methanol under conditions that did not prevent inhalation of the vapour, this latter factor thereby making appraisal difficult. Several sources report the case of a worker who accidentally spilled methanol on his clothes and shoes, who continued to wear the soaked garments and in whom blindness developed within several days.[3 5,40] Browning[49] notes that here too there was the possibility of effects from inhalation. (See also conclusions deduced from animal skin absorption experiments noted from ref. 50 in Section 18.4.)

Browning states that cases of acute poisoning by inhalation are rare. ACGIH[51] note a fatal case of occupational methanol intoxication by inhalation when a female worker was exposed to the vapour for 12 hours. A subsequent study of the process revealed concentrations ranging from 4,000 – 13,000 ppm. ACGIH[51] also note the belief of one authority (in 1937) that intake of 8 g methanol could seriously effect the eyes. ACGIH note that such a dose could result from inhalation of 800 ppm to 1,000 ppm for eight hours.

Browning[49] states that other acute symptoms reflect the strong irritancy that methanol has for mucous membranes, causing conjunctivitis and bronchopneumonia. The conjunctivitis has been known to progress to corneal destruction (1912 report) and the bronchopneumonia was fatal (in a case quoted in 1878).[49]

Rowe and McCollister in ref. 5 observe that as inhalation is the most likely route of exposure to methanol in industrial operations it is surprising that this route has not been associated with significant industrial illness, given the wide use and high volatility of the material.

Ingestion of methanol has been responsible for most of the non-occupational acute toxicity reported in humans. The methanol is often in admixture with ethanol (possibly as a denaturant) or other material. Apart from being ingested in mistake for ethanol, or through its toxicity not being appreciated, methanol-containing fluids may be purposely ingested by compulsive alcoholics for whom its toxicity is no deterrent.

The lethal dosage of methanol in humans varies considerably. This could be due in part to the difficulty in obtaining exact information from disoriented patients. Another factor could be the different amounts of ethanol that might be consumed with the methanol. Ethanol is preferentially metabolized and there is a longer latent period before the appearance of poisoning compared with methanol-only ingestions. Where methanol alone has been consumed, earlier and more severe symptoms are reported to be shown by patients who had ingested more of the substance than those who had consumed less. Also, death occurred more quickly in those who drank more methanol. In addition, it has also been suggested that insofar as formate metabolism may be folate dependent, different degrees of nutritional deficiency, such as may be observed in debilitated and inebriated persons, may also be a factor.[42] The usual fatal dose is between 30 and 250 ml[35] but the smallest dose reported is 15 ml of 40% solution ($\equiv$ 6 ml absolute methanol) while survival has followed the ingestion of 500 – 600 ml.[34] The lowest dose recorded as having toxic effect is 4 ml.[34]

The initial effects of methanol ingestion are due to CNS depression. There then follows an asymptomatic latent period occurring about 8 – 24 hours after ingestion. There are large individual differences in the duration of the latent period and symptoms of poisoning may be delayed for up to three days.[42]

Schneck[35] stresses that it is important to note that metabolism of methanol in humans is slow and that one third of an ingested dose can remain in the body unaltered for 48 hours with some remaining as long as a week. It is also noted that it is likely that the symptomless latent period terminates when formaldehyde/formic acid metabolites rise to significant levels and then exert their toxicity.[35] Then headache, dizziness, weakness and nausea are reported, followed in more severe cases by intense vomiting and excruciating abdominal and muscular pain. The abdominal pain can be so intense as to cause patients to throw themselves out of bed. Disorientation and severe mental disintegration may be exhibited. With more severe cases, classic respiratory difficulties of metabolic acidosis are shown, *viz*, severe dyspnoea characterized by marked increases in both depth and rate of respiration (Kussmaul breathing),[42] though Goodman and Gilman[52] state that Kussmaul respiration is not common despite the severe acidosis.

Tephly and McMartin[42] add that with the onset of respiratory problems there occur visual defects ranging from blurred vision to complete loss of vision. Ref. 41 notes that toxic optic retrobulbar neuritis is a specific effect of methanol poisoning and may result in permanent blindness due to optic atrophy.

These symptoms and coma may occur in rapid sequence.[41] In fact, coma can develop with amazing rapidity in relatively asymptomatic subjects.[52] Death may be sudden or it may occur after many hours of coma. It occurs in inspiratory apnoea, with terminal opisthotonosis and convulsions[52] Patients are usually blind prior to death though often a patient may not die and be left partially or totally blind.[42].

Litovitz[34] states that in the absence of a history of methanol ingestion, the initial diagnosis is difficult as evidenced by the diverse initial diagnoses that have been made, such as cholera, diabetic ketoacidosis, pancreatitis, meningitis, subarachnoid haemorrhage and nephrolithiasis. The delay in the onset of symptoms that the latent period engenders is deceptive, frequently leading to delayed initiation of therapy.[34] Litovitz[34] observes that the single most helpful diagnostic clue is the presence of visual changes but these are not uniformly noted. Cloudy, blurred, misty or indistinct vision is typical. Yellow spots and central scotomata may also be noted. In place of permanent, complete blindness developing, there may be irreversible but less severe visual abnormalities with complete resolution and, if visual function returns, it generally does so in the first six days.[34] Refs. 34, 42 and 48 give further details of symptoms of visual impairment.

Becker[40] summarizes the pathological findings. The primary site of ocular injury is in the optic nerve head and the intraorbital portion of the optic nerve rather than in the retinal ganglia. Haemorrhages into portions of the brain are an important aspect of methanol poisoning and necrotic areas in the putamen have been shown. Damage to liver, pancreas and kidneys have been described but these are not specific.

Litovitz[34] observes that even severely poisoned patients can be expected to survive if they are treated intensively within a few hours of ingestion, and Becker[40] states that it is essential that a blood methanol level be determined as soon as possible if methanol poisoning is suspected. Methanol levels in excess of 50 mg/dl are probably a clinical indication for haemodialysis and ethanol treatment. With levels below 50 mg/dl, ethanol treatment should be begun or continued and the test repeated. Ethanol treatment should not be delayed pending initial test results if clinical suspicion of methanol poisoning is high.[40] Andrews and Snyder in ref. 43 state that a review of papers that have extensively considered methanol concentration in the blood and the clinical outcome of poisonings indicates that an initial blood level in excess of 100 mg/100 ml would be required for irreversible effects such as visual disturbances.

Although the half-life of blood methanol is quite long in poisoning cases, it has been found with human volunteers who ingested small [*sic*] amounts (1 – 5 ml) of methanol that under these milder conditions the blood methanol half-life is only about three hours and that peak blood levels were around 10 mg/100 ml. It can then be calculated that, with a TLV of 200 ppm (261 mg/m³), a 100% absorption of vapours and a respiratory volume of 10 m³ in an eight hour workday, then a daily body burden of 2,610 mg is achieved.

If one assumes as a worst circumstance that this body burden is absorbed within the first few minutes of the workshift and that the methanol distributes with total body water, then the worst-case peak blood methanol level of a 70 kg person with 70% water content can be calculated as 5.3 mg/100 ml. This level is about one twentieth of that associated with acute irreversible toxic effects. With a half-life of three hours, the blood level methanol would have fallen to negligible levels by the start of the next working shift 24 hours later. It seems unlikely therefore that continuous exposure to 200 ppm methanol vapour will cause ocular toxicity.[43]

Following studies on the use of methanol in duplicating machines, the US Advisory Center on Toxicology published, in 1959, human tolerance values for exposure to methanol vapour. Ref. 53 notes that, 'the guiding principles used in determining these values were to prevent accumulation of methanol and its metabolites in the tissues and to be somewhat conservative'. For exposures up to 24 hours the estimated tolerance values are given as follows.

Duration	*Estimated tolerance value (ppm)*
Single but not repeated exposure:	
1 h	1000
8 h	500
24 h	200
Single or repeated exposures:	
1 h out of every 24 h	500
Two 1 h exposures every 24 h *or* One 2 h exposure every 24 h	200

18.4 Animal Acute Toxicity[4,5,42,50,54,55]

Rowe and McCollister in ref. 5 observe that methanol is a mild eye irritant and quote the following data, *viz*, undiluted methanol caused conjunctival redness in all of six rabbits and moderate corneal opacity in three of six. A 50% aqueous solution caused minimal to no effects and a 25% aqueous solution caused no adverse effects.

McCord[50] in experiments on rats, rabbits and monkeys found that, with skin application (on clipped but not shaved abdomens and with concurrent inhalation and evaporation of methanol precluded) the lowest amount regarded as responsible for death was $0.5 \, cm^3$ per kilogram of weight in one monkey (of eight) though clinical findings in this individual were atypical.

It was concluded that the threshold of danger following skin absorption is somewhere below 0.5 ml per kilogram of animal weight applied four times daily, this quantity producing illness in monkeys within 24 hours (during which time four applications of the specified amount were made) with eventual death. 1.3 ml/kg applied four times per day was found to produce death within 48 hours. With all species tested, methanol was found present in brain, lungs, heart muscle, skeletal muscle, liver, spleen, pancreas, kidneys, blood and urine after exposure to skin absorption (and in the inhalational experiments of this study referred to below). McCord concludes that, extrapolating from the results on monkeys, approximately one ounce (31 ml) of methanol, repeatedly in contact with the human body, under conditions favourable to retention, constitutes a practical threat to well-being.

Rowe and McCollister[5] quote various LD_{50} single-dose, oral toxicity values for the rat, ranging from 6.2 to 13.0 g/kg. Cooper and Felig[54] give the LD_{50} value for monkeys as $7-9$ g/kg and note this to be well above the minimal lethal dose of 3 g/kg reported by previous workers who had, in addition, found 78% mortality at 6.0 g/kg, which level was a no-effect one in the Cooper and Felig study.

With the foregoing the differences in metabolism between primate and non-primate species should be borne in mind, and this factor makes it difficult to assess the human implications of a mass of data on the results of single exposures of non-primates to methanol vapour (*e.g.*, refs. 5 and 4). In experiments with rats, rabbits and monkeys, McCord[50] found that four hours exposure to 40,000 ppm was lethal to all the animals so exposed.

Rowe and McCollister in ref. 5 note that the pathological changes found in the tissues of animals exposed to methanol by inhalation or by ingestion are similar. Hyperaemia of the choroid, and oedema of the ocular tissue (leading to local anoxia[55]) with early signs of degeneration of the ganglionic cells of the retina and nerve fibres have been found in the eyes of the dog. In other studies, it was found that the vessels of the choroid were markedly congested, that the entire retina was oedematous, that the ganglion cells were degenerated and that occasionally there were degenerative changes and fibrosis of the optic nerve. The lungs of various species exposed to methanol vapour experienced haemorrhage, oedema, congestion and pneumonia. Livers and kidneys have shown congestion, albuminous and fatty degeneration and fatty infiltration. Hearts have shown cardiac dilation and myocardial degeneration. Degenerative injuries of the CNS have been described. Pinpoint haemorrhages and congestion of the gastric mucosa were believed by some early (1914) observers to be characteristic of poisoning by inhalation of methanol.

Tephly and McMartin[42] note that despite the ingestion of lethal doses of methanol, non-primate species generally do not develop significant metabolic acidosis and that the effect of methanol in non-primate animals is manifested almost exclusively as a CNS depression such as that observed with other aliphatic alcohols. They note, in addition, that no impairment of vision has been observed in methanol-poisoned non-primate animals and quote a paper whose authors point out that, although there have been reports of the production of clinical visual impairment in such animals, these [visual impairment] claims were often based on several common sources of confusion not related to the typical visual disturbances seen in humans. Tephly and McMartin further note that, in well-conducted studies of ocular effects of methanol in rabbits, chickens, dogs and rats, there was neither evidence of visual disturbance nor any changes in the appearance of the fundus. [But see previous paragraph regarding ocular tissue changes in dogs.]

18.5 Human Sub-acute and Chronic Toxicity Including Reproductive Effects[3,5,6,43,45,49,51,56]

General

OSHA[3] states that prolonged or repeated exposure may cause symptoms such as blurred vision, contraction of visual fields and, sometimes, complete blindness. Browning[49] adds conjunctivitis, headache, giddiness, insomnia and gastrointestinal disturbances.

Ref. 6 notes that reports on the effects of prolonged exposure to methanol involve occupational exposure. The exposure levels are often poorly characterized and the studies are frequently confounded by the presence of other chemicals.

Andrews and Snyder in ref. 43 refer to a study of office workers who were exposed to methanol in the vicinity of duplicating machines. Complaints were solely of frequent and recurrent headaches. Exposure levels were reported to range from 15 to 375 ppm with most being in the 200 – 375 ppm range. Duplicating fluids were changed in favour of less methanol. However, the effect, if any, on workers' headaches was not recorded and it is unclear if the headaches were attributable to methanol or to other components of the duplicating fluids. ACGIH[51] states that this study contributed to NIOSH's recommendation of a TWA of 200 ppm for methanol.

Ref. 45 notes that one of 33 persons occupationally exposed to vapour concentrations between 1,000 and 10,000 mg/m³ [approx. 760 – 7,600 ppm] for ten years suffered temporarily reduced visual acuity though the study does not make it clear if this effect is as that seen with oral poisoning. Eye irritation has been reported by people using spirit duplicating machines after 25 minutes of exposure to 1,245 – 1.441 mg/m³ [approx. 950 – 1,100 ppm]. In another study involving spirit duplicating machines,[56] methanol vapour concentrations found ranged from 365 – 3,080 ppm and, in addition to headaches, found "more ominous symptoms", consistent with methanol toxicity, such as blurred vision.

Ref. 45 and Rowe and McCollister in ref. 5 note that general physical, ocular and haematologic examinations of 19 workers in a shirt factory who had been repeatedly exposed to 22 – 25 ppm methanol and to 40 – 45 ppm acetone over periods ranging from nine months to two years, revealed no significant abnormalities.

Browning[49] cites the case of a 27 year-old man who had been employed for four years in a factory department in which nicotinic acid was crystallised from methanol. For two years he had complained of weakness and numbness of his hands and arms plus disturbance of vision, which last was diagnosed as slight astigmatism and corrected by spectacles. While cleaning a tank in which the two substances had been heated he renewed his gas mask filter but inadvertently chose one intended for ammonia. During his five hour task he felt giddy at times but at the end had no symptoms. The next morning he vomited but continued working and then suddenly developed spots before the eyes and dimness of vision. He also complained of a sweetish taste and of loss of appetite. Blood and urine examinations showed nothing abnormal but ophthalmic examination showed papilloedema in both eyes. Visual acuity became normal after five weeks of treatment with eye drops, ointments, vitamin B, *etc*. After a further five weeks, vision became irregular and his urine contained formic acid, but later this (qualitative) test was negative. It has been stated that the amount of methanol inhaled must have been small but the man had been in the habit of cleaning his hands with methanol.[49]

It has been pointed out that the slow elimination of methanol facilitates its build up in blood and tissue with repeated exposures.[51]

Ref. 45 concludes that, from available data, it can be established that headaches occur at exposures higher than about 260 mg/m^3 [approx. 200 ppm] and that eye irritation and damage to vision result from exposures above about 1,000 mg/m^3 [approx. 760 ppm].

Carcinogenicity to Man

Ref. 6 states that the carcinogenic potential of methanol has not been investigated. No relevant human data have been found.

Reproductive Effects in Man

No data on reproductive effects in humans were found, but the ability of methanol to pass through the placenta in sheep and guinea pigs (see Section 18.2) should be borne in mind regarding possible effects on offspring.

18.6 Animal Sub-acute and Chronic Toxicity Including Reproductive Effects[1,5,6,43,57–59]

General

Andrews and Snyder in ref. 43 state that animal studies involving repeated long-term exposure to methanol are sparse and that, in view of the species difference in biotransformation, it is debatable whether or not rodent studies would be meaningful.

Rowe and McCollister in ref. 5 note that repeated oral doses of 3 – 6 g/kg to rhesus monkeys for 3 to 20 weeks resulted in ultrastructural abnormalities of hepatocytes of a nature indicating alteration of RNA metabolism of hepatic cells.

Ref. 6 notes a study in which two monkeys received a total dose of 17 and 18 g/kg divided over nine days. Their electroretinograms revealed reduced a- and b-waves indicating a decreased retinal activity.

Also noted in ref. 6 is work in which, in order to produce the human methanol intoxication syndrome in rhesus monkeys, an initial oral dose of 2 g/kg was used followed by supplemental oral doses for three to seven days. The monkeys developed signs of methanol poisoning such as metabolic acidosis, accumulation of formic acid in the blood, coma and visual impairment as evidenced by optic disc oedema, decreased light reflex and pupillary dilation. No changes were seen in the cerebrospinal fluid pressure, the retinal vasculature and the retinal ultrastructure. It was considered that the ocular lesion was essentially a toxic optic neuropathy, probably due to accumulation of formate in the blood.

Ref. 6 observes that data on the repeated dermal administration of methanol are not known. (See also *Carcinogenicity to Animals*.)

Carcinogenicity to Animals

Sax[1] states that there is no direct evidence of carcinogenicity of methanol. Ref. 6, while noting that the carcinogenic potential has not been investigated, also notes that some mouse skin painting studies in which the application of 100 µl of methanol daily for thirty five weeks, or 50 µl three times per week for fourteen months did not produce skin tumours.

Reproductive Effects in Animals

Methanol was administered by inhalation at concentrations of 20,000 ppm, 10,000 ppm and 5,000 ppm to groups of 15 pregnant Sprague–Dawley rats (for days 7 – 15 of gestation at 20,000 ppm and for days 1 – 19 for the lower concentrations). Dams were sacrificed on day 20 (copulation = day 0). 20,000 ppm but not the lower dosages produced an irregular gait in the dams following the exposure and a weight reduction during the first few days of exposure. There was a high incidence of malformations in the foetuses from the 20,000 ppm group (p < 0.001), predominantly, extra or rudimentary cervical ribs and urinary or cardiovascular defects being observed. Similar malformations were seen in the 10,000 ppm group's foetuses but the incidence was not significantly increased above control values. No adverse effects were noted in the 5,000 ppm group's foetuses.[6,57]

Two groups of ten Long–Evans rats were given drinking solutions of 2% methanol instead of distilled water on gestational days 15 – 17 or 17 – 19, with the average daily intake on these days being 2.5 g methanol/kg. No maternal toxicity was apparent by weight gain, gestational duration and daily fluid intake. Litter size, birth weight and infant mortality did not differ among the two groups and the control group. Postnatal development such as growth and date of eye opening were unaffected. Methanol was detected in placenta and in maternal and foetal brain and blood. Pups took longer than controls to begin suckling on postnatal day 1. On postnatal day 10 they took longer to locate nesting material from their own cages. Offspring from the days 17 – 19 exposed dams were significantly more active on day 15 than controls (circular alley). The authors[58] conclude that the data suggest that prenatal methanol exposure induces behavioural abnormalities early in life that are unaccompanied by overt toxicity.

Ref. 6 notes that no significant teratogenic effects were observed when a methanol/formaldehyde mixture was given by gavage to mice on days 6 – 15 of gestation (approx. 60 – 75 mg methanol/kg/day).

No increase in the incidence of abnormal sperm was observed when five daily doses (5 g/kg total) of methanol were administered orally to mice.[59]

18.7 Mutagenicity[6,59 – 62]

Ref. 6 observes that most of the *in vitro* tests cited therein were negative. These negative results were from; a human lymphoblasts mutation assay; a sister chromatid exchange test, CHO cells; cell transformation assays using mouse embryo fibroblasts and Syrian hamster embryo cells; a DNA-repair test in *E. coli* using strains WP2, WP67 and CM871; a *Salmonella typhimurium* plate incorporation assay using strains TA98, TA100, TA1535, TA1537, TA1538 and a *Schizosaccharomyces pombe* forward mutation assay.

Two positive *in vitro* findings are reported in refs. 60 and 61. In the first of these, in the presence of a metabolic activation system, 7.9 mg methanol/ml caused significant increases in mutation frequencies in L5178Y mouse lymphoma cells.[60] In the second study, 12.5% v/v methanol induced mutations in *Saccharomyces cerevisiae* which reverted spontaneously and readily. Ref. 6 notes that the methanol concentrations used in the foregoing two studies were rather high.

In an *in vivo* study,[59] the results of bone marrow cytogenic analysis indicated a dose-related response for structural aberrations, especially centric fusions in mice treated with three daily intraperitoneal doses of methanol of between 75 – 300 mg/kg total dose. The same report[59] states that the urine from mice, to which five daily oral doses (5 g/kg total) were administered, showed no mutagenic activity and no increase in the incidence of abnormal sperm was observed.

In another *in vivo* study,[62] oral administration of ^{14}C-labelled methanol to rats resulted in covalent binding to haemoglobin. Binding exhibited a linear dose relationship between 10 and 100 μmole/kg. The same report[62] states that the oral administration in mice of 1 g/kg methanol increased the incidence of chromosomal aberrations in bone marrow cells, particularly aneuploidy, sister chromatid exchanges and the incidence of micronuclei in polychromatic erythrocytes.

Ref. 6 observes that the methanol metabolite, formaldehyde, has been shown to be mutagenic.

18.8 Summary

[Methanol is generally of fairly low acute toxicity, though oral doses to humans of as low as 6 ml have occasionally proved lethal, and 4 ml has caused toxic effects. It can be absorbed through the skin or by inhalation, and sub-lethal doses, besides initially depressing the central nervous system, may cause metabolic acidosis, may affect the liver, and often cause delayed partial or complete blindness which may be permanent. Methanol is readily absorbed and distributed in the body, but only slowly metabolized and then excreted. Metabolites include formate, which is thought to be the cause of the visual damage; formate formation may be reduced by timely treatment with ethanol. The undiluted liquid is irritating to eyes, and may cause skin defatting on prolonged or repeated contact. The vapour does not irritate eyes or the respiratory tract below about 760 ppm, while acute toxic effects including visual damage can follow inhalation of 800 ppm for eight hours; repeated exposure to 200 ppm is unlikely to cause toxicity, though levels above this may cause toxic effects including headache. No significant additional long-term toxic effects in man have been reported, and no reports have been traced showing carcinogenicity, or adverse reproductive effects except at maternally toxic doses.]

19. MEDICAL / HEALTH SURVEILLANCE

A decision on the need for, and content of, medical surveillance should be based on an assessment of the possibility and extent of exposure in the work operation. In addition, medical examination may be directed to identifying any pre-existing or newly arising condition in the individual workers which might either be aggravated by subsequent exposure, or might confuse any subsequent medical assessment in the event of excessive exposure or an illness not related to exposure. A particular aspect to be considered is the identification of sensitive subjects not adequately protected by the control limit in operation. A professional medical judgement is required on continuance of employment in the specified process. The following information is relevant in the case of this solvent.

Pre-employment Medical Examination

A complete medical and occupational history should be taken and a physical examination carried out. Particular attention is necessary to the skin, eyes, liver and kidneys. Pre-existing skin disease should be reviewed. As methanol may affect the liver, a history of alcoholism may require a decision on fitness for employment on the process. Liver function testing may be adjudged necessary. Because a recognized toxic effect of methanol is optic atrophy and blindness, an ophthalmological examination is wise to identify those with pre-existing eye disease and as a baseline for future health assessments, notably in the event of excessive exposure.

Periodic Medical Examination

Medical examination on a regular basis or in the event of excessive exposure, splashes in the eye, or for any employee who develops ocular symptoms whilst working with methanol will consist of similar clinical evaluation as for the pre-employment medical examination. It has been reported that under constant exposure the concentration of methanol in an end-of-shift urine sample correlates with the intensity of exposure but it is recommended that where concentration in the air varies substantially during a shift, a whole-shift urine sample is required.

20. OCCUPATIONAL EXPOSURE LIMITS

[The Committee felt that a time-weighted average limit of 200 ppm was appropriate for methanol. In the light of available information the Committee saw no real evidence for the STEL of 250 ppm frequently adopted. Data in these reports indicate a possible STEL of 500 ppm. However, in the light of the serious systemic effects of methanol the 250 ppm STEL often adopted may be prudent. This substance may produce serious toxic effects by absorption through the skin (see Introduction).]

REFERENCES

1. Sax, N.I., Ed. 'Hazardous Chemicals Information Annual', No.1. Van Nostrand Reinhold Information Services, New York, 1986.
2. Council of Europe. 'Dangerous Chemical Substances and Proposals Concerning their Labelling'. ('The Yellow Book'.) 4th ed. Maisonneuvre, 1978.
3. Occupational Safety & Health Administration. Material safety data sheets from the Occupational Health Services database, OSHA, Washington D.C., 1986.
4. National Institute for Occupational Safety & Health. 'Registry of Toxic Effects of Chemical Substances (RTECS)'. DHHS (NIOSH) Publication No.84-101-6. US Department of Health & Human Services, April, 1986.
5. Clayton, G.D. and F.E. Clayton, Eds. 'Patty's Industrial Hygiene and Toxicology'. 3rd ed. rev. Wiley Interscience, 1982.
6. Mulder, D.E. and J.T.J. Stouten. 'Review of literature data on methanol'. NOTOX Toxicological Research & Consultancy C.V., The Netherlands, 1986.
7. Verschueren, K. 'Handbook of Environmental Data on Organic Chemicals'. 2nd ed. Van Nostrand Reinhold, 1983.
8. Graedel, T.E. 'Chemical Compounds in the Atmosphere'. Academic Press, 1978.
9. Grayson, M., Exec. Ed. 'Kirk-Othmer Concise Encyclopedia of Chemical Technology'. 3rd ed. John Wiley & Sons, 1985.
10. Lowenheim, F.A. and M.K. Moran. 'Faith, Keyes and Clark's Industrial Chemicals'. 4th ed. John Wiley & Sons, 1975.
11. Hawley, G.G., Ed. 'The Condensed Chemical Dictionary'. 10th ed. Van Nostrand Reinhold, 1981.
12. Grayson, M., Exec. Ed. 'Kirk-Othmer Encyclopedia of Chemical Technology'. 3rd ed. John Wiley & Sons, 1979.
13. Grasselli, J.G. and W.M. Ritchey, Eds. 'Atlas of Spectral Data and Physical Constants for Organic Compounds'. 2nd ed. CRC Press, 1975.
14. National Institute for Occupational Safety & Health. 'NIOSH Manual of Analytical Methods'. 2nd ed. DHEW (NIOSH) Publication No.77-157-A. US Department of Health, Education & Welfare, 1977.
15. Tyras, H. and J. Stufka-Olczyk. *Chemia Analityczna (Warsaw)*, **29**, No.3(1984):281.
16. Drugov, Y.S. and G.V. Murav'eva. *Zhurnal Analiticheskoi Khimii*, **37**, No.7(1982):1302.
17. Diaz-Rueda, J. *et al. Applied Spectroscopy*, **31**, No.4(1977):298.
18. Upadhyav, S. and V.K. Gupta. *Analyst*, **109**, No.11(1984):1427.
19. Verma, P. and V.K. Gupta. *Talanta*, **31**, No.5(1984):394.
20. Druyan, E.A. *Gigiena i Sanitariya*, **50**, No.9(1985):57.
21. Sweger, D.M. and J.C. Travis. *Applied Spectroscopy*, **33**, No.1(1979):46.
22. Smith, N.B. *Clinical Chemistry*, **30**, No.10(1984):1672.

23. Bock, J.L. *Clinical Chemistry*, **28**, No.9(1982):1873.
24. Kozu, T. and Y. Yumoto. *Eisei Kagaku*, **30,** No.3(1984):156.
25. Commission of the European Communities. 'Environmental Chemicals Data and Information Network (ECDIN)' Databank. Commission of the European Communities Joint Research Centre, ECDIN Group, I-21020 Ispra (Varese), Italy, 1986.
26. Marsden, C. and S. Mann, Eds. 'Solvents Guide'. 2nd ed. rev. Cleaver-Hume Press, 1963.
27. International Labour Office. 'Encyclopaedia of Occupational Health and Safety'. 3rd rev. ed. ILO, 1983.
28. Bretherick, L. 'Handbook of Reactive Chemical Hazards'. 3rd ed. Butterworths, 1985.
29. National Institute for Occupational Safety & Health/Occupational Safety & Health Administration. 'NIOSH/OSHA Occupational health guideline for methyl alcohol'. DHHS (NIOSH) Publication. US Department of Health & Human Services/US Department of Labor, 1978.
30. The Royal Society of Chemistry. Laboratory Hazards Data Sheet No.25: 'Methanol', 1984.
31. National Fire Protection Association. 'Fire Protection Guide on Hazardous Materials'. 9th ed. NFPA, Massachusetts, USA, 1986.
32. Bretherick, L., Ed. 'Hazards in the Chemical Laboratory'. 4th ed. Royal Society of Chemistry, UK, 1986.
33. Warren, P.J., Ed. 'Dangerous Chemicals Emergency Spillage Guide'. 1st ed. Jensons (Scientific) Ltd., Leighton Buzzard, UK, 1985.
34. Litovitz, T. Pediatric Toxicology 33 No.2(1986):311. 'The alcohols: ethanol, methanol, isopropanol, ethylene glycol'.
35. Schneck, S.A. 'Methyl alcohol'. In 'Handbook of Clinical Neurology' 36(1979):351. Eds. P.J. Vinken and G.W. Bruyn. Elsevier, Amsterdam.
36. Windholz, M. *et al.*, Eds. 'The Merck Index'. 10th ed. Merck & Co., 1983.
37. Reynolds, J.E.F. and A.B. Prasad, Eds. 'Martindale, The Extra Pharmacopoeia'. 28th ed. The Pharmaceutical Press, 1982.
38. Horsley, L.H. 'Azeotropic Data III'. Advances in Chemistry Series No.116, American Chemical Society, 1973.
39. Weast, R.C., Ed.-in-Chief. 'CRC Handbook of Chemistry and Physics' ('The Rubber Handbook'). 67th ed. CRC Press, 1986-1987.
40. Becker, C.E. *The Journal of Emergency Medicine*, 1(1983):51. 'Methanol poisoning'.
41. Last, J.M., Ed. 'Maxcy-Rosenau Public Health and Preventive Medicine'. 11th ed. Appleton-Century-Crofts, New York, 1980.
42. Tephly, T.R. and K.E. McMartin. Food Science & Technology 12(1984):111. 'Methanol metabolism and toxicity'.
43. Klaassen, C.D. *et al.*, Eds. 'Casarett and Doull's Toxicology: The Basic Science of Poisons'. 3rd ed. Macmillan, 1986.
44. Ruth, J.H. *American Industrial Hygiene Association Journal*, 47(1986):A-142. 'Odor thresholds and irritation levels of several chemical substances: a review'.
45. Swedish Criteria Group for Occupational Standards. *Arbete och Halsa* 1985:32. (Ed. P. Lundberg). 'Scientific basis for Swedish occupational standards VI'.
46. Apol, A.G. Health Hazard Evaluation Report No.HETA-81-177, 178-988. Hazard Evaluations & Technical Assistance Branch, National Institute for Occupational Safety & Health, 1981.
47. Moriarity, A.J. Proceedings – International Symposium on Alcohol Fuel Technology: Methanol and Ethanol (Wolfsburg, FRG, 1977),(1978):8-1(1). 'Toxicological aspects of alcohol fuel utilization'.
48. Grant, W.M. 'Toxicology of the Eye'. 2nd ed. Charles C. Thomas, Illinois, USA, 1974.
49. Browning, E. 'Toxicity and Metabolism of Industrial Solvents'. Elsevier, 1965.
50. McCord, C.P. *Industrial & Engineering Chemistry*, **23**, No.8(1931):931. 'Toxicity of methyl alcohol (methanol) following skin absorption and inhalation'.
51. American Conference of Governmental Industrial Hygienists. 'Documentation of the threshold limit values for chemical substances in the workroom environment'. ACGIH, 1986.

52. Goodman, L.S. and A. Gilman, Eds. 'The Pharmacological Basis of Therapeutics'. 5th ed. MacMillan, 1975.

53. Posner, H.S. *Journal of Toxicology & Environmental Health*, 1(1975):153. 'Biohazards of methanol in proposed new uses'.

54. Cooper, J.R. and P. Felig. *Toxicology & Applied Pharmacology* 3(1961):202. 'The biochemistry of methanol poisoning – II. Metabolic acidosis in the monkey'.

55. Tyson, H.H. and M.J. Schoenberg. *Journal of the American Medical Association*, 63(1914):915. 'Experimental researches in methyl alcohol inhalation'. (Abstract.)

56. Frederick, L.J. *et al. American Industrial Hygiene Association Journal*, **45,** No.1(1984):51. 'Investigation and control of occupational hazards associated with the use of spirit duplicators'.

57. Nelson, B.K. *et al. Teratology*, **29,** No.2(1984):48A. 'The teratogenic effects of methanol administered by inhalation to rats'. (Abstract.)

58. Infurna, R. *et al.* Toxicologist 1(1981):32. 'Developmental toxicology of methanol'. (Abstract.)

59. Chang, L.W. *et al. Environmental Mutagenesis*, 5(1983):381. 'The evaluation of six different monitors for the exposure of formaldehyde in laboratory animals'. (Abstract.)

60. McGregor, D.B. *et al. Environmental Mutagenesis*, **7,** Suppl., 3(1985):10. 'Optimisation of a metabolic activation system for use in the mouse lymphoma L5178Y tk$^+$ tk$^-$ mutation system'. (Abstract.)

61. Lund, P.M. and B.S. Cox. *Genetical Research* (Cambridge University Press) 37(1981):173. 'Reversion analysis of (psi-) mutations in *Saccharomyces cerevisiae*'.

62. Pereira, M.A. *et al. Environmental Mutagenesis*, 4(1982):317. 'Battery of short-term tests in laboratory animals to corroborate the detection of human population exposures to genotoxic chemicals'. (Abstract.)

CONTENTS

1. CHEMICAL ABSTRACTS NAME

benzene, nitro-
(10th Collective Index)

2. SYNONYMS AND TRADE NAMES[1-4]

essence of mirbane
essence of myrbane
mirbane oil
mononitrobenzene
NCI-C60082
nitrobenzeen (Dutch)
nitrobenzen (Polish)
nitrobenzene
nitrobenzène (French)
nitrobenzol
nitrobenzolo (Italian)
OHS16590
oil of mirbane
oil or myrbane
U169
UN 1662

3. CHEMICAL ABSTRACTS SERVICES REGISTRY NUMBER

98-95-3

4. NIOSH NUMBER

DA6475000

5. CHEMICAL FORMULA

$C_6H_5NO_2$
(Molecular weight 123.12)

6. STRUCTURAL FORMULA

7. OCCURRENCE[5-9]

Graedel *et al.*[5] indicate that the occurrence in the atmosphere of nitrobenzene arises from the latter's industrial use. Ref. 6 notes a statement that it may form spontaneously in the atmosphere from the photochemical reaction of benzene with oxides of nitrogen.

Ref. 6 also notes 1976 figures indicating that the greatest loss during production (then estimated as a loss of 3,600 tonnes annually) occurred in the effluent wash.

Graedel *et al.*[5] state that virtually nothing is known of the reactions of atmospheric nitro compounds, particularly in the condensed phase (as constituents of aerosol particles).

Nitrobenzene has been found as a micropollutant in one of 14 treated-water sites sampled in Great Britain (at an upland reservoir).[7]

Gomolka and Gomolka[8] when investigating the ability of activated sludge to degrade nitrobenzene in municipal wastewater found that a concentration of 300 g/m^3 was biodegradable by an adapted, activated biomass, and in a flow-through system it was degradable at 400 g/m^3 concentration. Concentrations exceeding 10 g/m^3 inhibit the [final] nitrification process.

In the USA, the EPA decided in 1984 that as the release of nitrobenzene to air and water are limited, additional testing of the material for environmental effects was not justified.[9]

8. COMMERCIAL AND INDUSTRIAL NITROBENZENE[10,11]

Commercial grades of nitrobenzene noted in the literature include: 'Technical', this being an undistilled product containing small amounts of un-nitrated hydrocarbons and traces of *m*-dinitrobenzene, nitrophenol and water.[10]

Also noted are two purer grades:

Oil of Mirbane[10]:

Purity[a] %	< 99.5
Colour	clear, light yellow to brown
Distillation range,[b] first drop, °C	> 210
Drypoint,[c] °C	< 215
Moisture, %	< 0.1
Dinitrobenzene, %	< 0.1
Acidity, as HNO_3, %	< 0.001

[a]Determined by freezing point.
[b]95% boiling within 1 °C, including the true boiling point [210.8 °C].
[c]Temperature at which no liquid remains.

Double distilled nitrobenzene[11]:

Purity, %	⩾99.8%
Colour	clear, light yellow to brown
Freezing point, °C	⩾5.13

Distillation range*d* (first drop), °C	$\geqslant 207$
Dry point,*e* °C	212
Moisture, %	< 0.1
Acidity (as HNO_3), %	< 0.001

*d*95% boiling at 207 – 210 °C
*e*Temperature at which no liquid remains.

9. SPECTROSCOPIC DATA[12]

Infrared, Raman, ultraviolet, both ^1H and ^{13}C NMR and mass spectral data for nitrobenzene have been tabulated.[12]

10. MEASUREMENT TECHNIQUES[13–17]

The method recommended for the determination of nitrobenzenes in workshop atmospheres is the NIOSH[13] silica gel trapping method for collection and concentration, followed by solvent extraction of the silica gel and a gas chromatographic (gc) analysis of the extract. This method determines nitrobenzene, nitrotoluene and 4-chloronitrobenzene.

The collection tube is 7 cm long with a 4 mm internal diameter. It contains a total of 225 mg of silica gel (20 – 40 mesh) divided into a front section of 150 mg and a rear section of 75 mg separated by a small plug of urethane foam. Air is sampled at a flow-rate of 0.01 – 1 l per minute by means of a small pump for a length of time depending on the particular gc analytical method. The entire apparatus is portable so that it may be carried in a pocket with the sampling tube in the breathing zone (normally a coat lapel). The apparatus can also be static.

Methanol is used to extract the nitrobenzene, which is separated on one of two gc systems, *viz.*

(a) A 3 m × 2 mm stainless steel column packed with 10% FFAP on Chromosorb WHP (100 – 120 mesh) for an air sample of 10 – 150 l.
(b) A 3 m × 2 mm stainless steel column packed with 5% SE-30 on Chromosorb WHP (100 – 120 mesh) for an air sample of 1 – 30 l.

Flame ionization detection is used in both systems. For full details of the method see ref. 13.

The sampling method uses a small portable apparatus with no liquids and is therefore easy to handle and maintain. The analytical method is relatively specific with the additional advantage of flexibility of operating conditions.

Air samples are normally taken by means of suction using an electric pump capable of operating at low flow-rates (*i.e.*, 10 – 300 ml per minute). If higher flow-rates are used, breakthrough can occur (*i.e.*, the nitrobenzene is not completely adsorbed and some of it passes through the adsorbent). It is also convenient either to establish the trapping device in a static position, or to have it attached to the user in such a way that the breathing zone is sampled (this usually means attachment to the worker's lapel).

It should be noted that suitable personal samplers may be commercially available.

Tenax-GC has been used to trap nitrobenzene plus benzene and chlorinated hydrocarbons from air.[14] The organic compounds were thermally desorbed and swept by helium on to a 50 m long capillary column coated with SP-2100 for gc analysis. The column was temperature-programmed from -90 to $+140\,°C$ and detection was by flame ionization. The limit of detection for nitrobenzene was 10 ppt.

Nitrobenzene and 4-chloronitrobenzene were collected from dye industry workplace air samples by passing the samples through ethanol.[15] The compounds were reduced to the corresponding anilines and coupled with sodium 1,2-naphthaquinone-4-sulphonate and the resulting dye was extracted into carbon tetrachloride. The extract was concentrated and the dyes were separated by ascending paper chromatography using 0.01N sodium hydroxide saturated with n-butanol as mobile phase. The separated dyes were extracted with carbon tetrachloride and spectrophotometrically determined at 450 nm. The method was used for total amounts of nitrobenzene in air of $10 - 100\,\mu g$.

Piezoelectric detectors have found some use in the determination of nitrobenzene. A silica crystal resonator was used as an indicator for such a detector which was used to determine toluene and ethyl acetate as well as nitrobenzene.[15]

Air was passed at a flow-rate of 94 ml per minute through quartz piezoelectric crystals operated at 15 MHz.[16]

Activated charcoal, quadrol, tetrabase and polyoxyethylene glycols, 400 and 750 were tried as coatings for the piezoelectric crystals and activated charcoal proved to be the most sensitive for the determination of nitrobenzene in the range $0.7 - 7.6$ ppm in air.[17]

11. CONDITIONS UNDER WHICH NITROBENZENE IS PUT ON THE MARKET

Production[11,18,19]

Nitrobenzene was first synthesised in 1834 by treating benzene with fuming nitric acid. Commercial production commenced in 1856.[11]

Nitrobenzene is manufactured commercially by the direct nitration of benzene with 'mixed' or 'nitrating acid', (approx. by weight, 60% sulphuric acid, 30% nitric acid and 10% water, though these proportions may be altered to suit individual processes). The reactor is of acid-resistant cast-iron or steel.

Dunlap in ref. 11 observes that most if not all producers now use continuous processes in place of earlier batch ones. A continuous process generally offers lower capital costs and more efficient labour usage.

Two phases are formed in the reaction mixture and the reactants are distributed between them. The rate of reaction is therefore controlled by mass transfer between the phases as well as by chemical kinetics. The interfacial area between the phases is kept as high as possible by vigorous agitation. Internal cooling coils control the temperature of what is a highly exothermic reaction.[11]

In a typical batch process, nitrating acid (56 – 60 wt% sulphuric acid, 27 – 32 wt% nitric acid and 8 – 17 wt% water) is added slowly below the surface of benzene in a reactor. Adjustment of feed rate and cooling maintains the temperature of the mixture at 50 – 55 °C until towards the end of the reaction when it can be raised to about 90 °C to promote completion. The reaction time varies but usually is two to four hours.

The reaction mixture is fed into a separator where the spent acid settles to the bottom to be drawn off and refortified. Crude nitrobenzene is drawn from the top, and may be used directly for aniline manufacture. To obtain a purer product, the crude nitrobenzene is washed with water, then with dilute sodium carbonate solution and finally with water again. If desired, the product may be further purified by distillation. Usually, a slight excess of benzene is used to ensure little or no remaining nitric acid in the spent acid which latter is either recovered or used as cycle acid to start subsequent runs after steam stripping to remove un-nitrated material. The nitrobenzene yield, based on the benzene used, is 95 – 98 wt%.[11,18]

The basic sequence of operations for a continuous process is the same as for a batch process though, for a given rate of production, the nitrators are smaller in a continuous process. A 115 l continuous nitrator has roughly the same production capacity as a 5,700 l batch reactor. In addition, a continuous process typically utilizes a lower nitric acid concentration and, because of the rapid and efficient mixing in the smaller reactors, higher reaction rates obtain. The nitrator can be a stirred cylindrical reactor with internal cooling coils or a cascade of such reactors. It also can be designed as a tubular reactor involving turbulent flow in which case the reaction mixture generally is pumped through the reactor in a recycle loop and a portion of the mixture is withdrawn and fed to the separator. A slight excess of benzene ensures maximum possible uptake of nitric acid and the minimizing of dinitrobenzene formation. The nitrating acid for continuous processes is about 56 – 65 wt% sulphuric acid, 20 – 26 wt% nitric acid and 15 – 18 wt% water. The temperature is maintained at 50 – 100 °C by adjustment of the cooling. Reaction times of 10 – 30 minutes are typical.

Separation of the two layers after the reaction mixture has been sent to a decanter takes 10 – 20 minutes, and washing is carried out in tanks similar in appearance and size to the nitrator. With high speed mechanical agitation, washing is completed in ten minutes. If the sulphuric acid is then concentrated and re-used, the losses amount to only 1% – 2%.[11,18]

Dunlap in ref. 11 notes that an adiabatic nitration process has been developed in which very low levels of by-product dinitrobenzene appear and in which the necessity of removing heat of reaction disappears; this heat can be used to reconcentrate the sulphuric acid. The concept is applicable to both continuous and batch processes. The nitrating acid consists of 60 – 70 wt% sulphuric acid, 5 – 8.5 wt% nitric acid and not less than 25% of water. Again a slight excess of benzene is used. The initial temperature is 60 – 75 °C and this rises to 105 – 145 °C at the end of the reaction. Reaction times are 0.5 – 7.5 minutes and vigorous agitation is required.

In an azeotropic process, the need to reconcentrate the sulphuric acid is partially or completely eliminated. The nitration takes place at 120 – 160 °C and excess water is distilled off as an azeotrope with benzene. After being separated from the product, sulphuric acid is recycled without needing to be concentrated. An excess of benzene is required to provide for the azeotropic distillation. After the azeotrope is condensed, water is separated and the benzene is returned to the nitrator. In the reactor, temperature control requires partial or complete vaporization of the recycled benzene. This need for vaporization is avoided in a duplex process in which the nitration is carried out in an azeotropic first stage which is followed by a lower temperature, mixed acid, second stage. All or part of the benzene resulting from the azeotrope of the first stage is taken to the second stage for nitration without the need for this benzene being first vaporized.

$$C_6H_6 + HNO_3 \rightarrow C_6H_5NO_2 + H_2O$$

Nitrobenzene is supplied in road and rail tankers, drums, cans and bottles.[19]

Uses[11,18–21]

Nitrobenzene is produced on a large scale only by aniline manufacturers.[18]

Approximately 97 – 98% of US nitrobenzene production is converted to aniline which in turn is further reacted in a wide range of industrial syntheses.[11] Nitrobenzene is a good organic solvent. It dissolves aluminium chloride, and thereby finds use in Friedel–Crafts reactions which nitrobenzene itself does not undergo (but see Section 14). It is used in the refining of certain lubricating oils and in the production of benzidine and chloronitrobenzenes.[19] It has found use as an ingredient of metal polishes and shoe polishes and in modifying esterification of cellulose acetate.[20] It has also been used in perfumeries and soaps and as an insect repellent.[21]

In the US, nitrobenzene production rose from 73,600 tonnes in 1960 to over 248,000 tonnes in 1970. Production fluctuated in the 1970s but had risen to over 301,000 tonnes in 1983.[11,19] EC production in 1983 was 500,000 tonnes and its consumption was 505,000 tonnes.[19] Japanese production in 1983 is given as 100,000 tonnes.[19]

12. STORAGE, HANDLING AND USE PRECAUTIONS[22,23]

Nitrobenzene is of moderate flammability and the vapour can form explosive mixtures in air.

[As with other flammable materials the organization of storage facilities has to take into account two separate requirements. These are that the stored material be protected from a fire elsewhere on the premises and that the premises be protected from a fire involving the stored materials. On an industrial scale, sufficient physical separation and water spray curtains are often used effectively. On a laboratory scale, such materials need to be stored in fire resistant, thermally insulated cupboards.

It is important to appreciate that a non-combustible barrier is not synonymous with a fire barrier. A fire barrier has to protect against both conducted and radiant heat. For example, a metal cupboard is useless on both these counts.]

Care must be taken with storage process and drying plant to ensure that hazard-free conditions exist regarding fire and explosion risks and exposure of personnel to nitrobenzene. The material should be kept in clearly marked containers, stored in a ventilated space and kept separate from incompatible chemicals. The floor of the storage area should be impermeable and have adequate drainage for spilt liquid, though not to the public drains. Sills or ramps should be at storeroom openings. Detached storage is preferable with protection against physical damage, intense heat and freezing. However, ref. 22 states that nitrobenzene storage areas should not be heated.

Nitrobenzene is toxic by ingestion, vapour inhalation and skin absorption. It should be handled only with adequate ventilation. Process enclosure, local exhaust ventilation and personal protective equipment should be used, as appropriate, to keep personnel exposures below permissible limits. Personal protective equipment should consist of impervious aprons or overalls, gloves, face shields (eight inch minimum) and such other protective clothing as is necessary to prevent skin contact with the liquid. (Ref. 23 states that butyl rubber, PVA* or Viton (a synthetic fluoro-rubber) may provide protection.) Any clothing that could have become contaminated should be kept in closed containers pending disposal or laundering in a safe manner. Emergency eye-wash and water-drench facilities should be available plus the facility for washing the body with soap or mild detergent and water. [Note that personnel should not be allowed to enter areas containing potentially hazardous concentrations of nitrobenzene unless all relevant legislation and official guidelines have been strictly adhered to. Suitable self-contained breathing apparatus should be used.] Such exposed personnel should be able to communicate with each other and with those (*i.e.*, more than one person) monitoring the work from outside the hazardous area. Monitors should be suitably equipped to effect rescue (possibly with lifelines to exposed personnel) and should themselves ensure, so far as is possible, that one of their number remains outside the hazard area in the event of an emergency.

Vessels that have contained nitrobenzene should be drained and purged before any welding or cutting. Plant such as drying ovens into which nitrobenzene might be introduced and heated should be fitted with flameproof exhaust ventilation to prevent the formation of a flammable mixture with air. Note that exhaust ventilation should be sited to avoid recirculation of nitrobenzene and that any discharge to the atmosphere should not have a nitrobenzene concentration that constitutes a toxic hazard.

N.B. the caveat in Section 13 (final paragraph).

---

*Polyvinyl alcohol coated gloves are available in the U.K. but it should be noted that the PVA coating may be impaired by contact with water.

13. FIRE HAZARDS[1,3,22,24]

Nitrobenzene is a flammable liquid though it does not ignite readily.[1] It presents a moderate fire hazard when exposed to heat, flame or oxidizing agents. It has a boiling point 210.8 °C, specific gravity 1.2 and approximately 0.2% water solubility. It has a flash point of 88 °C (closed cup), an auto-ignition temperature of about 480 °C and a lower flammability limit in air of 1.8% v/v at 93 °C (see Section 17).

Fire-exposed containers of nitrobenzene can explode when heated and, if feasible, should be cooled with flooding amounts of water applied from as far as possible. Fire-control water should be bunded for later disposal.

Suitable fire extinguishants for small fires are carbon dioxide, dry chemical powder, alcohol-resistant foam or water spray. For larger fires, water spray, fog or alcohol-resistant foam are suitable. Water spray should be applied carefully over the higher boiling liquid (excessive frothing is a hazard) in order to blanket the fire (nitrobenzene being denser and of low solubility). In addition, solid jets of water may scatter burning chemical. The fire should be fought from upwind from the maximum distance in order to avoid breathing toxic vapours of nitrobenzene and its combustion products. Water spray can help control these vapours but personnel exposed to these hazards must have appropriate safety clothing and equipment (see Section 12).

Ref. 24 notes that where a fire is fed by an uncontrolled flow of combustible liquid the decision on how or if to fight it will depend on the size and type of fire anticipated and must be carefully considered. This may call for special engineering judgement, particularly in large-scale applications.

[Notwithstanding the foregoing review of various authorities noted in Sections 12 and 13, it is essential that managers and others responsible for the planning, implementation and overseeing of personnel and plant safety should be familiar with the legal constraints and official guidelines applicable to them and that they liaise with their local emergency services in the planning of plant and storage facilities and in the preparation of contingency plans for dealing with fires and other emergencies. Managers should regularly monitor staff knowledge of, and ability to implement, emergency procedures and should ensure that equipment provided for use in emergencies is regularly inspected and maintained.]

14. HAZARDOUS REACTIONS[25]

Bretherick[25] notes several hazardous reactions that have occurred between nitrobenzene and alkalis, either alone or in the presence of additional compound(s), *viz.*:

An explosion occurred when a mixture of nitrobenzene, flake sodium hydroxide and a little water were heated in an autoclave.

During the technical-scale preparation in a 6 m³ vessel of a warm solution of nitrobenzene in methanolic potassium hydroxide (flake 90%), the accidental omission of most of the methanol led to an accelerating exothermic reaction which eventually ruptured the vessel. Laboratory investigation showed that, with no methanol or with the full amount (3.4 vol.) of methanol present, no exothermic reaction occurred but that it did so in presence of only a little methanol. The residue was largely a mixture of azo- and azoxy-benzene.

On heating nitrobenzene with finely powdered anhydrous potassium hydroxide though, violent conversion to (mainly) 2-nitrophenol is reported, and the accidental substitution of nitrobenzene for aniline as diluent during large-scale fusion of benzanthrone with potassium hydroxide caused a violent explosion.

The drying of wet nitrobenzene in the presence of flake sodium hydroxide (probably by distilling off the nitrobenzene–water azeotrope) led to the separation of finely divided solid base which remained in contact with the hot, dry nitrobenzene under essentially adiabatic conditions. A violent explosion ensued which blew out a valve and part of the 10 cm steel pipeline. These conditions were simulated in a subsequent investigation and led to thermal runaway and explosion.

Mixtures of nitrobenzene and aluminium chloride are thermally unstable and may lead to thermal decomposition. Simulation of an incident in which a 4,000 l vessel was ruptured suggested that the pressure in the reactor probably increased from 3 to 40 bar in five seconds.

Studies on the thermal degradation of nitrobenzene–aluminium chloride addition compounds formed in Friedel–Crafts reactions show that it is characterized by a slow, multi-step decomposition reaction above 90 °C, which self-accelerates with high exothermicity producing azo- and azoxy-polymers.

Addition of aluminium chloride to a large volume of recovered nitrobenzene containing 5% phenol caused a violent explosion. Experiments showed that mixtures containing all three substances reacted violently at 120 °C.

The unpredictable violence of the Skraup synthesis of quinolines (which usually has nitrobenzene as the oxidant) is often attributed to a lack of stirring and to inadequate temperature control. In a synthesis (that had been doubled in scale), sulphuric acid was added to a stirred mixture of aniline, glycerol, nitrobenzene, ferrous sulphate and water. The ambient temperature was unusually high (reaction contents 32 °C), and the reaction went out of control soon after the addition of what was an accidental excess of acid. A 150 mm rupture disc blew out first, followed by the manhole cover of the vessel. The violence of the reaction was attributed to the increased scale and the high ambient temperature. Experiment showed that a critical temperature of 120 °C was attained immediately on addition of excess acid under these conditions.

Concentrated aqueous solutions of urea perchlorate dissolve nitrobenzene to give high velocity explosives.

Mixtures of tetranitromethane with nitrobenzene are high explosives of high sensitivity and detonation velocity and are spark-detonatable.

The combination of sodium chlorate with nitrobenzene is a powerful explosive.

A series of mixtures of nitrobenzene with nitric acid has been shown to possess high-explosive properties. An unexpected plant explosion occurred even when a substantial quantity of water was present. The detonation parameters of such mixtures have been studied.

Uncontrolled contact of nitrobenzene with peroxodisulphuric acid may cause an explosion.

Mixtures of nitrobenzene with dinitrogen tetraoxide, once used as liquid high-explosives (with the addition of carbon disulphide to lower the freezing point) had disadvantageously high sensitivity to mechanical stimulus.

A solution of phosphorus pentachloride in nitrobenzene is stable at 110 °C but begins to decompose with accelerating violence above 120 °C with evolution of nitrous fumes.

Nitrobenzene is rendered shock-sensitive by the addition of traces of potassium or potassium–sodium alloy.

When a wash of 5% sulphuric acid was used to remove amines from nitrobenzene, the latter became contaminated with a tarry emulsion that formed. On distillation, the hot, tarry, acidic residue attacked the iron vessel. Hydrogen evolution and an eventual explosion took place. It was later found that addition of the nitrobenzene to the diluted acid did not give emulsions, while the reverse addition did. The accident occasioned a final wash with sodium carbonate being added to the process.

During hazard evaluation of a continuous adiabatic process for the manufacture of nitrobenzene it was found that nitrobenzene with 85% sulphuric acid gave a violent exotherm above 200 °C, and with 69% acid a mild exotherm was given at 150 – 170 °C.

In the preparation of 3-nitrobenzenesulphonic acid by sulphonating nitrobenzene with oleum, a completed 270 l batch exploded violently after hot storage at approximately 150 °C for several hours. An exotherm develops at 145 °C and the acid itself decomposes at approximately 200 °C. In a similar incident, water leaking into the fuming sulphuric acid reaction medium caused an exotherm to over 150 °C and subsequent violent decomposition. Additional detail on evaluating the conditions for this synthesis is given by Bretherick.

15. EMERGENCY MEASURES IN THE CASE OF ACCIDENTAL SPILLAGE[1,2,26]

In the event of a spillage all possible sources of ignition should be eliminated. Non-essential personnel should be prohibited from entering the area. Those engaged in dealing with the emergency should wear appropriate apparel and, if necessary, breathing equipment (see Section 12). If feasible, closed spaces should be ventilated before being entered. If a fire hazard exists, containers of flammable materials should be removed to a safe area if this is a practicable, safe procedure.

Contact with spilled nitrobenzene must be avoided. In the event of contact with the skin, the affected area should be washed at once using soap (or mild detergent) and water. Contaminated clothing or shoes should not be worn again until freed from nitrobenzene.

Small quantities of spilled nitrobenzene can be absorbed on paper towels or vermiculite, dry sand, earth or similar. The disposal of spilled material must comply with local and national legislation but a possible means is the burning of nitrobenzene-soaked paper in a suitable location away from combustible materials.[26]

The burial of nitrobenzene-contaminated absorbent in a safe open area, possibly in sealed containers in a secured, sanitary, approved landfill, may be possible.[26]

Incineration of nitrobenzene may be effected by first pouring it onto sodium bicarbonate or a 90/10 sand/soda ash mixture, then mixing in heavy paper cartons and incinerating in a permitted place. The fire may be augmented with wood or paper. Alternatively an incinerator with an afterburner and an alkaline scrubber may be used. Such an incinerator may also be used to burn nitrobenzene after having dissolved the chemical in a flammable solvent.[2]

If the spillage is due to a leaking container the leak should be stopped if safely possible. Where there is appreciable nitrobenzene vapour, this can be reduced with water spray. In the case of a large spillage, the area ahead of the spill should be bunded pending later disposal of the chemical and downwind evacuation must be considered.[1] The appropriate water authority must be informed of any contamination of a drain, sewer, water course or land.

16. FIRST AID

General

The casualty should be removed from the danger of further exposure into the fresh air. Rescuers should ensure their own safety from inhalation and skin contamination where necessary. Conscious casualties should be asked for information about what has happened. Bear in mind that the casualty may lose consciousness at any time. Where necessary continued observation and care should be ensured. First aiders should take care not to become contaminated.

Inhalation

The conscious casualty should be kept at rest. The unconscious casualty should be placed in the recovery position and an open airway maintained. If breathing or heart beat stops, resuscitation should be commenced immediately. Medical aid should be obtained or the casualty removed to hospital immediately.

Skin contamination

Remove contaminated clothing. Wash with copious amounts of water, or soap and water, for at least 15 minutes. If necessary, seek medical aid. In the case of persistent skin irritation refer for medical advice.

Eye contamination

Irrigate with copious amounts of water for at least ten minutes but avoid further damage to the eye from the use of excessive force. Seek medical aid. Remember that contact lenses may be worn and that these may be affected by some solvents and may impede decontamination of the eye.

Ingestion

If the lips or mouth are contaminated rinse thoroughly with water. Do not induce vomiting. Give further supportive treatment as for inhalation. In the case of ingestion of significant amounts wash out the mouth with water. Obtain immediate medical attention. Contact a hospital or poisons centre at once for advice. Remember that symptoms may develop many hours after exposure so continued care and observation may be necessary. If signs of cyanosis appear, oxygen should be administered. Specific antidote (methylene blue by intravenous injection) may be administered by qualified medical personnel; the judgement for this rests on clinical assessment and on blood test for methaemoglobin level. The antidote may therefore be kept at work sites where nitrobenzene is handled and medical aid is available.

In all cases note information on the nature of exposure and give this to ambulance or medical personnel. Information for doctors is provided in some technical literature issued by manufacturers.

17. PHYSICO-CHEMICAL PROPERTIES

17.1 General[11,24]

Nitrobenzene is an oily liquid with a characteristic odour resembling that of bitter almonds. Its colour ranges from pale yellow to brown, depending on its purity.[11,24].

Nitrobenzene is miscible in all proportions with benzene and diethyl ether, and is readily soluble in most other organic solvents.[11] The solubility of nitrobenzene in water is recorded as 0.19% at 20 °C, and 0.8% at 80 °C.[11]

17.2 Melting Point[27]

5.7 °C.

17.3 Boiling Point[27,28]

The boiling point of nitrobenzene is given by Weast[27] as

210.8 °C.

Nitrobenzene will form binary azeotropes with some compounds, two of which are listed below[28]:

Wt % nitrobenzene	Second component	Wt % second component	Boiling pt azeotrope (°C)
41	Ethylene glycol	59	185.9
12 (vol%)	Water	88 (vol %)	98.6

17.4 Density/Specific Gravity[27,29]

The specific gravity of nitrobenzene (referred to water at 4 °C) is 1.205 at 15 °C,[29] and 1.2037 at 20 °C.[27]

17.5 Vapour Pressure[27,30]

Weast[27] tabulates the following temperatures corresponding with vapour pressures below atmospheric pressure for nitrobenzene:

Temperature in °C	Pressure in mmHg
44.4	1
84.9	10
115.4	40
139.9	100
185.8	400

The critical temperature of nitrobenzene is 482.8 °C.[30]

17.6 Vapour Density[24]

4.3 (relative to air = 1).

17.7 Flash Point[24]

88 °C, closed cup test.

17.8 Explosive Limits[24]

The lower limit of flammability of nitrobenzene in air has been determined as 1.8% v/v at 93 °C; no upper limit is stated.[24]

The National Fire Protection Association[24] give the ignition temperature of nitrobenzene as 482 °C. However, they point out that different test conditions (and also different definitions of "ignition temperature") can result in widely varying values being quoted. They therefore recommend that any value should be treated as an approximation.

17.9 Viscosity[27]

The viscosity of liquid nitrobenzene is reported as follows[27]:

Temperature in °C	Viscosity in cp
2.95	2.91
5.69	2.71
5.94	2.71
9.92	2.48

Temperature in °C	Viscosity in cp
14.94	2.24
20.00	2.03

17.10 Concentration Conversion Factors

At 25 °C and 760 Torr (1 atm),

1 ppm = 5.04 mg/m³, and
1 mg/m³ = 0.199 ppm.

18. TOXICITY

18.1 General[6,22,26,31,32]

Nitrobenzene is toxic by inhalation of vapour, by ingestion of the liquid or by absorption of vapour or liquid through the skin. Rapid absorption through the skin is frequently the main route of entry. Even a small amount absorbed from clothes or shoes may cause toxic symptoms due to methaemoglobinaemia whose onset is insidious and may cause death from anoxia. Nitrobenzene also affects the central and peripheral nervous systems causing, in some cases, excitement and tremors followed by severe depression, unconsciousness and coma. Repeated exposure may be followed by liver impairment up to yellow atrophy, haemolytic icterus and anaemia of varying degrees, with the presence of Heinz bodies in the red cells.[22] An allergic skin rash may occur from long-term exposure.[26] Ingestion of alcohol aggravates the toxic effects of nitrobenzene. Direct contact with the eyes or skin may cause mild irritation. It may produce dermatitis due to primary irritation or sensitization.[26]

The odour threshold of nitrobenzene is stated to range from about 0.005 ppm to about 2 ppm.[31] Several organizations recommend an exposure limit of 1 ppm TWA in which circumstances odour may not be an adequate warning to some persons of a hazardous concentration. The saturated vapour pressure of nitrobenzene at 25 °C is given by Timmermans[32] as 0.340 mmHg. This is equivalent to a concentration of about 450 ppm in the atmosphere.

Ref. 6 notes that the odour of nitrobenzene in water is detectable at concentrations as low as 30 μg/l.

18.2 Toxicokinetics[33-35]

Beauchamp *et al.*[33] observe that nitrobenzene is readily absorbed through human skin and lungs. Under conditions precluding simultaneous absorption through the skin, human subjects breathing air containing 5 – 30 μg nitrobenzene per litre [approx. 1 – 6 ppm] retained approximately 80%. This percentage remained relatively constant (87 – 73%) over a period of six hours, indicating that steady state conditions were achieved. Thus humans exposed to 10 mg/m³ [2 ppm] may absorb

18.2 – 24.7 mg in six hours. For skin absorption, it was found that, in six hours, 8 – 18 mg were absorbed through the skin of naked subjects exposed to a 10 mg/m³ [2 ppm] vapour concentration, with only minor, and probably statistically insignificant, decreases in absorption when subjects were clothed.

In two studies of the absorption of liquid nitrobenzene through the skin, there was some discrepancy between the results of each. In one study, the nitrobenzene was applied at 15 mg/cm² on the forearm and the initial rates of absorption varied between 0.5 to 2.5 mg/cm²/hr. The rate decreased over a period of several hours to 0.1 to 0.2 mg/cm²/hr. In the other study absorption through the skin was estimated by monitoring ¹⁴C excreted in the urine after application of labelled nitrobenzene at 4 μg/cm² on the forearm. It appeared here that only 1.5% of the applied dose was absorbed *via* the skin in 24 hours. Beauchamp *et al.* observe that the discrepancies between the two studies may arise from the very different doses per unit area.[33]

Studies using human volunteers and low inhalational doses of nitrobenzene have detected only *p*-nitrophenol as a metabolite and it has been suggested that monitoring the urinary excretion of this compound be used to estimate worker exposure to low levels of nitrobenzene (*e.g.*, up to 1 ppm).

p-Aminophenol has also been found as a metabolite in humans but only after more severe exposure. However, as the acid hydrolysis used to liberate the *p*-aminophenol from its conjugates would also have converted *p*-acetamidophenol to aminophenol, the actual identity of the metabolite in question is uncertain. Beauchamp *et al.* note that both *p*-nitrophenol and *p*-aminophenol were found in the urine of a woman who became ill after having been exposed for 17 months to paint containing nitrobenzene. The ratios of *p*-aminophenol to *p*-nitrophenol in the urine ranged from 0.5 to 1. The urinary metabolites disappeared eight weeks after exposure ended while changes such as methaemoglobinaemia subsided earlier, in approximately five weeks. A similar ratio of excreted metabolites was found in rats exposed to 25 ppm nitrobenzene vapour for eight hours.[33]

Beauchamp *et al.*[33] observe that a number of metabolites were identified in the urine following the administration of nitrobenzene to rabbits by gavage at 150 – 200 mg/kg. About 56% of the dose was accounted for within two days, the major product being *p*-aminophenol. All phenolic compounds were excreted as conjugates such as the glucuronide or as sulphate. In another study, randomly labelled ¹⁴C-labelled nitrobenzene at 250 mg/kg was administered by stomach tube to rabbits, and the metabolites were measured in expired air, urine and faeces. Nearly 70% of the radioactivity was recovered by these routes within 4 – 5 days after dosing. Again, the major product was *p*-aminophenol (31% of dose) excreted in the urine. The other urinary metabolites found were *m*-nitrophenol (9% of dose), *p*-nitrophenol (9%), *m*-aminophenol (4%), *o*-aminophenol (3%), 4-nitrocatechol (0.7%), aniline (0.3%), *p*-nitrophenylmercapturic acid (0.3%), *o*-nitrophenol (0.1%), nitroquinol (0.1%), unchanged nitrobenzene (<0.1%). Radioactivity recovered from air within 30 hours was from carbon dioxide (1.0% of dose) and nitrobenzene (0.5%). Both intravenous and dermal administration of nitrobenzene to rats gave similar results, with *p*-aminophenol accounting for 19% of the dose within 48 hours and *p*-nitrophenol 7%.[33]

As nitrobenzene does not of itself react to form methaemoglobin it must be metabolized *in vivo* to a suitably active material, presumably by mixed-function oxidase activity in the liver.[34] Beard and Noe in ref. 35 observe that the causative agents in cyanosis due to aromatic nitro compounds are probably derivatives created by biochemical redox processes. These derivatives form haemoglobin complexes thus hindering oxygen transport to the tissues. The phenylhydroxyl-amines are probably the most potent cyanosis producers in the redox series. It has been suggested that detoxification by rearrangement of the phenylhydroxylamine to less toxic *o*- and *p*-aminophenols probably is responsible for restoration of the oxygen transport balance.[34,35]

Beauchamp *et al.*[33] state that *in vivo* metabolism of nitrobenzene in humans and in animals is not well characterized but it is clear that both oxidation (hydroxylation) and reduction occur to form the major excretory products of *p*-nitrophenol and *p*-aminophenol.

These reviewers cite studies indicating a possible role being played by bacterial reductases. Following the intraperitoneal administration of nitrobenzene at 200 mg/kg to normal rats, 30 – 40% of the blood haemoglobin was converted to methaemoglobin, while with germ-free rats, whose gut contents contained only 1 – 2% of the nitro-reductase activity of the normal controls, no measurable methaemoglobin formation occurred up to seven hours after administration. In both types of rat, tissues such as liver, kidney and gut wall contained comparable levels of nitro-reductase.[33]

On the other hand studies of microsomal mixed function oxidase activity in rats pretreated with inducers or inhibitors led to the postulation of an important role for cytochrome P450 dependent nitro reductase in the formation of a methaemoglobin-causing metabolite. Beauchamp *et al.* note the apparent contradiction with the germ-free rat study but observe that it seems possible that the inducers decreased the duration of methaemoglobinaemia by increasing the rate of detoxification of a metabolite produced by gut flora. Beauchamp *et al.* also note that the effect of inhibitors on microfloral nitro-reduction had not been assessed in the inducer/inhibitor study and that it was possible that decrease in nitrobenzene-induced methaemoglobinaemia, which had been observed following protective administration of 3-amino-1,2,4-triazole, SKF-525A and carbon tetrachloride before the nitrobenzene, was due to alteration or inhibition of microfloral nitro-reductase.[33]

In vitro experiments show that mammalian reductases produce intermediates which probably include the nitro anion radical. Molecular oxygen can oxidize this back to the parent compound with concomitant production of a superoxide anion (O_2^-). Both nitro and superoxide anions are potentially toxic but Beauchamp *et al.*[33] observe that the *in vitro* conditions of these observations may not compare with actual *in vivo* ones, that nitrosobenzene, phenylhydroxylamine and aniline, other intermediates in the reduction of nitrobenzene, are potentially toxic and have been found in animals treated with the substance, and that there is evidence that these metabolites originate, at least in part, from the activity of bacterial reductases.

Beauchamp *et al.* also note that there are few data on the human metabolism of nitrobenzene after reasonable exposures under controlled conditions and so it is not clear how comparable the production of reduced metabolites in animals is to the same situation in humans.[33]

18.3 Human Acute Toxicity[1,26,33-41]

Browning[36] and OSHA[26] observe that the outstanding effect of nitrobenzene is its capacity to cause the formation of methaemoglobin. There is risk of death from respiratory failure if the methaemoglobinaemia is severe. In man the onset of this condition is insidious, cyanosis being grossly recognizable only when the methaemoglobin concentration reaches 15% or more. At up to about 40% the individual may feel perfectly well and have no complaints. Above 40% there usually is weakness and dizziness and at up to 70% there may be ataxia, dyspnoea on mild exertion, tachycardia, nausea, vomiting and drowsiness.

Vapour concentrations near 40 ppm have caused intoxication in workers, while at concentrations averaging 6 ppm, headache, vertigo, cyanosis and sulphaemoglobinaemia have been caused.[1,26] OSHA[1] states that inhalation of 6 ppm may cause headache, vertigo, cyanosis and sulphaemoglobinaemia; that higher concentrations may cause weakness, nausea, vomiting, drowsiness, dyspnoea on mild exertion, tachycardia, lethargy, giddiness, anoxia, ataxia, semi-stupor, vertigo, hypotension and numbness, and that severe exposure may affect the CNS, resulting in excitement, tremors, severe depression, unconsciousness, coma and death, 200 ppm being immediately dangerous to life and health. OSHA also notes that the onset of symptoms of poisoning may be insidious, with a possible delay of several hours, and that jaundice, pain on urination and anaemia may occur later. OSHA also notes that ocular discolouration, brownish colour of the vessels of the fundus and conjunctiva, has been reported due to methaemoglobinaemia. Hamilton and Hardy[37] state that the cyanosis associated with nitrobenzene poisoning can be alarming but that, even at methaemoglobin conversion levels approximating 75%, recovery without specific therapy has been the rule.

Nitrobenzene may be absorbed through the intact skin from contaminated shoes or clothing or from the directly applied liquid which itself may cause smarting of the skin and first degree burns.[1] As noted in Section 18.2, the degree of absorption of nitrobenzene vapour through the skin approaches that of inhalational absorption of the same aerial concentration.

Beauchamp *et al.* observe, regarding methaemoglobinaemia, that considerable individual variation exists and that the relationship between the absorbed dose of nitrobenzene and the severity of the response in man is not clear-cut. Infants though appear to be particularly sensitive to nitrobenzene-induced methaemoglobi-naemia.[33] Levin[38] notes that cases have been reported in infants in which cyanosis followed the wearing of recently marked clothing and in which nitrobenzene was isolated from the stamping ink. Zeitoun[39] reports on cases of nitrobenzene poisoning in newborn infants who had been rubbed with what were found to be mixtures of 2 – 10% nitrobenzene and 90 – 98% cotton-seed oil.

Nitrobenzene has been used in shoe dyes as well as laundry marking ink. Levin[38] notes several reports of poisoning from freshly dyed shoes in which nitrobenzene was found to be the toxic agent.

There have been several reports of peroral poisoning in humans. Beauchamp *et al.*[33] provide the following precis of several cases of acute poisoning though the amount or concentration of nitrobenzene involved is not always known.

Following light (but unspecified) peroral exposure, there was an increase of Heinz bodies to 60% on the fourth day. This was not proportional to the decrease in RBC as the latter went from $4 - 3.1 \times 10^6$ at one week with the Heinz bodies disappearing. The relatively small RBC reduction was not due to regeneration as reticulocytes amounted to only 5%.

Following peroral intake of about 100 g by a male there were leucocytosis (32,000) with transient plasmocytosis, a moderate decrease in RBC and an increase in nucleated RBC (21 per 100 WBC). Beauchamp *et al.* note the possibility of a direct or indirect effect of nitrobenzene on proliferative WBC response.

On autopsy, following the lethal peroral intake of about 100 ml by a female, the brain membrane was soft, the brain matter plethoric and there were petechial haemorrhages in the white matter. There were haemorrhages on the upper and lower extremities and shallow haemorrhages under the pleura. On incision, the lung exuded a significant amount of foamy, bloody fluid. Beauchamp *et al.* note though that whether this is truly an effect depends on the case history and the effect may be agonal.

In other acute peroral cases are noted headache and dizziness and, in an autopsy, oedema of the upper CNS.

Transient leucopenia appeared in another case of peroral intake (amount unknown) by a female. Leucopenia (2,600) and a slight decrease in haemoglobin (79%) were present five weeks after poisoning. After three months the WBC count was 5,800 and the haemoglobin was 84% of normal. The report on this case also noted the neurotoxic effects of headache, hyperalgesia and paraesthesia in the feet. In another acute case of peroral intake of an unknown amount of nitrobenzene, there was loss of cognition, confusion, hyperaesthesia in the lower limbs and 3 mg % hyperbilirubinaemia (1.9 mg % indirect).[33]

Browning[36] notes the case of poisoning by an industrial product later found to consist of 3% nitrobenzene, 95% mineral oil and the remainder a chlorinated solvent believed to be trichloroethylene. An (unstated) amount was ingested in error and the toxic effects were dominated by intense cyanosis and semi-consciousness. The effects were not immediate but supervened in about $0.5 - 1$ hour with violent headache, loss of consciousness and very dark cyanosis. There was moderate anaemia (3,137 red blood corpuscles [*sic*]) but a normal white cell and differential count. The blood was dark brown in colour and methaemoglobinaemia was diagnosed. Treatment was with methylene blue and vitamin C. Improvement in the symptoms followed but temporary anuria occurred with final recovery on the seventh day.

Beard and Noe in ref. 35 note that most references advocate the use of methylene blue intravenously if the methaemoglobin level reaches 40% or if the clinical signs and symptoms are severe or there is coma or stupor. They note though that the total dose should not exceed 7 mg/kg as toxic effects such as dyspnoea, precordial pain, restlessness, apprehension, red cell haemolysis and 'changes in the electrocardiogram (reduction in the height or even reversal of the T wave, frequently with lowering of the R wave)' may be caused. Also, some investigators question the efficacy of methylene blue treatment for methaemoglobinaemia. The use of vitamin C also apparently has had questionable results.

Von Oettingen[40] quotes reports indicating that the lethal peroral dose of nitrobenzene can vary widely. Twenty drops have been estimated as a fatal dose while recovery is reported following a dose of 400 g. In most cases death is due to respiratory failure. In addition, ref. 34 quotes an observation that all chemical agents which generate methaemoglobin have additional toxic effects that may make profound contributions to the toxic syndrome. Some, such as nitrobenzene, have central and prominent cardiac effects that in some species, including man, appear to be the proximal cause of death [but *v.s.*].

Von Oettingen[40] notes that there have been numerous cases of nitrobenzene having been used as an abortifacient. It appears not to be known when and for what reason the substance acquired a reputation as such an agent. In 1895 four cases were instanced of which three resulted in abortion, but generally the desired effect is not accomplished. One observer believed nitrobenzene to be at least as toxic to the mother as to the foetus and another noted that in 24 cases collected from the literature, twelve ended fatally [no dosages are noted].

Beauchamp *et al.*[33] note that when nitrobenzene was used as a preservative in cutting oils, contact dermatitis, associated with exposure to the substance, was frequently reported. OSHA[26] notes that nitrobenzene may produce dermatitis due to primary irritation or sensitization and that it is mildly irritating to the eyes.

ACGIH[41] notes a report that 200 ppm is the maximum concentration that can be inhaled for one hour without serious poisoning, the observers reporting this considering 1 – 5 ppm to be a safe level for daily exposure.

18.4 Animal Acute Toxicity[3,33,42,43]

Methaemoglobinaemia has been experimentally induced by nitrobenzene in a number of animal species. Cats, rabbits, pigeons, rats, dogs and mice have been studied, with mice having been found more resistant than the other species.[33] In a study by Smith *et al.*.[42] intraperitoneal administration to mice of 0.1 – 1.0 mmol [0.0123 – 0.123 g]/kg resulted in approximately 5% methaemoglobin formation. Five mmol [0.615 g]/kg produced slight methaemoglobinaemia, while 10 mmol [1.23 g]/kg was lethal in 40 minutes.

Studies with animals suggest that nitrobenzene and/or its metabolites are toxic to bone marrow, Beauchamp *et al.*,[33] noting that several investigators have reported decreased ^{59}Fe utilization in bone marrow following nitrobenzene exposure. For example, a 28% decrease in iron utilization by bone marrow was indicated following subcutaneous administration of 32 mg/kg nitrobenzene to male and female rats.

Neurotoxicity has been observed in animals to which nitrobenzene has been administered. Dogs inhaling 'saturated air' [presumably *ca.* 400 – 450 ppm] for 5.5 hours experienced nystagmus and paralysis. It was reported that there were no changes of any region of the brain except for the Purkinje cells of the cerebellum. These were reduced in number, showed chromatolytic degeneration, swollen cells and an absence of tigroid bodies (Nissl bodies).[33]

Beauchamp *et al.*[33] also note a study in which an intraperitoneal administration of 8.1 mmole [1 g]/kg (lethal dose) of nitrobenzene to female mice caused loss of righting reflex and comatose tremor. Some animals displayed running movements while in coma.

Intraperitoneal administration of 0.01 – 0.03 g/kg to cats produced marked opisthotonos, running movements and hind limb paralysis.[33]

Bond *et al.*[43] performed experiments to ascertain the possible deleterious effects of nitrobenzene in different tissues of male Fisher-344 rats. Single oral doses of 50 – 450 mg/kg were given. At the time of sacrifice (this varied according to the particular response-study), 25 tissues were removed for examination. The report states that histopathological changes consistently involved only the liver and testes. One rat receiving 450 mg/kg had a microscopic cellular lesion. Hepatic centrolobular necrosis appeared inconsistently in rats subjected to various doses, while hepatocellular nucleolar enlargement was consistently detected in rats given doses as low as 110 mg/kg. The authors state that these data suggest that nucleolar enlargement was independent of cell death and subsequent regeneration.

The report states that testicular lesions were confined to the seminiferous tubules. They consisted of necrosis of the primary and secondary spermatocytes with the appearance of multinucleated giant cells between one and four days after administration of 300 mg/kg of nitrobenzene. Necrotic debris and decreased numbers of spermatozoa were seen in the epididymis as early as three days after administration.

Bond *et al.* observe that the nitrobenzene-induced methaemoglobinaemia [this had also been manifest] did not appear to be solely responsible for the formation of early lesions in the liver, testes or brain of the rat since sodium nitrate administration, at doses which produced methaemoglobinaemia equivalent to that of nitrobenzene, did not produce any histopathological changes. The authors conclude therefore that the liver and testicular damage observed in the study were probably due to a direct effect of nitrobenzene or its metabolites.

The oral LD_{50} for the rat is noted as 640 mg/kg.[3]

Beauchamp *et al.*[33] note an acute toxicity study in which female rabbits were dosed by gavage with 180 – 370 mg/kg. The eventual death of 50% of the animals resulted from doses of greater than 200 mg/kg and occurred from two to six days post-dosage when this was 250 mg/kg.

18.5 Human Sub-acute and Chronic Toxicity Including Reproductive Effects[1,9,19,33,40,41,44,45]

General

Beauchamp *et al.*[33] note the case of a female patient who had worked with nitrobenzene for 17 months but who was apparently exposed to high concentrations (level unstated) for a period of two months. Neurotoxic and hepatotoxic effects were observed. Neurotoxic effects noted were headache, nausea, vertigo, paraesthesia in the legs and hyperalgesia in the back of the hands and in the feet. Hepatotoxic effects included an enlarged and tender liver, icterus, and altered serum chemistry. Total serum bilirubin was not elevated but a slight increase in conjugated bilirubin with icterus was reported. Bromsulphthalein retention was markedly elevated, a cephalin-flocculation test was positive and serum cholesterol was slightly reduced. Serum enzymes, GOT, GPT and alkaline phosphatase were unchanged.

OSHA[1] observes that persons with glucose-6-phosphate dehydrogenase deficiency or blood disorders may be at an increased risk from exposure.

Apart from hyperbilirubinaemia noted in Section 18.3, decreased prothrombin activity has also been reported amongst alterations in the liver function. In respect of neurotoxicity, polyneuritis was reported in a case history of inhalational and possibly dermal exposure to nitrobenzene (unquantified) for several years.[33]

ACGIH[41] note that, according to one observer (in 1957), poisoning by nitrobenzene is probably more common than is generally realized, although most cases are mild. Apart from the primary effect of methaemoglobin formation, in sub-acute and chronic poisoning anaemia is the leading feature as well as symptoms such as vertigo, headache and vomiting [already noted].[41]

ACGIH[41] report that nitrobenzene vapour averaging 6 ppm was found in the air of a [chemical] plant making nitro-aromatic compounds. No intoxication, but 'one or two' cases of headache were reported. Small amounts of methaemoglobin and sulphaemoglobin and some Heinz bodies were found in the blood of workers there. Previous concentrations of nearly 40 ppm had resulted in intoxication of workers.

In another study noted by ACGIH,[41] researchers who had studied the excretion of *p*-nitrophenol by nitrobenzene workers, and had estimated the daily absorption of nitrobenzene to be 80 mg or less, considered 35 mg per day to be the maximum allowable absorption. Other workers were exposed at 1 – 6 ppm with no ill effects reported. In a paper by one of these researchers (Piotrowski[44]), the daily dose corresponding to the American TLV (1 ppm averaged over an eight hour shift) is estimated as approximately 35 mg (about one third of this being *via* skin absorption).

ACGIH[41] note the recommendation by one observer for a TLV of 5 ppm (at that time the same as for aniline) rather than of 1 ppm, on the basis of the fact that nitrobenzene's 'urinary BTLV', derived from levels of metabolites associated with the occurrence of cyanosis, was higher than that of aniline. However, ACGIH point out that "nevertheless, both the cyanogenic and anaemiagenic potentials of nitrobenzene were listed as considerably greater than those for aniline, in fact the overall potential for producing the blood effects was second only to that of dinitrobenzene, among thirteen substances evaluated". The ACGIH Committee recommended that their TLV-TWA of 1 ppm be retained but the STEL [of 2 ppm[45]] be deleted pending additional toxicological data and industrial hygiene experience becoming available. Notice of the intended change in the STEL was published in 1984.

Reproductive and Carcinogenic Effects in Man

Beauchamp *et al.*[33] stated (1982) that there was inadequate information on the carcinogenic or teratogenic potential of nitrobenzene in man (and animals) and that there was no knowledge of the potential for nitrobenzene adversely to affect fertility and/or reproduction in humans (or animals).

ECDIN[19] recommends caution with nitrobenzene with regard to pregnant women as methaemoglobinaemia may induce abortion.

Von Oettingen[40] notes a (1933) report which includes menstrual disturbances among the effects of prolonged exposure to sub-acute toxic concentrations of nitrobenzene.

Ref. 9 (published 1984) notes that the EPA had decided not to proceed with testing nitrobenzene (under the US Toxic Substances Control Act) for certain health and environmental effects, and that the EPA has determined that, as the release of nitrobenzene to air and water are limited, additional testing of nitrobenzene for environmental effects is not justified. With regard to human health effects, industry and Federal testing agencies are producing data that EPA believes will be sufficient to determine or predict the potential for nitrobenzene to produce oncogenic, mutagenic, teratogenic or reproductive effects.

18.6 Animal Sub-acute and Chronic Toxicity Including Reproductive Effects[6,33,40,43,46]

General

Ref. 6, referring to the sub-chronic and chronic effects of nitrobenzene on animals notes a 1927 study demonstrating *in vivo* production of methaemoglobin in dogs, cats and rats, but not in guinea pigs or rabbits [but *v.i.*].

Refs. 33 and 6 note a later (1958) chronic toxicity study (ref. 46) on rabbits that received a subcutaneous dose of 0.7 ml/kg/day of nitrobenzene for three months (the maximum acute non-lethal dose), and which were further studied up to the 24th week. Of the eight rabbits in the study, three died at week 4, two at week 6, one at week 11, one at week 14 and one was sacrificed at week 24. A decrease in erythrocyte number and haemoglobin content were observed early in the exposure. These values then increased during the three months but did not return to normal levels. Urinary excretion of detoxification products was variable in the early stages of the exposure but subsequently all the detoxification reactions, *viz.* reduction, hydroxylation and assimilation, were depressed. Bone marrow showed increased megakaryocytes, and spleens showed marked congestion of sinusoids, increased megakaryocytes, hypertrophy and hyperplasia of reticuloendothelial cells, increased yellow-brown pigmentation within macrophages, hypertrophy and hyaline degeneration of vascular walls. Degenerative changes in the adrenal cortex were also reported. The five animals dying within six weeks exhibited fascicular degeneration and disorganization of cellular cords. Such changes were less pronounced in those animals which survived longer. Slight pathological changes also occurred in other organs, *e.g.*, the liver [of generally doubtful significance in the absence of adequate control data].

In another study in which rabbits received 0.1 ml/kg intravenously every other day for five days, haemorrhage, congestion, oedema, emphysema, atelectasis, alveolar wall hypertrophy and adenomatous hyperplasia of the bronchial epithelium were shown as well as hepatotoxic effects.[33]

In a (1972) study in which nitrobenzene was administered to guinea pigs at 200 mg/kg subcutaneously every other day for six months hyperplasia of the adrenal cortex was reported[33]

Reproductive Effects in Animals

Beauchamp *et al.*[33] found that no studies designed to assess the effects of nitrobenzene on reproduction had been reported. Testicular effects on the rat have been noted in Section 18.4 in the study by Bond *et al.*,[43] and the one-time use of nitrobenzene as a human abortifacient has been noted in Section 18.3 in the review by Von Oettingen.[40]. Beauchamp *et al.*[33] note two limited reports on the adverse effects of nitrobenzene on pregnant albino rats but note also that due to the number of animals being small and the exposures not being controlled, interpretation of the findings was made impossible.

Carcinogenicity to Animals

Beauchamp *et al.*[33] found that no studies of sufficient duration had been reported which assess nitrobenzene's potential for carcinogenicity.

18.7 Mutagenicity[33]

Beauchamp *et al.*[33] observe that nitrobenzene together with eight putative metabolites have been examined for genotoxicity in a variety of test systems and they present summaries from about three dozen papers. They note that the only effect with a clear genetic basis is where exposure of *Drosophila* to nitrobenzene vapour produced an increase in sex-linked recessive lethal mutations. Beauchamp *et al.* note though that the report on this study contained limited experimental details and results and that evaluation of the work was difficult. They also state that there are defects in some studies wherein nitrobenzene was not shown to be mutagenic in the Ames *Salmonella*/microsome assay. Various metabolites of nitrobenzene, such as phenylhydroxylamine, *p*-nitrosophenol, *p*-nitrophenol, *p*-aminophenol, acetanilide and aniline have been shown to produce genotoxic effects, including both gene and chromosomal mutations. Negative results with these compounds also have been reported, but these reviewers note that independent evaluation of results is not possible in all cases. The need for more data to assess the genotoxicity of nitrobenzene is noted. Also noted is that although certain metabolites of nitrobenzene appear to be more genotoxic than nitrobenzene itself, the relative importance of oxidative and reductive metabolism in generating genotoxic products is not clear from the available data.[33]

18.8 Summary

[Nitrobenzene is readily absorbed by inhalation, percutaneously or by ingestion, and is toxic by all three exposure routes. The main primary acute toxic effect is methaemoglobinaemia, with its normal secondary symptoms, together with central and peripheral nervous system effects, while prolonged over-exposure may also cause liver damage and anaemia. Infants are particularly sensitive to the compound. The toxic effects may be potentiated by ethanol ingestion. The liquid is mildly irritating to eyes and skin, and skin sensitization may occur. Adverse reproductive effects only appear likely following exposure to amounts toxic to either sex adult, on the basis of limited data. No evidence of, or adequate tests for, carcinogenicity have been found, and available evidence on mutagenicity is inconclusive. The main metabolite following low exposures is *p*-nitrophenol, but *p*-aminophenol conjugates and other metabolites may appear following toxic exposures, with only slow excretion. In severe poisoning, methaemoglobinaemia may be treated with methylene blue. Exposure to 40 ppm nitrobenzene in air causes severe poisoning, while headache and dizziness with mild haematological changes have occurred with chronic exposure to 6 ppm. Daily absorption by all routes (including skin) should not exceed 35 mg/man.]

19. MEDICAL / HEALTH SURVEILLANCE

A decision on the need for, and content of, medical surveillance should be based on an assessment of the possibility and extent of exposure in the work operation. In addition, medical examination may be directed to identifying any pre-existing or newly arising condition in the individual workers which might either be aggravated by subsequent exposure, or might confuse any subsequent medical assessment in the event of excessive exposure or an illness not related to exposure. A particular aspect to be considered is the identification of sensitive subjects not adequately protected by the control limit in operation. A professional medical judgement is required on continuance of employment in the specified process. The following information is relevant in the case of this solvent.

Pre-employment Medical Examination

A clinical history should be taken and a general physical examination carried out. The urine should be examined for glucose and urea, and the blood for anaemia. Workers with anaemia should be investigated. As they tolerate methaemoglobin poorly, workers with ischaemic heart disease should be excluded from work with nitrobenzene.

Periodic Medical Examination

Clinical examination to exclude cyanosis may be conducted at appropriate intervals. Quarterly examinations of blood for methaemoglobin levels and annually for haemoglobin level are recommended.

[It should be noted that regulatory requirements exist in some states for pre-employment and periodic medical examinations of workers employed in manufacture of nitrobenzenes.]

20. OCCUPATIONAL EXPOSURE LIMITS

[The Committee felt that a time-weighted average limit of 1 ppm was appropriate for nitrobenzene. The Committee could find no evidence to set a STEL and hence recommends a provisional STEL of three times the TWA, *i.e.*, a STEL of 3 ppm.

This substance may produce serious toxic effects by absorption through the skin (see Introduction).]

REFERENCES

1. Occupational Safety & Health Administration. Material safety data sheets from the Occupational Health Services database, OSHA, Washington D.C., 1986.
2. Sax, N.I., Ed. 'Hazardous Chemicals Information Annual', No.1. Van Nostrand Reinhold Information Services, New York, 1986.
3. Sax, N.I. 'Dangerous Properties of Industrial Materials'. 6th ed. Van Nostrand Reinhold, 1984.
4. Council of Europe. 'Dangerous Chemical Substances and Proposals Concerning their Labelling'. ('The Yellow Book'.) 4th ed. Maisonneuvre, 1978.
5. Graedel, T.E. *et al.* 'Atmospheric Chemical Compounds'. Academic Press, 1986.
6. Safe Drinking Water Committee. 'Drinking Water and Health', Vol. 4. Board on Toxicology & Environmental Health Hazards, Assembly of Life Sciences, National Research Council. National Academy Press, Washington D.C., 1982.
7. Fielding, M. *et al.* 'Organic Micropollutants in Drinking Water'. Technical Report TR 159. WRC Environmental Protection, Water Research Centre, UK, 1981.
8. Gomolka, E. and B. Gomolka. *Acta Hydrochimica et Hydrobiologica*, **7,** No.6(1979):605. 'Ability of activated sludge to degrade nitrobenzene in municipal wastewater'.
9. Hazards Review 6 Nos.3/4(1984):8. 'Dichloromethane and nitrobenzene; decision to withdraw proposed rules'.
10. Standen, A., Exec. Ed. 'Kirk-Othmer Encyclopedia of Chemical Technology'. 2nd ed. John Wiley & Sons, 1967.
11. Grayson, M., Exec. Ed. 'Kirk-Othmer Encyclopedia of Chemical Technology'. 3rd ed. John Wiley & Sons, 1979.
12. Grasselli, J.G. and W.M. Ritchey, Eds. 'Atlas of Spectral Data and Physical Constants for Organic Compounds'. 2nd ed. CRC Press, 1975.
13. National Institute for Occupational Safety & Health. 'NIOSH Manual of Analytical Methods'. 3rd ed. DHHS (NIOSH) Publication No.84-100. US Department of Health & Human Services, 1984.
14. Kebbekus, B. and J.W. Bozzelli. *Journal of Environmental Science & Health*, **A17,** No.5(1982):713.
15. Dangwal, S.K. *American Industrial Hygiene Association Journal*, **42,** No.7(1981):557.
16. Mierzwinski, A. and Z. Witkiewicz. *Chemia Analityczna (Warsaw)*, **30,** No.3(1985):429.
17. Sanchez-Pedreno, J.A.O. *et al. Analytica Chimica Acta*, 182(1986):285.
18. Lowenheim, F.A. and M.K. Moran. 'Faith, Keyes and Clark's Industrial Chemicals'. 4th ed. John Wiley & Sons, 1975.
19. Commission of the European Communities. 'Environmental Chemicals Data and Information Network (ECDIN)' Databank. Commission of the European Communities Joint Research Centre, ECDIN Group, I-21020 Ispra (Varese), Italy, 1986.
20. Hawley, G.G., Ed. 'The Condensed Chemical Dictionary'. 10th ed. Van Nostrand Reinhold, 1981.
21. Reynolds, J.E.F. and A.B. Prasad, Eds. 'Martindale, The Extra Pharmacopoeia'. 28th ed. The Pharmaceutical Press, 1982.
22. *Safety Practitioner*, **3,** No.9(1985):14. 'Hazard Data Bank, Sheet No. 69: Nitrobenzene'.
23. Keith, L.H. and D.B. Walters, Eds. 'Compendium of safety data sheets for research and industrial chemicals'. VCH Publishers, 1985.
24. National Fire Protection Association. 'Fire Protection Guide on Hazardous Materials'. 9th ed. NFPA, Massachusetts, USA, 1986.
25. Bretherick, L. 'Handbook of Reactive Chemical Hazards'. 3rd ed. Butterworths, 1985.
26. National Institute for Occupational Safety & Health/Occupational Safety & Health Administration. 'NIOSH/OSHA Occupational health guideline for nitrobenzene'. DHHS (NIOSH) Publication. US Department of Health & Human Services/US Department of Labor, 1978.

27. Weast, R.C., Ed.-in-Chief. 'CRC Handbook of Chemistry and Physics' ('The Rubber Handbook'). 67th ed. CRC Press, 1986-1987.
28. Horsley, L.H. 'Azeotropic Data III'. Advances in Chemistry Series No.116, American Chemical Society, 1973.
29. Windholz, M. *et al.*, Eds. 'The Merck Index'. 10th ed. Merck & Co., 1983.
30. Dreisbach, R.R. 'Physical Properties of Chemical Compounds'. Advances in Chemistry Series No.15, American Chemical Society, 1955.
31. Ruth, J.H. *American Industrial Hygiene Association Journal*, 47(1986):A-142. 'Odor thresholds and irritation levels of several chemical substances: a review'.
32. Timmermans, J. 'Physico-Chemical Constants of Pure Organic Compounds'. Elsevier, 1950.
33. Beauchamp, R.O., jr. *et al. CRC Critical Reviews in Toxicology*, **11**, No.1(1982):33. 'A critical review of the literature on nitrobenzene toxicity'.
34. Doull, J. *et al.*, Eds. 'Casarett and Doull's Toxicology: The Basic Science of Poisons'. 2nd ed. Macmillan, 1980.
35. Clayton, G.D. and F.E. Clayton, Eds. 'Patty's Industrial Hygiene and Toxicology'. 3rd ed. rev. Wiley Interscience, 1982.
36. Browning, E. 'Toxicity and Metabolism of Industrial Solvents'. Elsevier, 1965.
37. Hamilton, A. and H.L. Hardy. 'Industrial Toxicology'. 3rd ed. Publishing Sciences Group, Massachusetts, USA, 1974.
38. Levin, S.J. *Journal of the American Medical Association*, **89**, No.26(1927):2178. 'Shoe-dye poisoning – relation to methemoglobin formation'.
39. Zeitoun, M.M. *The Journal of Tropical Pediatrics*, **5**, No.3(1959):73. 'Nitrobenzene poisoning in infants due to inunction with false bitter almond oil'.
40. Von Oettingen, W.F. Public Health Bulletin No.271, US Public Health Service, Washington D.C., 1941. 'The aromatic amino and nitro compounds, their toxicity and potential dangers: a review of the literature'.
41. American Conference of Governmental Industrial Hygienists. 'Documentation of the threshold limit values for chemical substances in the workroom environment'. ACGIH, 1986.
42. Smith, R.P. *et al. Biochemical Pharmacology*, 16(1967):317. 'Chemically induced methemoglobinemias in the mouse'.
43. Bond, J.A. *et al. Fundamental & Applied Toxicology*, **1**, No.5(1981):389. 'Induction of hepatic and testicular lesions in Fischer-344 rats by single oral doses of nitrobenzene'.
44. Piotrowski, J.K. Biological Monitoring & Surveillance of Workers Exposed to Chemicals (Proceedings of the International Course, Espoo, Finland, 1980),(1984):165. (Eds. A. Aitio *et al.*). 'Phenol, aniline and nitrobenzene'.
45. American Conference of Governmental Industrial Hygienists. 'Documentation of the threshold limit values for chemical substances in the workroom environment'. 4th ed. ACGIH, 1982.
46. Yamada, Y. *Kobe Journal of Medical Sciences*, 4(1958):227. 'Studies on the experimental chronic poisoning of nitrobenzene'.

CONTENTS

1. CHEMICAL ABSTRACTS NAME

pyridine
(10th Collective Index)

2. SYNONYMS AND TRADE NAMES[1,2]

azabenzene
azine
NCI-C55301
OHS19990
pyridin (German)
piridina (Italian)
pirydyna (Polish)
U196
UN 1282

3. CHEMICAL ABSTRACTS SERVICES REGISTRY NUMBER

110-86-1

4. NIOSH NUMBER

UR8400000

5. CHEMICAL FORMULA

$N(CH)_5$
(Molecular weight 79.10)

6. STRUCTURAL FORMULA

7. OCCURRENCE[3-6]

Pyridine is one of nearly 150 heterocyclic nitrogen compounds known to occur in the atmosphere where it is found at picogram/m³ levels.[3,4] Sources of atmospheric pyridine given by Graedel *et al.*[4] are; acrylonitrile and other chemical manufacture, coffee processing, sewage treatment, polymer combustion, coke ovens and tobacco smoke. It is reported to occur naturally in wood oil, the leaves and roots of *Atropa belladonna* and in other plants, *e.g.*, coffee and tobacco.[5]

Graedel[3] notes that the occurrence of the nitrogen heterocycles in the atmosphere is almost entirely in the aerosol phase, suggesting that gas phase chemistry can be ignored. Work on the photochemical fate of pyridine in aqueous solutions following UV radiation is noted by Jori *et al.*[6] The photo-decomposition kinetics and yield appeared to be pH sensitive, though in all cases the final products were glutaconic aldehyde and ammonia.

Jori *et al.*[6] also note studies showing that pyridine is degradable by activated sludge and by other methods, but they also state that as regards its degradation directly in soil and water there is only limited information available. They note a 1914 report that numerous micro-organisms in the soil were capable of degrading pyridine that had been used as a soil antiseptic. Also noted is a 1954 study in which the persistence of pyridine in river water samples was determined with 1 ppm additions of pyridine. It was found that, as the microflora of the water became increasingly adapted to the pyridine, the degradation took progressively less time (1 – 3 days).[6]

8. COMMERCIAL AND INDUSTRIAL PYRIDINE[7,8]

Ref. 7 states that pyridine is available in various technical grades distinguished as 20 °C, 2 °C, *etc.* according to distillation range. Ref. 8 states that 1 °C pyridine is at least 99.5% pure with a maximum water content of 0.1%, with most 2 °C pyridine actually having a boiling range of less than 1 °C and a purity of more than 99.8%.

Pyridine is also available in other high purity grades, *e.g.*, spectrophotometric, medicinal.

9. SPECTROSCOPIC DATA[9]

Infrared, Raman, ultraviolet, both ^1H and ^{13}C NMR, and mass spectral data have been tabulated for pyridine.[9]

10. MEASUREMENT TECHNIQUES[10-12]

The method recommended for the determination of pyridine in workshop atmospheres is the NIOSH[10] activated charcoal trapping method for collection and concentration, followed by solvent extraction of the charcoal and a gas chromatographic (gc) analysis of the extract.

The collection tube is 7 cm long with a 4 mm internal diameter. It contains a total of 150 mg of activated charcoal, (20 – 40 mesh) divided into a front section of 100 mg and a rear section of 50 mg separated by a small plug of urethane foam. Air is sampled at a flow-rate of 1 l per minute or less for 100 minutes or less by means of a small pump. The entire apparatus is portable so that it may be carried in a pocket with the sampling tube in the breathing zone (normally a coat lapel). The apparatus can also be static.

Dichloromethane is used to extract the pyridine, which is separated on a 10 ft × 0.125 in stainless steel column packed with 5% Carbowax 20M on Chromosorb W-AW DMCS using flame ionization detection. For full details of the method see ref. 10.

The sampling method uses a small portable apparatus with no liquids and is therefore easy to handle and maintain. The analytical method is relatively specific with the additional advantage of flexibility of operating conditions.

Air samples are normally taken by means of suction using an electric pump capable of operating at low flow-rates (*i.e.*, 30 – 300 ml per minute). If higher flow-rates are used, breakthrough can occur (*i.e.*, the pyridine is not completely adsorbed and some of it passes through the adsorbent). It is also convenient either to establish the trapping device in a static position, or to have it attached to the user in such a way that the breathing zone is sampled (this usually means attachment to the worker's lapel).

It should be noted that suitable personal samplers may be commercially available.

There has been little interest in the determination of pyridine in air. In an examination of the air from a fume cupboard outlet of a laboratory using pyridine and other organic compounds, the sample was led through a tube divided into three sections containing 100, 800 and 800 mg activated charcoal.[11] Pyridine was extracted with carbon disulphide and the extract was analysed by gc using flame ionization detection.

Pyridine can be associated with malodour problems, *e.g.*, in animal fat rendering plants. Air samples from such a plant were collected in poly(vinyl fluoride) bags and then adsorbed on Chromosorb 103.[12] Acetone was used to extract the pyridine, with 90 – 95% recovery, and was determined by gc-mass spectrometry. Tenax-GC was also used to adsorb pyridine and was similarly determined after thermal desorption with greater than 96% recovery.

11. CONDITIONS UNDER WHICH PYRIDINE IS PUT ON THE MARKET

Production[13–15]

Pyridine is obtainable from coal tar and by synthesis. The noncondensable gas from the pyrolysis of coal includes ammonia and most of the usable pyridine bases formed during the coking. The basic compounds are removable with sulphuric acid and on neutralization of the acid solution the pyridine bases are freed. They can be combined with bases produced by acid-base separation of the appropriate boiling range material from fractionation of coal tar and the whole can be then further refined by fractional distillation, pyridine being one of the resultant products.[13]

Goe in ref. 13 notes that, in the US, the isolation of pyridine bases from coal tar is now practised only to a very small extent given the availability of synthetic material. In this, a mixture of acetaldehyde, formaldehyde and ammonia undergoes vapour-phase catalytic reaction to yield a mixture of pyridine and 3-methylpyridine.

The inclusion of methanol in the feed achieves an increased yield. The reaction is carried out at about 350 – 550 °C. The catalyst is usually alumina or silica–alumina which may bear a di- or trivalent metal in the form of oxide, halide or phosphate.

$$CH_3CHO + CH_2O + NH_3 \rightarrow \underset{\text{pyridine}}{C_5H_5N} + \underset{\text{3-methylpyridine}}{C_6H_7N}$$

Goe also observes that acrolein, a derivative of which may be an intermediate in the above reaction, yields mostly 3-methylpyridine and lesser amounts of pyridine when reacted under similar conditions.[13]

Pyridine may be separated from 3-methylpyridine (boiling point 143.9 °C) by fractionation to yield a high purity product.

Pyridine is marketed in bottles, drums, barrels and in tankers.[14] EC production of pyridine in 1982 was 9 kilotonnes whereas its consumption was 22 kilotonnes.[15]

Uses[5,16-19]

ILO[16] notes that the use of pyridine for the preparation of biologically active derivatives is facilitated by the availability of high purity, synthesis material. Other uses of pyridine are; the manufacture of explosives, agrochemicals, pharmaceuticals, paints, rubbers, dyestuffs, polycarbonate resins and water-repellent finishes for textiles. It has for long been used as a denaturant for ethanol and finds similar use in anti-freeze mixtures. It is a useful solvent in the chemical industry especially in conditions where its basic properties may facilitate reaction (*e.g.*, in acylations).[16-18] Furia and Bellanca[5] note pyridine's use as a flavour in non-alcoholic beverages, confectionery, baked goods and ice cream, ices and the like. Furia[19] notes its use in seafood flavours, smoke flavours and chocolate.

12. STORAGE, HANDLING AND USE PRECAUTIONS[18,20-24]

Pyridine is a flammable, water-miscible liquid of boiling point 115.5 °C and a vapour density about 2.7 times that of air.

Care must be taken with storage, process and drying plant to ensure that hazard-free conditions exist regarding fire and explosion risks and exposure of personnel to pyridine. Marsden and Mann[20] state that iron, mild steel and aluminium are suitable for plant and containers. Discolouration of the solvent may result from its contact with copper. Small quantities should be stored and carried in safety cans. Glass containers should be used only for small amounts for laboratory use.[18]

[As with other flammable solvents the organization of storage facilities has to take into account two separate requirements. These are that the stored material be protected from a fire elsewhere on the premises and that the premises be protected from a fire involving the stored materials. On an industrial scale, sufficient physical separation and water spray curtains are often used effectively. On a laboratory scale, flammable solvents need to be stored in fire resistant, thermally insulated cupboards.

It is important to appreciate that a non-combustible barrier is not synonymous with a fire barrier. A fire barrier has to protect against both conducted and radiant heat. For example, a metal cupboard is useless on both these counts.]

Storage tanks in the open should be bunded and there should be a ramped sill at the doorways of storerooms in order to prevent egress of spilled material. Such storerooms should be well ventilated, flammable liquid stores, free from source of ignition. Pyridine should be stored away from chemicals with which it is incompatible (see Section 14). The construction and siting of process buildings and plant should be such as to prevent the spread of escaping liquid throughout the building or site. Large items of equipment should be earthed and tools or equipment used should be of the non-sparking/flameproof type, notwithstanding the hygroscopic nature of the solvent.

OSHA[21] recommends that engineering controls such as process enclosure, general dilution ventilation, local exhaust ventilation and the use of personal protective equipment should be used, as appropriate, to keep personnel exposures at or below permissible limits. Care should be taken to prevent recirculation of pyridine exhaust due to the proximity of air conditioning intakes or windows. Skin contact with pyridine should be prevented by personnel being required to use, as appropriate, impervious clothing, gloves, splash proof safety goggles, face shields (eight inch minimum) and such other protective clothing as may be appropriate.

Note that pyridine attacks some forms of plastics, rubber and coatings. Ref. 22 notes that ordinary rubber gloves may be penetrated by the solvent and recommends the use of heavy rubber. Polyethylene gloves may be suitable.[23]

Adequate emergency eye-wash facilities should be available as should facilities for washing pyridine from the skin (including quick-drenching facilities). Contaminated clothing and footwear should not be re-worn unless and until it can be completely freed from the solvent.

[Personnel should not be allowed to enter areas containing potentially hazardous concentrations of pyridine unless all relevant legislation and official guidelines have been strictly adhered to. Suitable respiratory protective equipment should be used.] Such exposed personnel should be able to communicate with each other and with those (*i.e.*, more than one person) monitoring the work from outside the hazardous area. Monitors should be suitably equipped to effect rescue (possibly with lifelines to exposed personnel) and should themselves ensure, so far as is possible, that one of their number remains outside the hazard area in the event of an emergency.

Pyridine vapour in high concentration may cause narcosis with dizziness, drowsiness, headache, and possible unconsciousness. A concentration of 3,600 ppm is immediately dangerous to life or health.[24]

N.B. the caveat in Section 13 (final paragraph).

13. FIRE HAZARDS[21,22,24,25]

Pyridine is a highly flammable, water-miscible liquid of boiling point 115.5 °C, flash point 20 °C (closed cup) and lower and upper explosive limits in air of 1.8% and 12.4% v/v. The auto-ignition temperature is about 480 °C (see Section 17). The vapour has a density about 2.7 times that of air and, if in sufficient quantity, may therefore flow to an ignition source and then 'flash back'. Pyridine burns with a smoky flame. When heated to decomposition, highly toxic cyanide fumes are produced while combustion products may include oxides of carbon and of nitrogen.

Containers of pyridine can explode when heated and, if feasible, should be cooled with flooding quantities of water from as far as possible until well after the fire is extinguished. Immediate withdrawal from the area is imperative in the case of rising sound from a safety venting device or in the case of discolouration of a storage tank due to the fire. As combustion of pyridine may produce toxic vapours fire-fighting personnel should keep upwind of the hazard.

Small fires may be fought with dry chemical, carbon dioxide, vaporizing liquid, water spray, alcohol-resistant foam or by dilution with plenty of water. Large fires can be attacked with water spray, fog or alcohol-resistant foam. Jets of water may spread burning liquid notwithstanding its water miscibility.

Ref. 25 notes that where a fire is fed by an uncontrolled flow of combustible liquid the decision on how or if to fight it will depend on the size and type of fire anticipated and must be carefully considered. This may call for special engineering judgement, particularly in large-scale applications.

[Notwithstanding the foregoing review of various authorities noted in Sections 12 and 13, it is essential that managers and others responsible for the planning, implementation and overseeing of personnel and plant safety should be familiar with the legal constraints and official guidelines applicable to them and that they liaise with their local emergency services in the planning of plant and storage facilities and in the preparation of contingency plans for dealing with fires and other emergencies. Managers should regularly monitor staff knowledge of, and ability to implement, emergency procedures and should ensure that equipment provided for use in emergencies is regularly inspected and maintained.]

14. HAZARDOUS REACTIONS[22,24,26]

Bretherick[26] notes the following reactions.

Maleic anhydride decomposes exothermally, evolving carbon dioxide in the presence of (*inter alia*) pyridine at temperatures above 150 °C. With pyridine, a concentration even below 0.1% is effective and decomposition is rapid.

The solid produced by the action of bromine trifluoride on pyridine in carbon tetrachloride ignites when dry.

A highly explosive by-product is formed where pyridine is used as an acid-acceptor in reactions involving trifluoromethyl hypofluorite.

The chromium trioxide–pyridine complex is a powerful oxidant and its preparation is hazardous unless due precautions are taken. Lack of really efficient stirring led to flash fires as the oxide was added to the pyridine at -15 to -18 °C. Reversed addition of pyridine to the oxide is extremely dangerous, ignition usually occurring. Bretherick notes that a later preparation specifies temperature limits of $10-20$ °C to avoid an excess of unreacted trioxide; that a safe method of preparing the complex in solution has been described and that the preparation and use of solutions of the isolated complex in dichloromethane or acetic acid have been detailed. Solution of the oxide is not smooth. It first swells then suddenly dissolves exothermally in the pyridine. This hazard may be eliminated without loss of yield by dissolving the trioxide in an equal volume of water before adding it to ten volumes of pyridine. Pulverising chromium trioxide before use is not recommended as its rate of reaction with organic compounds is thereby increased to a hazardous level.

Pyridine incandesces on contact with fluorine.

Pyridine is attacked violently by liquid nitrogen tetraoxide.

OSHA[24] notes the following additional incompatibilities.

Explosive reactions occur with perchloric acid and perchromates.

Silver perchlorate forms a shock-sensitive solvated salt with pyridine.

There is vigorous or violent reaction with sulphuric acid, oleum, chlorosulphonic acid, nitric acid, other strong acids and strong oxidizers and with β-propiolactone.

Ref. 22 notes that pyridine reacts violently with nitromethane after several hours delay.

15. EMERGENCY MEASURES IN THE CASE OF ACCIDENTAL SPILLAGE[21,22,24]

All possible sources of ignition should be eliminated and, if possible, any leak should be stopped if this can be done without risk. The aim should be to isolate the material, *e.g.*, by closing doors to the spillage area, and by ventilation *e.g.*, by opening windows. Non-essential personnel should leave the area and those engaged in dealing with the hazard should wear appropriate apparel and breathing equipment (see Section 12).

Very small spills may be mopped up with plenty of water and run to waste (a sewer) diluting greatly [at least one hundred-fold] with running water,[22] or else such spillage may be absorbed on paper, sand or vermiculite and either allowed to evaporate in a fume cupboard with a flameproof extractor until the whole of the ducting is clear of vapour or, if feasible, burnt in a suitable location away from combustible materials.[21] Larger spills can be collected and atomized in a suitable combustion chamber equipped with an appropriate effluent gas cleaning device.[21] Conveyance of collected material should be in suitable containers such as covered buckets. Water-spray may be used to reduce a hazardous concentration of vapour.[24] The disposal of contaminated water though should not engender a hazard elsewhere.

Large spills may need to be bunded far ahead of the spill for later disposal.[24] Disposal of whatever quantity must comply with local legislation and should not put others at risk. In the event of a likely need for the attendance of the fire services or of there being an environmental hazard, the appropriate authorities, including the water authority, must be informed.

16. FIRST AID

General

The casualty should be removed from danger of further exposure into the fresh air. Rescuers should ensure their own safety from inhalation and skin contamination where necessary. Conscious casualties should be asked for information about what has happened. Bear in mind that the casualty may lose consciousness at any time. Where necessary continued observation and care should be ensured. First aiders should take care not to become contaminated.

Inhalation

The conscious casualty should be kept at rest. The unconscious casualty should be placed in the recovery position and an open airway maintained. If breathing or heart beat stops, resuscitation should be commenced immediately. Medical aid should be obtained or the casualty removed to hospital immediately.

Skin contamination

Remove contaminated clothing. Wash with copious amounts of water, or soap and water, for at least 15 minutes. If necessary, seek medical aid. In the case of persistent skin irritation refer for medical advice.

Eye contamination

Irrigate with copious amounts of water for at least ten minutes but avoid further damage to the eye from the use of excessive force. Seek medical aid. Remember that contact lenses may be worn and that these may be affected by some solvents and may impede decontamination of the eye.

Ingestion

If the lips or mouth are contaminated rinse thoroughly with water. Do not induce vomiting. Give further supportive treatment as for inhalation. In the case of ingestion of significant amounts wash out the mouth with water. Provided the patient is conscious, give 200 – 300 ml of warm water to drink. Obtain immediate medical attention. Contact a hospital or poisons centre at once for advice. Remember that symptoms may develop many hours after exposure so continued care and observation may be necessary.

   In all cases note information on the nature of exposure and give this to ambulance or medical personnel. Information for doctors is provided in some technical literature issued by manufacturers. Antidotes may be available for retention at the place of work for use by persons competent to treat casualties.

17. PHYSICO-CHEMICAL PROPERTIES

17.1 General[25,27,28]

Pyridine is a flammable, colourless liquid with a nauseous and penetrating odour.[25,27] Merck[27] describes it as having a sharp taste.

Pyridine is miscible with water[27] and a wide range of organic solvents, *e.g.*, acetone, benzene, carbon tetrachloride, diethyl ether and ethanol.[28] (Miscibility with the organic solvents was determined by shaking together equal volumes of the two solvents and noting the absence of an interfacial meniscus after the mixture had settled.[28]) Pyridine is a good solvent for both organic and inorganic compounds.[27] It is a weak base, and will form salts with strong acids[27]

17.2 Melting Point[13]

The freezing point of pyridine is quoted as

$-41.6\,°C$[13].

17.3 Boiling Point[28,29]

The boiling point of pyridine is given by Weast[28] as

115.5 °C.

Details are listed below of some binary azeotropes formed by pyridine with some common organic solvents and with water[29]:

Wt% pyridine	Second component	Wt% second component	Boiling pt azeotrope (°C)
47 – 49	Acetic acid	51 – 53	138 – 141
56.1	n-Octane	43.9	109.5
48.5	Tetrachloroethylene	51.5	112.85
57 – 59	Water	41 – 43	93.6 – 94

17.4 Density/Specific Gravity[20,27]

The specific gravity of pyridine is 0.987 at 20 °C,[20] and 0.9780 at 25 °C[27] (both referred to water at 4 °C).

17.5 Vapour Pressure[13,28]

The vapour pressure of pyridine below atmospheric pressure is tabulated by Weast[28] as follows:

Temperature in °C	Vapour pressure in mmHg
−18.9	1
13.2	10
38.0	40
57.8	100
95.6	400

The critical temperature and critical pressure of pyridine are recorded as 346.8 °C,[28] and 55.58 atm [*ca.* 42200 mmHg][13] respectively.

17.6 Vapour Density[25]

2.7 (relative to air = 1).

17.7 Flash Point[25]

20 °C, closed cup test.

17.8 Explosive Limits[25]

The limits of flammability of pyridine in air at normal atmospheric temperature and pressure are given by the NFPA[25] as:

Lower limit: 1.8% v/v
Upper limit: 12.4% v/v.

The NFPA[25] give the ignition temperature of pyridine as 482 °C. However they point out that different test conditions (and also different definitions of "ignition temperature") can result in widely varying values being quoted. They therefore recommend that any value should be treated as an approximation.

17.9 Viscosity[13,28]

The viscosity of liquid pyridine is 0.974 cp at 20 °C,[28] and 0.878 cp at 25 °C.[13]

17.10 Concentration Conversion Factors

At 25 °C and 760 Torr (1 atm),

1 ppm = 3.24 mg/m^3, and
1 mg/m^3 = 0.31 ppm.

Pyridine

18. TOXICITY

18.1 General[13,20,21,30-33]

Reinhardt and Brittelli in ref. 30 state that pyridine is absorbed by inhalation, ingestion and percutaneously. It is a mild mucous membrane and eye irritant.[21,31] Most of the effects observed in man are transient, involving the CNS and gastrointestinal tract. Despite its relatively large industrial and some medicinal use, reports of injurious effects of pyridine in man have been relatively uncommon. Kidney and liver injury and death have resulted though from ingestion of the substance for up to two months, this being the result of using pyridine as an anti-convulsant in the treatment of epilepsy.[30,31].

Pyridine has a narcotic effect on experimental animals.[30] Browning[31] states that it has proved less toxic in animals than in human beings. Skin irritation may result from prolonged or repeated contact with the vapour or liquid. Corneal necrosis can be caused by the liquid.[21] Pyridine is reported to be a photosensitizer.[30]

Ruth[32] gives the odour threshold range of pyridine as approximately 0.009 mg/m³ [approx. 0.003 ppm] to 15.0 mg/m³ [approx. 4.6 ppm]. It has a vapour pressure of 20 mmHg at 24.8 °C[20] [equivalent to approx. 26,000 ppm in air]. Pyridine has an odour that is objectionable at 10 ppm to unacclimatized individuals. A test on 12 human subjects gave an olfactory threshold of 0.21 mg/m³ [0.063 ppm].[33] Reinhardt and Brittelli in ref. 30 point out though that the odour and mild irritant properties of pyridine are not objectionable enough to prevent toxic exposures to the vapour and that skin exposures may cause irritation or systemic intoxication. Olfactory fatigue can occur and acclimatized individuals may soon be able to tolerate what would otherwise be obnoxious concentrations.[30]

N.B. Pyridine from coal tar contains methyl derivatives,[13] and observations on toxic effects may have resulted from the use of crude, technical material.

18.2 Toxicokinetics[6,8,30,34]

Jori *et al.*[6] note reports that pyridine is absorbed through the lungs, the gastrointestinal tract and the skin, and is eliminated through the lung, skin, faeces and urine as free base and as metabolites. They add that, depending on the animal species, from 50 – 70% of intraperitoneally administered pyridine is recovered in the 24 hour urine collection.

Santodonato *et al.*[8] note that specific data on the distribution and accumulation of pyridine had not been located by them in the available literature. From the acute toxic effects of the solvent they assume that its distribution and possibly accumulation occurs in the CNS, the liver and the kidneys with much being excreted in the urine unchanged. These reviewers note that the extent of faecal excretion has not been determined.

Jori *et al.*[6] and Reinhardt and Brittelli in ref. 30 note that the metabolic fate of pyridine is not completely known or understood. Metabolites derived by hydroxylation, methylation and oxidation have been identified but with no information on their distribution and half-lives. The possibility of 2-pyridone formation has been suggested (as a similar compound is a metabolite of nicotinamide) but its occurrence as a pyridine metabolite has not been reported.[34]

N-methylation, producing the N-methylpyridinium ion, has been observed in animal species to which pyridine has been administered orally or intraperitoneally. Variation between species regarding the amount of pyridine found in the urine has been shown. Of the administered doses, 40% appeared as N-methylpyridinium ion in the cat, 30% in the guinea pig, 26% in the hamster and the gerbil, 21% in the rabbit, 12% in the mouse, 5% in the rat and 9% in man (this value referring only to two subjects). The route of administration appeared to make no difference. Methylation of pyridine in the rat became even lower with increasing dose indicating that this metabolic pathway is easily saturated in this species.[6] Williams[34] notes a study in which traces of 3-hydroxypyridine were excreted by rabbits given 0.25 g pyridine/kg orally.

Jori *et al.*[6] note that pyridine can be oxidized to nitrogen by pyridino-N-oxidase, giving rise to a pharmacologically active metabolite of its own. Pyridino-N-oxidase is NADPH dependent and is located in the liver and lung microsomal fraction. A study showed its activity to be species dependent and to decrease through the series of rabbit, hamster, rat, guinea pig and mouse. Under the conditions of the study only 2% of pyridine's metabolism was accountable for in this way. The activity could be enhanced by pretreatment of animals with inducing agents such as phenobarbital but not with those of the 3-methylcholanthrene type. These data indicate that N-oxidation of pyridine is mediated by cytochrome P450 and not P448 which is responsible for N-oxidation of amides and aromatic tertiary amines.[6]

Jori *et al.*[6] conclude that knowledge of the kinetics and metabolism of pyridine is totally inadequate and that information on metabolic fate is based either on very old publications or on recent studies which were not specifically designed to study pyridine metabolism. Most data were obtained *in vitro*, are incomplete, and lack *in vivo* confirmation.

18.3 Human Acute Toxicity[24,30,31,33,35]

Browning[31] notes a fatal case of poisoning reported in 1893. This followed the accidental ingestion of half a cupful of commercial pyridine. Initially there was vomiting. Later there was slight cyanosis, a choking sensation, precordial and abdominal pain and a raised temperature, pulse and respiration. There was no abnormality of the urine. Delirium and acute congestion of the lungs preceded death which occurred 43 hours after the ingestion. Autopsy revealed that the chief lesions were inflammatory and were situated in the respiratory tract, the oesophagus and the stomach. There were a few fatty patches in the liver.

Browning[31] also notes a case of acute narcosis in a man who had been employed in cleaning a tank-waggon that had contained pyridine (no further details were given).

Another case of acute poisoning is noted by Reinhardt and Brittelli in ref. 30. In this case a woman had been decontaminating a pyridine spillage for 15 to 20 minutes. Symptoms did not occur for ten hours and then intensified until the third day. They included speech disorders and what was reported as 'rather diffused cortical affliction' which receded after thiamine therapy. There were no upper respiratory symptoms.

Grant[35] notes that pyridine is irritating to the skin, and its vapour is irritating to the eyes and lids, as well as to the respiratory tract.

In a human experiment, $0.098 - 0.11$ mg/m^3 [$0.0295 - 0.033$ ppm] temporarily affected sensitivity of the eye to light.[33] [The practical significance of this unconfirmed finding is unclear.]

OSHA[24] states that a vapour concentration in air of 3,600 ppm is immediately dangerous to life and health.

18.4 Animal Acute Toxicity[21,30]

Reinhardt and Brittelli in ref. 30 state that pyridine has a narcotic action when administered to experimental animals. Toxic doses cause weakness, ataxia, unconsciousness and salivation irrespective of the route of administration. Animals that survive the acute episode generally recover.

Exposure to pyridine vapour produces symptoms of moderate mucous membrane irritation.[30] OSHA[21] notes that a 40% solution in the eye of a rabbit caused corneal necrosis. Reinhardt and Brittelli[30] note that one drop of undiluted pyridine in the eye of a guinea pig caused corneal damage.

Reinhardt and Brittelli in ref. 30 also note that pyridine and some of its derivatives have been the subject of range-finding studies. Results for pyridine are given below. For the skin absorption test in guinea pigs, undiluted pyridine was applied to the intact skin under a rubber cuff which was left in place for 24 hours. All animals were observed for two weeks. The vapour concentrations given are calculated values. The ranges quoted represent no deaths at the lower dose but 100% mortality at the higher dose.

Rats – oral administration : $0.8 - 1.6$ g/kg
Rats – i.p. administration : $0.8 - 1.6$ g/kg
Mice – oral administration : $0.8 - 1.6$ g/kg
Mice – i.p. administration : $0.8 - 1.6$ g/kg
Guinea pigs – percutaneous administration : $1.0 - 2.0$ ml/kg

Inhalational toxicity, rats – 4,000 ppm over six hours caused the death of three out of three animals. Also noted by Reinhardt and Brittelli are the deaths of five out of six rats exposed to 4,000 ppm for four hours; of two out of three rats exposed to 3,600 ppm for six hours and of six out of six rats exposed to 23,000 ppm for 1.5 hours. The inhalation of saturated vapour [approx. 26,000 ppm] for 30 minutes is quoted as the maximum for no deaths in the rat.[30]

18.5 Human Sub-acute and Chronic Toxicity Including Reproductive Effects[6,17,21,30,31]

General

Reinhardt and Brittelli in ref. 30 observe that effects centred on the CNS and gastrointestinal tract are those most often observed in man and that these have most often resulted from repeated or intermittent exposures to pyridine vapour. Symptoms include headache, dizziness or giddiness, nervousness, insomnia, mental dullness, nausea and anorexia. These reviewers also note that, in some cases, lower abdominal or back discomfort with urinary frequency has been observed.

Various reviewers, *e.g.*, Reinhardt and Brittelli in ref. 30 and Jori *et al.*[6] note that persons exposed to pyridine vapour at an average concentration of 125 ppm for four hours/day for 1 – 2 weeks exhibited the foregoing transient symptoms, but without any associated evidence of liver or kidney damage.

Serious liver and kidney injury (one case being fatal) though have resulted from pyridine poisoning following a course of administration of the substance for up to two months to control epileptic convulsions.[6,30] Browning[31] in her account of this occurrence states that the short-term oral administration of 0.6 ml of pyridine, three or four times/day was found to cause minor complaints of anorexia, nausea, occasional vomiting and headache, faintness, weakness and mental depression. The blood picture showed a tendency to a fall in leucocytes and an increase in eosinophils. The urine contained a few erythrocytes and, in one case, a few hyaline casts, but on the whole the toxic effects were minimal. Treatment was continued at a higher dosage (1.85 and 2.4 ml) in two cases, with much more unfavourable results. One of these cases was fatal. In this there had been rapid development of anuria and jaundice with evidence of parenchymal destruction of the liver.

The second was a case of convulsive disorder due to an astrocytoma, with some residual signs of brain disease. This case then received 10 minims [0.6 ml] of pyridine three times/day. Ten days later there was evidence of hepatorenal injury with weakness of the limbs, nausea, stupor, confusion and disorientation. There was eventual recovery following a blood transfusion, with a return to the condition that existed before the pyridine medication.

Browning[31] also describes cases of occupational exposure (though without exposure or attributability data). In one case a chemist, who had worked with pyridine for six months, suffered from disturbance of equilibrium, facial paralysis and attacks of loss of consciousness. In another case the symptoms after two years' exposure were paralysis of the ocular muscles with nystagmus, facial paralysis, hemiparesis with anaesthesia to heat and paraesthesia of the left side of the face, right-sided excessive perspiration, cerebellar ataxia, bladder paralysis, difficulty of hearing and neuralgic headache. Similar symptoms were observed elsewhere, with the addition of marked loss of weight, noises in the ears and pain on moving the arms. Browning notes that one observer considered these symptoms similar to Wernicke's pseudo-encephalitis but that another remarked that some of the multiform symptomatology of pyridine poisoning does not conform exclusively to injury of the medulla and that it resembled more closely the manifestations of tabes dorsalis.[31]

Browning notes the report by this last observer of a man who had worked with pyridine for 13 years. He had complained of nervousness and sleeplessness. Over the last two years he had suffered attacks of giddiness, pain in the right calf and chest, weakness of the legs and tremor. There was then a sudden attack of anuria and a catheter specimen provided two litres of reddish urine with a trace of albumin. Operation revealed a gross diverticulum of the bladder which was believed to be secondary to the urine retention.[31]

ACGIH[17] notes that large doses of pyridine act as a heart poison whereas smaller doses stimulate the bone marrow to increase production of blood platelets. ACGIH also quotes the consideration of one observer that the most important effect of pyridine inhalation was chronic poisoning, centering in the liver, kidney and bone marrow.[17]

ACGIH[17] notes a report of chronic poisoning with mild symptoms of CNS injury in an establishment where pyridine vapour concentrations ranged from 6 – 12 ppm.

ACGIH proposed that a TWA of 5 ppm and a STEL of 10 ppm should be low enough to prevent systemic effects from exposure to pyridine provided skin absorption is not permitted. However, they recognize that the proposed level may not eliminate complaints about odour.[17]

OSHA[21] states that skin irritation may result from prolonged or repeated contact with the vapour or liquid.

Carcinogenicity and Reproductive Effects in Man

No reports suggesting adverse reproductive effects or carcinogenicity of pyridine in man have been traced.

18.6 Animal Sub-acute and Chronic Toxicity Including Reproductive Effects[6,8,31,33,36,37]

General

Browning[31] notes that the repeated oral administration of pyridine to rabbits at a dosage of 44 mg/kg for 5 – 7 days had no toxic effect. 165 mg/kg caused the death of one animal out of four on the sixth day. Also noted is an 1887 report in which the administration to dogs of 0.1 g pyridine as a 5% solution daily for 35 days caused only occasional vomiting and diarrhoea.

Baxter and Mason[36] reported that six young male rats were fed on a basal diet containing 0.1% added pyridine. The diet was low in casein and choline but not so low that significant lesions were produced during the experimental period by the diet alone. The rats, which had been gaining weight on a stock diet over the eight days prior to the start of the basal diet and pyridine regimen, then gradually lost weight. Five rats died during the 14th day to 32nd day of the pyridine treatment and the single survivor was sacrificed on the 32nd day. The livers and kidneys revealed acute lesions and some of the livers exhibited well marked cirrhosis.

Another group of rats in this study received 0.164% in the diet of the pyridine metabolite, methylpyridinium ion (as chloride), this dosage being equivalent to 0.1% pyridine. All six animals survived and showed continual growth. Two were sacrificed on the 32nd day and were found to have essentially normal livers and kidneys.

Continuous inhalational exposure of rats for two months at 1 mg/m³ [0.3 ppm] caused minor effects on CNS function and blood protein composition, while 0.1 mg/m³ [0.03 ppm] was without effect.[33]

Carcinogenicity to Animals

Jori *et al.*[6] and Santodonato *et al.*[8] note a study in which rats were subcutaneously injected with pyridine twice weekly for one year. Four dosage regimens were used, 3, 10, 30 and 100 mg/kg. The animals were killed at 18 months. Body weight, mortality, incidence of tumours and tumour location were the same as in controls (including testicular tumours which arose in 100% of controls and in 95% of treated rats).

Baxter[37] reported that hepatic nodules (in addition to cirrhosis and fatty degeneration) occurred in some rats exposed to pyridine in their diet at 0.1% – 0.29% for up to four months, followed by up to two months observation. The lesions were not invasive and no metastases were seen, indicating their benign nature in this relatively short experiment in which the diet was deficient in choline and casein.

Reproductive Effects in Animals

No reports of tests for adverse reproductive effects of pyridine in mammals have been found.

18.7 Mutagenicity[6,8]

Santodonato *et al.*[8] note that pyridine was not mutagenic in an extensive assay using *Salmonella typhimurium* strains TA1535, TA1537, TA1538, TA1536 and TA100, both with and without a metabolic activation system (from seven rat tissues). Testing was at 1, 10 and 100 µg/plate. An NTP study reported that assays in *S. typhimurium* tester strains TA98, TA100, TA1535 and TA1537 were negative following pyridine incubation with and without added metabolic activation, and another study proved negative in tester strain TA1537.

A significant increase in forward mutation frequency was seen in *S. typhimurium* strain TM677 with microsomal activation. The minimum dose with mutagenic effect (calculated from the dose response curve) was 6 mM. For comparison, benzo(*a*)pyrene was mutagenic at 4 µM. However, in another study using this strain, no mutagenesis was reported. No chromosomal aberrations were observed in Chinese hamster cells exposed to pyridine at concentrations of 1, 2, 5 and 50 mM (this last level being the maximum with no lethal effects on these cells). In another Chinese hamster cell test, sister-chromatid exchange was noted. This though was at the limit of significance and was not dose-related.[6,8].

Santodonato *et al.*[8] note that, in a spot test in *Escherichia coli*, no increase in reverse mutants to Trp⁺ were produced following pyridine exposure.

18.8 Summary

[Pyridine is readily absorbed by inhalation, percutaneously or by ingestion. Acute toxic effects are primarily on the central nervous system, including temporary narcosis and headache, together with gastrointestinal disturbance particularly following ingestion, while repeated over-exposure may additionally cause kidney and liver damage. The liquid is strongly irritating to eyes, and may cause skin irritation on prolonged or repeated contact, perhaps with photo-sensitization, while high vapour concentrations may also cause irritation to eyes and perhaps skin. No evidence of adverse reproductive effects or carcinogenicity has been found, and most mutagenicity tests were negative. Absorbed pyridine is rapidly excreted as free base or metabolites. Repeated human inhalational exposure to 125 ppm in air caused marked CNS effects, and mild symptoms are reported following long-term exposure to 6 – 12 ppm.]

19. MEDICAL / HEALTH SURVEILLANCE

A decision on the need for, and content of, medical surveillance should be based on an assessment of the possibility and extent of exposure in the work operation. In addition, medical examination may be directed to identifying any pre-existing or newly arising condition in the individual workers which might either be aggravated by subsequent exposure, or might confuse any subsequent medical assessment in the event of excessive exposure or an illness not related to exposure. A particular aspect to be considered is the identification of sensitive subjects not adequately protected by the control limit in operation. A professional medical judgement is required on continuance of employment in the specified process. The following information is relevant in the case of this solvent.

It is unlikely that workers will be continuously exposed to pyridine as the odour is so repellent, being detectable at about 1 ppm and objectionable at 10 ppm. Although nasal detection of the odour may recede, an unpleasant taste may persist. If exposure at potentially hazardous levels may occur at work, or in the event of accidental excessive exposure, medical considerations are as follows.

Pre-employment Medical Examination

A complete history and physical examination should be carried out. Pre-existing conditions which might be aggravated by exposure to pyridine should be assessed. Particular attention should be paid to the liver, kidneys and central nervous system, and appropriate tests of function should be performed if necessary. Alcoholism should be noted and assessed.

Periodic Medical Examination

The same medical assessment should be made as in pre-employment examinations. Assays of pyridine in urine and exhaled air, and assay of the metabolite N-methyl pyridine in urine, have been proposed but are not in routine use. As skin sensitization and photo-sensitization have been reported, particular attention should be paid in the examination to the skin.

20. OCCUPATIONAL EXPOSURE LIMITS

[The Committee felt that a time-weighted average limit of 5 ppm was appropriate for pyridine. The Committee could find no evidence to set a STEL and hence recommends a provisional STEL of three times the TWA, *i.e.*, a STEL of 15 ppm. This substance may produce serious toxic effects by absorption through the skin (see Introduction).]

REFERENCES

1. Sax, N.I. 'Dangerous Properties of Industrial Materials'. 6th ed. Van Nostrand Reinhold, 1984.
2. National Institute for Occupational Safety & Health. 'Pocket Guide to Chemical Hazards'. 5th printing. DHEW (NIOSH) Publication No.78-210. US Department of Health & Human Services, 1985.
3. Graedel, T.E. 'Chemical Compounds in the Atmosphere'. Academic Press, 1978.
4. Graedel, T.E. *et al.* 'Atmospheric Chemical Compounds'. Academic Press, 1986.
5. Furia, T.E. and N. Bellanca, Eds. 'Fenaroli's Handbook of Flavor Ingredients'. 2nd ed. CRC Press, 1975.
6. Jori, A. *et al. Ecotoxicology & Environmental Safety*, **7**, No.3(1983):251. 'Ecotoxicological profile of pyridine'.
7. 'Hawley's Condensed Chemical Dictionary'. 11th ed. rev. by N.I. Sax and R.J. Lewis. Van Nostrand Reinhold, 1987.
8. Santodonato, J. *et al.* 'Monograph on Human Exposure to Chemicals in the Workplace: Pyridine'. Report No.SRC-TR-84-1119. National Cancer Institute, USA, 1985.
9. Grasselli, J.G. and W.M. Ritchey, Eds. 'Atlas of Spectral Data and Physical Constants for Organic Compounds'. 2nd ed. CRC Press, 1975.
10. National Institute for Occupational Safety & Health. 'NIOSH Manual of Analytical Methods'. 2nd ed. DHEW (NIOSH) Publication No.77-157-A. US Department of Health, Education & Welfare, 1977.
11. Snell, J. *Pollution Engineering*, **14**, No.6(1982):36.
12. Barnes, R.D. *et al. Analyst*, 106(1981):412.
13. Grayson, M., Exec. Ed. 'Kirk-Othmer Encyclopedia of Chemical Technology'. 3rd ed. John Wiley & Sons, 1979.
14. Hawley, G.G., Ed. 'The Condensed Chemical Dictionary'. 10th ed. Van Nostrand Reinhold, 1981.
15. Commission of the European Communities. 'Environmental Chemicals Data and Information Network (ECDIN)' Databank. Commission of the European Communities Joint Research Centre, ECDIN Group, I-21020 Ispra (Varese), Italy, 1986.
16. International Labour Office. 'Encyclopaedia of Occupational Health and Safety'. 3rd rev. ed. ILO, 1983.
17. American Conference of Governmental Industrial Hygienists. 'Documentation of the threshold limit values for chemical substances in the workroom environment'. ACGIH, 1986.
18. *Safety Practitioner*, **4**, No.8(1986):44. 'Hazard Data Bank, Sheet No. 80: Pyridine'.
19. Furia, T.E., Ed. 'Handbook of Food Additives'. 2nd ed. CRC Press, 1980.
20. Marsden, C. and S. Mann, Eds. 'Solvents Guide'. 2nd ed. rev. Cleaver-Hume Press, 1963.
21. National Institute for Occupational Safety & Health/Occupational Safety & Health Administration. 'NIOSH/OSHA Occupational health guideline for pyridine'. DHHS (NIOSH) Publication. US Department of Health & Human Services/US Department of Labor, 1978.
22. The Royal Society of Chemistry. Laboratory Hazards Data Sheet No.8: 'Pyridine', 1983.

23. Keith, L.H. and D.B. Walters, Eds. 'Compendium of safety data sheets for research and industrial chemicals'. VCH Publishers, 1985.

24. Occupational Safety & Health Administration. Material safety data sheets from the Occupational Health Services database, OSHA, Washington D.C., 1985.

25. National Fire Protection Association. 'Fire Protection Guide on Hazardous Materials'. 9th ed. NFPA, Massachusetts, USA, 1986.

26. Bretherick, L. 'Handbook of Reactive Chemical Hazards'. 3rd ed. Butterworths, 1985.

27. Windholz, M. *et al.*, Eds. 'The Merck Index'. 10th ed. Merck & Co., 1983.

28. Weast, R.C., Ed.-in-Chief. 'CRC Handbook of Chemistry and Physics' ('The Rubber Handbook'). 67th ed. CRC Press, 1986-1987.

29. Horsley, L.H. 'Azeotropic Data III'. Advances in Chemistry Series No.116, American Chemical Society, 1973.

30. Clayton, G.D. and F.E. Clayton, Eds. 'Patty's Industrial Hygiene and Toxicology'. 3rd ed. rev. Wiley Interscience, 1982.

31. Browning, E. 'Toxicity and Metabolism of Industrial Solvents'. Elsevier, 1965.

32. Ruth, J.H. *American Industrial Hygiene Association Journal*, 47(1986):A-142. 'Odor thresholds and irritation levels of several chemical substances: a review'.

33. Kristesashvili, T.S. *Gigiena i Sanitariya*, **30,** No.11(1965):3. 'Data to substantiate the maximum permissible concentration of pyridine in the atmosphere'.

34. Williams, R.T. 'Detoxication Mechanisms'. 2nd ed. Chapman & Hall, 1959.

35. Grant, W.M. 'Toxicology of the Eye'. 2nd ed. Charles C. Thomas, Illinois, USA, 1974.

36. Baxter, J.H. and M.F. Mason. *Journal of Pharmacology & Experimental Therapeutics*, 91(1947):350. 'Studies of the mechanisms of liver and kidney injury – IV. A comparison of the effects of pyridine and methyl pyridinium chloride in the rat'.

37. Baxter, J.H. *American Journal of Pathology*, 24(1948):503. 'Hepatic and renal injury with calcium deposits and cirrhosis produced in rats by pyridine'.

CONTENTS

1. CHEMICAL ABSTRACTS NAME

benzene, methyl-
(10th Collective Index)

2. SYNONYMS AND TRADE NAMES[1,2]

antisal 1a
methacide
methane, phenyl-
methyl-benzene
methylbenzene
methylbenzol
NCI-C07272
phenylmethane
tolueen (Dutch)
toluen (Czech)
toluene
toluéne (French)
toluol
toluole
toluolo (Italian)
tolu-sol
U220
UN 1294

3. CHEMICAL ABSTRACTS SERVICES REGISTRY NUMBER

108-88-3

4. NIOSH NUMBER

XS5250000

5. CHEMICAL FORMULA

C_7H_8
(Molecular weight 92.13)

6. STRUCTURAL FORMULA

$$CH_3$$

7. OCCURRENCE[3-6]

Toluene is found as a component of high-flash aromatic naphthas which are produced from crude oil by primary distillation. In the past it has been recovered from by-products in the coal tar industry.[3]

Graedel[4] observes that the only natural sources [of toluene in the atmosphere] known with certainty are the combustion of vegetation and emission from volcanoes, with neither of these sources being considered significant.

Most emission of toluene to the atmosphere arises from human activity. Internal combustion engine exhaust contains high concentrations of aromatic compounds, particularly since alternatives to lead additives (to achieve high octane ratings) have been sought. Volatilization of toluene used for solvent purposes is not uncommon and emission to the atmosphere of toluene used in industrial processes also occurs.[4] Of the four largest sources of emission of toluene to the atmosphere, the largest is automobile use, followed in turn by industrial solvent use, coke ovens and toluene production industries.[5]

Toluene is the most prevalent aromatic hydrocarbon in the atmosphere with average levels ranging from 0.14 to 59 ppb.[5] Graedel[4] notes that is has been detected in urban areas at concentrations as high as 129 ppb. Toluene has also been detected in drinking water and in the flesh of edible fish. A concentration of 19 ppb has been detected in a drinking water supply. In surface water and treated wastewater effluents, toluene levels have been detected generally below 10 ppb.[5]

Apart from some unquantifiable sources of exposure mentioned below, the estimated quantities of toluene taken in by the general public are between a trace and 94 mg/week by inhalation, depending on location (rural, urban or relevant industrial) and 0.0 to 0.75 mg/week from food and water. In addition there may be occupational exposure (up to 18,000 mg/week) and cigarette smoking, which can provide 14 mg/week from 140 cigarettes.[5]

Unquantifiable exposures that the foregoing does not take into account are inhalational and dermal exposure during the re-fueling of vehicles with toluene-containing fuel, and exposure to solvents or other toluene-containing consumer products. EPA observes that no quantitative estimate of either the number of people exposed or the extent of exposure can be provided for these sources although consumer usage could contribute significantly to total exposure.[5]

US environmental release for the period *circa* 1980 amounted to approximately 1,100 kilotonnes/year of which virtually all was released to the atmosphere. The make-up of atmospheric emission was 0.34% from production of toluene, 34.2% from industrial usage, 64.5% from 'inadvertent' sources (mainly automobile emissions) and 1% from coke production.[5] Figures for non-atmospheric emissions were incomplete [but could be of the order of 1 to 2 kilotonnes to water, 0.3 to 0.6 kilotonnes to land and 0.1 kilotonnes to publicly owned treatment works].

Toluene itself only absorbs light of wavelengths below 295 nm of which the solar spectrum in the troposphere contains little. However, a charge-transfer complex between toluene and molecular oxygen absorbs light of wavelengths up to at least 350 nm and it has been proposed that it is the photolysis of this complex that may be responsible for some of the observed photochemical reactions of toluene.[5]

Toluene is removed from the atmosphere apparently entirely through free radical chain processes. Of the free radicals in the atmosphere, potential initiators are hydroxyl ($^\cdot$OH), atomic oxygen (O), and peroxy radicals ($HO_2^\cdot$ or $RO_2^\cdot$, where R is an alkyl or acyl group). An additional reactive species is ozone.

Rate constants for the reaction of the above species with toluene indicate that the reaction with the hydroxyl radical is the only important reaction. Thus concentrations of hydroxyl radicals, which are a function of solar intensity and which are zero at night, will determine the chemical lifetime of atmospheric toluene. Ref. [6] provides the following table.

Species	Estimated average daytime annual concentration (ppm)	Rate constant $(ppm^{-1} min^{-1})$	Rate of toluene removal $(ppm min^{-1})$	Fraction of hydroxyl rate
Hydroxyl radical	4×10^{-8}	9.5×10^3	3.7×10^{-4}	1
Atomic oxygen	3×10^{-9}	1.1×10^2	3.3×10^{-7}	1×10^{-3}
Ozone	3×10^{-2}	5×10^{-7}	1.5×10^{-8}	5×10^{-5}
Peroxy radical	1×10^{-4}	2.5×10^{-7}	2.5×10^{-11}	4×10^{-8}

The daytime hydroxyl radical concentration in the above table is given as the best estimate for an annual average. This though has considerable uncertainty and might be totally inapplicable to high latitudes in winter when longer lifetimes would be expected, or in summer with concomitant shorter lifetimes expected. In addition the actual daily values may vary by as much as a factor of 10 upward or 50 or more downward, depending on the local solar intensity, temperature, and composition of trace gas chemicals in the air.[6] These factors make an estimate, cited by ref. 5, of an atmospheric residence time of toluene due to hydroxyl radical reactions as being 1.9 days, and liable to considerable variation.

The reaction products formed from toluene under simulated atmospheric conditions are not known with certainty.[5] A study has given the gaseous reaction products as *o*-cresol, *m*- and *p*-nitrotoluenes, benzyl nitrate and benzaldehyde, with *o*-cresol and benzaldehyde as the major components, each being about 8% of the total product yield.[5]

It is assumed that the reaction of toluene is *via* hydroxyl radical addition to the ring or by abstraction of hydrogen from the methyl group. From the amounts of reaction products formed it appears that the addition mechanism is of much greater significance than that of abstraction. Ring fragmentation products, such as acetylene, acetaldehyde and acetone also are formed under simulated atmospheric conditions, these three products totalling much less than 1%. Unknown amounts of formaldehyde and formic acid are also formed.[5]

In addition to the foregoing, photolysis of toluene in polluted atmospheres, containing NO_x, yields ozone and fairly high amounts of peroxyacetyl nitrate (PAN) (5 – 30% nitrogen yield) and peroxybenzoyl nitrate (PBzN) (0 – 5% nitrogen yield).[6] PBzN is much less stable than PAN and would not accumulate in the atmosphere to the extent that PAN does. However, it is a much more powerful lacrimator than PAN even at very low levels. These peroxy compounds are strong eye irritants and oxidizing agents and may induce plant damage in susceptible species.

8. COMMERCIAL AND INDUSTRIAL TOLUENE[5,7]

Toluene, unseparated from the benzene–toluene–xylene (BTX) mixture produced by catalytic reforming, is used for gasoline blending.[5] This also obtains where toluene is present in BTX from pyrolytic cracking and from coke-oven operation (see Section 11).

Hoff in ref. 7 states that [in the US] toluene is marketed mostly as nitration and industrial grades whose generally accepted quality standards are given by ASTM D841 and D362 respectively. The ASTM standards do not specify the actual toluene content but EPA[5] notes that the 1 °C distillation range nitration grade is of 99.5 – 100% purity and that the 2 °C distillation range industrial grade is of 98.5 – 99.4% purity. Non-fuel toluene (solvent grade) is of 90 – 98.4% purity.

EPA observes that commercial toluene may contain benzene as an impurity and that prior to the 1950s benzene was a common contaminant of toluene. All health effect studies involving toluene should specify therefore the quality of the toluene used.[5] The ASTM specifications for nitration and industrial grade toluene are:

Specification for Nitration Grade Toluene, ASTM D 841 – 80

Property	Specification	ASTM test method
S.G.(d_{20}^{20})	0.8690 – 0.8730	D 891
Colour	Not darker than 20 max. on the Pt-Co scale	D 1209
Distillation range at 101.3 kPa (1 atm)	Not more than 1 °C including 110.6 °C for any one sample	D 850
Paraffins	Not more than 1.5 wt%	D 851

Property	Specification	ASTM test method
Acid-wash colour	Not darker than no.2 colour standard	D 848
Acidity	No free acid, no evidence of acidity	D 847
Sulphur compounds	Free of H_2S and SO_2	D 853
Copper corrosion	Copper strip shall not show iridescence nor gray or black deposit or discoloration	D 849

Specifications for Industrial Grade Toluene, ASTM D 362-80

Property	Specification	ASTM test method
S.G.(d_{20}^{20})	0.860 – 0.874	D 891
Colour	Not darker than 20 max. on the Pt-Co scale	D 1209
Distillation range at 101.3 kPa (1 atm)	Not more than 2 °C from initial boiling point to dry point, including 110.6 °C	D 850, D 1078
Odour	Characteristic aromatic hydrocarbon odour as agreed on by buyer and seller	D 1296
Water	Not sufficient to show turbidity at 20 °C	
Acidity	Not more than 0.005 wt% (free acid calculated as acetic acid) 0.047 mg KOH (0.033 mg NaOH) per g of sample or no free acid; that is no evidence of acidity	D 847
Acid-wash colour	Not darker than no. 4 colour standard	D 848
Sulphur compounds	Free of H_2S and SO_2	D 853
Corrosion $\frac{1}{2}$ h at 100 °C	Copper strip shall not show greater discolouration than Class 2 in Method D 1616	D 1616
Solvent power	100 min kauri-butanol value	D 1113

9. SPECTROSCOPIC DATA[8]

Infrared, Raman, ultraviolet, both ^{1}H and ^{13}C NMR, and mass spectral data have been tabulated for toluene.[8]

10. MEASUREMENT TECHNIQUES[9-50]

The method recommended for the determination of toluene in workshop atmospheres is the NIOSH[9] activated charcoal trapping method for collection and concentration, followed by solvent extraction of the charcoal and a gas chromatographic (gc) analysis of the extract.

The collection tube is 7 cm long with a 4 mm internal diameter. It contains a total of 150 mg of activated charcoal (20 – 40 mesh) divided into a front section of 100 mg and a rear section of 50 mg separated by a small plug of urethane foam. Air is sampled by means of a small pump at a flow-rate of up to 200 ml per minute usually to obtain an optimum sample of 2 l, with a maximum of 8 l. The entire apparatus is portable so that it may be carried in a pocket with the sampling tube in the breathing zone (normally a coat lapel). The apparatus can also be static.

Carbon disulphide is used to extract the toluene, which is separated on a 3.0 m × 2 mm glass column packed with 10% OV-275 on Chromosorb W-AW (100 – 120 mesh) using flame ionization detection.

The method is designed to separate the following aromatic hydrocarbons: toluene, xylene, benzene, cumene, styrene, α-methylstyrene, vinyltoluene, *p*-t-butyltoluene, ethylbenzene and naphthalene. For full details of the method see ref. 9.

The sampling method uses a small portable apparatus with no liquids and is therefore easy to handle and maintain. The analytical method is relatively specific with the additional advantage of flexibility of operating conditions.

Air samples are normally taken by means of suction using an electric pump capable of operating at low flow-rates (*i.e.*, 30 – 300 ml per minute). If higher flow-rates are used breakthrough can occur (*i.e.*, the toluene is not completely adsorbed and some of it passes through the adsorbent). It is also convenient either to establish the trapping device in a static position, or to have it attached to the user in such a way that the breathing zone is sampled (this usually means attachment to the worker's lapel).

It should be noted that suitable personal samplers may be commercially available.

The UK Health and Safety Executive has produced a similar method to the NIOSH one.[10] Ten litres of air are drawn through an activated charcoal trap and adsorbed toluene is extracted with carbon disulphide. The extract is analysed by gc on a 2 m × 2 mm column packed with 10% Carbowax 20M on Chromosorb W (100 – 120 mesh) at a temperature of 70 °C and using flame ionization detection. Aliphatic petrol constituents, if present, can be separated from toluene by using a column of 30% NN-bis-(2-cyanoethyl)-formamide. The method can be used in the range 0.1 – 200 ppm toluene in air.

The UK Health and Safety Executive has also produced a method using Tenax-GC for trapping toluene.[11] Five litres of air are drawn through a 7 cm × 5 mm tube containing Tenax-GC and the trapped toluene is thermally desorbed at 250 °C. The toluene is separated by gc on a 2 m × 2 mm column packed with 10% polyoxyethylene glycol on Chromosorb W using flame ionization detection.

Wood-charcoal cigarette-filters in series have been used to adsorb toluene and other organic compounds.[12] This trapping technique involves the passage of air at a flow-rate of 2.5 ml per minute for eight days through two 25 mg cartridges in series.

Using this technique, toluene was subsequently extracted by carbon disulphide and determined by gc[13] on a 50 m capillary column coated with UCON LB 550X-33. The column was temperature programmed from 20 to 120 °C and detection was by flame ionization. The method was used for the range 1 to 210 μg toluene per cubic metre of air.

Organic gases, including toluene, *o*-xylene and methanol, were passed at up to 500 ml per minute (1 to 3 l sample) through a 7 cm × 4 mm tube containing 50 – 100 mg activated charcoal[14] Dimethylformamide was used to extract the toluene and gc was performed on a 3 m × 4 mm column packed with 10% Carbowax 20M on Chromosorb W (80 – 100 mesh) at a temperature of 120 °C and using flame ionization detection.

Having trapped toluene and xylene on activated charcoal, heptane was used as extractant.[15] Gc of the extract was made on a column of poly(ethanediol adipate) on Porolith at 100 °C and using flame ionization detection.

Activated charcoal was preferred to alumina as a trapping medium for toluene and xylene because the capacity of alumina was affected by moisture.[16] Following gc analysis, the detection limit for toluene was 10 ppb.

The major constituents of vapour from some industrial paint and varnish installations were found to be toluene and *m*-xylene.[17] The vapour was passed through activated charcoal and the organic compounds were extracted with pentane. The extract was analysed by gc on a 50 m capillary column coated with PFMS-4 using flame ionization detection. The column was temperature programmed from 50 to 90 °C at 10 °C per minute.

After collection of toluene and other organic compounds from workshop air on a tube of 20 cm × 4 mm containing activated charcoal, they were extracted and analysed by gc using 10% dioctyl phthalate on Chromosorb W (80 – 100 mesh) at 86 °C and using flame ionization detection.[18]

It can be seen that a popular method of collecting and analysing toluene from air is activated charcoal trapping followed by solvent extraction and gc determination. A variation is the use of thermal desorption. Having trapped toluene and other organic compounds, the collection tube was placed in a desorption chamber and heated to 190 °C and the vapour was swept on to a 5 m × 4 mm gc column packed with 10% SP-1000 on Chromosorb W AW-DMCS (100 – 120 mesh).[19] The column was temperature programmed between 100 and 160 °C and detection was by flame ionization (if butyl acetate was present, this co-eluted with toluene). The detection limit was 2 ppm. In a similar method, the heating block was at 250 °C.[20]

If the organic compound is not to be concentrated the detection needs to be ultra-sensitive. Air containing toluene and xylene was analysed directly by gc using mass spectrometric detection using m/e of 91.[21] The detection limit for toluene was 2 ppb.

Air containing toluene was drawn for 10 – 15 minutes at a flow-rate of 100 – 200 ml per minute through water contained in a glass ampoule.[22] The ampoule was broken open in the laboratory and gc was performed on the vapour on a 3 m × 2 mm column packed with 20% poly(ethanediol adipate) on Celite C-22. The column temperature was 100 °C and detection was by flame ionization. The working range was 1 – 50 mg per cubic metre of air.

Working atmospheres have been collected at a flow-rate of 0.5 ml per minute by passage through two bottles containing ethanol and the trapped toluene was determined spectrophotometrically at 261 nm.[23] The limit of detection was 10 mg per cubic metre of air for a 10 l sample.

In polythene production, the working atmosphere containing toluene and xylenes was sampled in a 2 l gas burette.[24] The burette was heated to 40 °C and the sample was analysed by gc on a 6 m × 4 mm column packed with 10% Carbowax–terephthalic acid on Chromosorb P (60 – 80 mesh). The column temperature was 120 °C and detection was by flame ionization. The detection limit for toluene was 1 ppm.

For the analysis of toluene, acetone, ethyl acetate and methanol in air, a tube containing 150 mg activated charcoal was used to trap the sample at a flow-rate of 11 ml per minute.[25] Carbon disulphide was used to extract the organic compounds and the extract was examined by infrared spectrometry using CsBr windows.

In an examination of the air in the driver's cab of a vehicle fuelled by gasoline–methanol, toluene and methanol were trapped on activated charcoal and then extracted with chlorobenzene.[26] The extract was analysed by gc on a 3 m × 3 mm column packed with 8% 1,2,3-tris-(2-cyanoethoxy)propane (TCEP) on Chromaton N using flame ionization detection. The detection limit for toluene was 0.5 – 1 µg per cubic metre of air.

The air from a laboratory fume cupboard outlet was examined for organic compounds, including toluene[27] The compounds were trapped on a three-sectioned activated charcoal tube, and extracted with carbon disulphide prior to gc analysis.

Air containing toluene and xylene was passed through a column of poly(phenylmethylsiloxane) on diatomite and the adsorbed compounds were analysed by gc on a 2 m × 3 mm column packed with 15% TCEP on Chromaton N AW-DMCS using flame ionization detection[28] The limit of detection for toluene was 0.01 mg per cubic metre of air.

In the production of cellulose lacquer film, solvents used include toluene and ethyl acetate[29] These were determined in the working atmosphere by trapping in ice-cold *o*-xylene and analysing by gc on 3 m × 3 mm column packed with 10% polyoxyethylene glycol 1500 on Chromaton N AW-HMDS at 80 °C and using flame ionization detection.

Air containing toluene and xylene was passed through 80% acetic acid at a flow-rate of 350 – 450 ml per minute.[30] After the addition of 40% KOH, the mixture was heated at 25 °C and the headspace was analysed by gc on a 1.25 m × 3 mm column packed with 10% nitrilotripropionitrile on Chromosorb P (60 – 80 mesh). The column temperature was 80 °C and detection was by flame ionization. The effective detection range for toluene was 0.05 – 5 mg per cubic metre of air.

A similar method involved passage of air containing toluene and xylene through 65 or 80% acetic acid cooled between 0 and −24 °C.[31] After neutralization, the headspace was analysed by gc on a 1.25 m × 3 mm column packed with 20% Apiezon L on Chromosorb W. The limit of detection for toluene was 8 µg per cubic metre of air.

Toluene vapour in air was ionized in a chamber subjected to radiation at 100 – 170 nm emitted by a hydrogen discharge tube.[32] A photo-ionization detector provided with a LiF window and a sensitive picoammeter measured the emitted current, this being proportional to the toluene concentration.

Tin oxide, doped with various impurities and acting as a metallic oxide semiconductor, was used to measure the change in its electrical conductivity as a result of the reaction of toluene with oxygen absorbed on the oxide surface.[33] The effective range for this unusual method was 50 – 100 ppm of toluene in air.

Solid state electrolytic sensors, each consisting of two electrodes to which a potential of 1.3 – 2.5 V was applied have been used to determine concentrations of air pollutants.[34] Toluene could be determined by this method at the ppm level.

Piezoelectric detectors have been used to measure toluene in air. A silica crystal resonator was used as an indicator for such a detector[35] and Carbowax 550 was evaluated as a coating material in the determination of toluene in the 30 – 300 ppm range.[36]

A crystal coated with Pluronic F-68 has been used in a portable piezoelectric detector in the examination of the working atmosphere in a Danish printing plant.[37] The optimum air-flow was 100 ml per minute, having removed water by using a Nafion permeation tube.

There has been considerable interest in the fate of inhaled toluene and consequently blood, urine and other body tissues have been examined for toluene and urine has also been analysed for the major metabolite, hippuric acid.

Blood samples were extracted with pentane and the extract analysed by gc[38] and headspace samples over heparinised blood were passed through Tenax-GC before thermal desorption and gc analysis on a 2 m × 0.125 in column packed with 0.1% SP-1000 on Carbopack C (80 – 100 mesh).[39] The column was temperature programmed from 70 to 200 °C and detection was by flame ionization. The detection limit was 7.5 µg toluene per litre of blood.

Water, methanol and ethylbenzene were added to blood in an ice-cooled bottle and the mixture was shaken for 20 minutes at 35 °C before sampling the headspace for gc analysis.[40] This was performed on a 1 m × 3 mm column packed with 20% dioctylsebacate on Uniport B (60 – 80 mesh) at 100 °C and using flame ionization detection. The method was used for toluene in blood concentrations of 0.5 – 500 µg per ml.

The blood and urine samples of persons exposed to toluene were mixed in a stoppered phial, heated to 60 °C for 30 minutes and the headspace subjected to gc.[41] Butanol was used as internal standard and chromatography was carried out at 145 °C on a 6 ft × 0.125 in column packed with 0.25% Carbowax 1500 on Carbopack C and using flame ionization detection.

Body tissues, *e.g.*, brain, liver, kidney, lung and blood, were examined for residues of toluene and xylene by placement in phials and heating them to 65 °C for 40 minutes and using the headspace for gc analysis[42] This was performed on a 50 m capillary column coated with OV-101 and using a temperature programme from 40 to 250 °C at 12 °C per minute. Flame ionization or mass spectrophotometric detection was applied, the method being used in the range 0.1 – 15.8 ppm for toluene.

Blood and other tissue containing toluene residues were extracted with methanol and the extract was imbibed by filter paper.[43] The toluene was desorbed in a stream of nitrogen onto Tenax-GC before gc analysis. This clean-up procedure was efficient in that it removed much of the debris which would otherwise have fouled the column.

In assessment of the levels of metabolites of toluene and xylene in the urine of exposed workers, the urine was acidified, saturated with salt and heptadecanoic acid in chloroform was added as internal standard[44] The methyl esters of the metabolic acids and internal standard were formed by treatment with diazomethane and the methyl ester of hippuric acid was measured by gc, thus giving a measure of toluene uptake. The chromatography was performed on a 1.5 m × 2 mm column packed with 3% OV-101 on Diatomite Q AW-DMCS (100 – 120 mesh) at 170 °C and using flame ionization detection.

Hippuric acid in urine was simultaneously extracted and methylated using 3-methyl-1-(p-tolyl)triazen solution in ethyl acetate and the ester was determined by gc.[45]

The urinary metabolites of toluene and xylene were determined after extraction using ethyl acetate after the addition of ammonium sulphate, 5% sulphuric acid and heptadecanoic acid as internal standard.[46] After evaporation of the extract, the trimethylsilyl derivatives were formed and analysed by gc on a 1.2 m × 2 mm column packed with 3% OV-1 on Gas Chrom Q (80 – 100 mesh). The column was temperature programmed up to 215 °C and detection was by flame ionization. The limit of detection for hippuric acid was 4 mg per litre.

A high performance liquid chromatographic (hplc) method was used to determine hippuric acid extracted from urine of workers exposed to toluene.[47] This was performed at 40 °C on a 15 cm × 4 mm column packed with Li Chrosorb RP-18 (5 μm) using water–methanol–acetic acid (40:10:0.1) as mobile phase at a flow-rate of 1.5 – 2.0 ml per minute with detection at 272.4 nm.

After extraction of hippuric acid from urine with butyl chloride/isopropanol, the extract was dried and dissolved in water/acetonitrile.[48] After the addition of 4-hydroxybenzoic acid as internal standard this solution was analysed by hplc on a 25 cm × 2.6 mm column packed with ODS SILX.

Chloroform was used to extract hippuric acid from urine of workers exposed to toluene and the extract was analysed by thin layer chromatography.[49] This was carried out on silica gel using chloroform–water–acetic acid (4:1:1) as mobile phase and 4-dimethylaminobenzaldehyde solution in acetic anhydride as locating agent. The limit of detection for hippuric acid was 6 μg per ml.

Isotachophoresis has been used to determine the urinary metabolites of toluene and styrene.[50] The sample was acidified and salt was added before diethyl ether extraction. Isotachophoresis was performed on the extract from 5 to 20 kV.

11. CONDITIONS UNDER WHICH TOLUENE IS PUT ON THE MARKET

Production[7,51,52]

Until World War I toluene was obtained as one of the products from coke ovens. Supplies from this source were insufficient for the large production of trinitrotoluene needed during the war, and they were augmented by toluene being obtained for the first time from petroleum sources. In this, narrow-cut naphthas, containing relatively small amounts of toluene, were subjected to thermal cracking and subsequent purification. Shortly after the end of the war, production from petroleum was discontinued but the advent of catalytic reforming and the need for large quantities of aviation fuel during World War II again made petroleum an important source.[7] In 1951, 50% of toluene was coal-derived, but by 1971 less than 2% came from this source.[51] Ref. 51 states that [in 1975] virtually all toluene was petroleum derived, about 84% coming from catalytic reformate, about 12% from pyrolysis gasoline and a small amount as a by-product of styrene manufacture. Hoff in ref. 7 states that about 90 – 95% of US toluene production is not isolated but is blended directly into the gasoline pool as a component of reformate and of pyrolysis gasoline.

Catalytic reforming (hydroforming) of $C_6 - C_9$ naphthas yields toluene along with benzene, xylenes and C_9 aromatics. These products can be extracted from the crude reformate and then separated by fractionation.[7]

Catalytic reforming involves catalytic dehydrogenation in the presence of hydrogen (which reduces coke formation). The hydroformer raw material should be rich in dimethylcyclopentane, methylcyclohexane, and ethylcyclopentane which are the three chief toluene synthesis compounds. Other naphthenes, such as cyclohexane and dimethylcyclohexane, yield benzene and xylene respectively. Paraffin hydrocarbons, of which n-heptane and dimethylhexane are typical, pass through the hydroformer essentially unchanged to contaminate the ultimate toluene product.[51]

Hoff in ref. 7 observes that before 1940, predominant were fixed-bed and fluidised-bed units, typically using a 10 – 15% molybdenum–aluminium oxide catalyst or similar catalysts promoted with 0.5 – 2% cobalt. In 1940 improved operation was obtained from a 0.3 – 0.6% platinum–aluminium oxide catalyst and since about 1970 further improvement has come from promoting this catalyst with up to 1% chloride, by using bimetallic catalysts containing 0.3 – 0.6% of both platinum and rhenium to retard deactivation, and by using molecular sieves as part of the catalyst base to gain activity.

The reaction is endothermic and most units comprise about three reactors with reheat furnaces in between. Operations are of three basic types, *viz.* semi-regenerative, cyclic and (since about 1971) continuous. With semi-regenerative operation, feedstocks and operating conditions are controlled so that the unit is on-stream for six months to two years before shutdown and catalyst regeneration. In cyclic operation, a swing reactor is employed so that one reactor can be regenerated while three are in operation. Regeneration may be as frequent as every 24 hours thus permitting continuous operation at high severity. With continuous operation, the catalyst is continuously withdrawn, regenerated and then fed back into the system.[7]

Crude petroleum and natural gasoline yield feedstock on distillation. A heart-cut between 90 – 115 °C may be taken from crude oil. Heart cutting of natural and straight-run gasoline may be arranged so as to yield the maximum obtainable amount of feedstock.[51] Naphthas from catalytic crackers and from coke stills are also used.[7]

In a process example given by ref. 51 a selected naphtha fraction feedstock is preheated in heat exchangers and then totally vaporized at 565 °C in a furnace. It is then joined with hydrogen-rich (70%) recycle gas at 575 °C and the mixture is passed through a reactor with a contact time of about 15 seconds. The dehydrogenation catalyst here is 10% molybdenum dioxide on alumina. The reaction takes place at a temperature of 535 – 575 °C and a pressure of 150 – 300 psi (approx. 10 – 20 atm).

The reaction gases pass through heat exchangers (which pre-heat the feedstock) and then to a gas–liquid condenser and separator. A large proportion of the separated gas is compressed at 75 – 200 psi and is recirculated to the furnace and reactor to increase the hydrogen concentration. The remaining gas and liquid pass to conventional absorption and stabilizing columns where fractionation produces fuel gas (butanes, *etc.*), gasoline and hydroformate containing about 21% toluene. This figure can be raised to 38% on recirculation of the one-pass hydroformate through the reactor. The hydroformate is then distilled with three cuts being taken, the first and last being of benzene and xylene respectively. The heart-cut contains about 65% toluene from two-pass hydroformate (or about 39% toluene from one-pass material).[51]

The aromatic hydrocarbons in reformates cannot easily be separated from non-aromatics of similar or close boiling points. The processes most widely used for extraction are azeotropic distillation, extractive distillation, solvent extraction and selective adsorption.

Azeotropic distillation involves adding *e.g.*, aqueous ethyl methyl ketone (10% water) to the reformate heart-cut in an azeotrope tower. This azeotrope loosely associates with the paraffins and naphthenes and carries them out from the top of the tower to a ketone recovery plant. Crude toluene is removed from the bottom of the tower from where it is passed to a flash tower where remaining ketone and some toluene are removed in the head cut. Ketone-free toluene is then washed with 98% sulphuric acid, water washed and caustic washed to remove small amounts of olefins. The washed toluene is then redistilled to produce nitration grade toluene.[51] Nitromethane, methanol or dioxane are also used in azeotropic distillation.

In extractive distillation the usual solvents are phenol, cresols, furfural, aniline and alkyl phthalates. Selective adsorption processes include the use of activated alumina or silica gel.[51] Hoff in ref. 7 states that the two processes now primarily in use are both liquid–liquid extractions using either sulpholane or tetraethylene glycol. Both these supplant the one-time use of diethylene glycol.

Reaction (typical):

$$C_6H_{11}CH_3 \rightarrow C_6H_5CH_3 + 3H_2$$
Methylcyclohexane   Toluene

One use of toluene is in the production of benzene by hydrodemethylation. By proper choice of feedstocks and the use of relatively severe operating conditions in the reformers, it is possible to produce streams high enough in toluene content to be directly usable for hydrodemethylation without the need for extraction.

If pyrolysis gasoline is used as a source of toluene it is usually first mixed with reformate and the mixture processed in a typical aromatics extraction unit. The pyrolysis gasoline is hydrotreated to eliminate dienes and styrene prior to processing.[7]

Toluene may be obtained by fractionation of the light oil formed by coal carbonization. The light oil, obtained by cooling and scrubbing the by-product coke-oven gas, and by rectification of tar condensate, is fractionally distilled in continuous or semicontinuous units. Of the four fractions usually taken, the second cut, from about 95 – 125 °C, consists of crude toluene. This cut is washed with 96% sulphuric acid to remove unsaturated compounds and then with 10% sodium hydroxide solution to remove the free acid. The washed toluene is then fractionally distilled in batch vacuum columns to yield refined toluene. One tonne of coal yields about two litres of toluene.

World production of [isolated] toluene in 1984 was over 7,300 kilotonnes to which the US contribution was 3,100 kilotonnes and that of the Japanese 810 kilotonnes.[52] EC production in 1983 was 780 kilotonnes and its consumption was 990 kilotonnes.

Toluene is marketed in tank cars, drums, cans and bottles.[51]

Uses[7.51.53]

About 90% of the toluene generated by catalytic reforming is blended into gasoline as a component of $> C_5$ reformate. Depending on the severity of the reforming operation, the octane numbers of such reformates typically lie in the range 88.9 – 94.5. Toluene itself has a blending octane number of 103 – 106.

Toluene usage in the US (in 1980) was as follows. Of the approximately 30 million tons used, 82.6% was used in gasoline without first being isolated and 9.4% was isolated prior to usage in gasoline. About 5.3% was converted to benzene, 1.3% was used for solvent purposes, 0.65% for toluene di-isocyanate production, 0.18% for benzoic acid production, 0.11% for benzyl chloride production, and 0.55% had other uses.

Di- and trinitrotoluenes are produced by the nitration of toluene. Ref. 51 observes that the use of these nitrotoluenes is now almost exclusively by the military as cheaper, competitive industrial explosives are now available.

Solvent use accounted for 379,000 tons of toluene in the US in 1980. About two thirds of this use was in paints and coatings, the remainder being used in adhesives, inks, pharmaceuticals and other formulations, and as an extractive solvent for various principles in plants.[7.53] Hoff in ref. 7 observes that the use of toluene as a solvent in surface coatings is declining and is expected to continue to decline primarily because of various environmental and health considerations; that it is being replaced by other solvents and by changes in product formulation.

12. STORAGE, HANDLING AND USE PRECAUTIONS[2.54 – 58]

Toluene is a flammable, non water-miscible liquid of boiling point 111 °C and a vapour density about three times that of air (see Section 17). Toluene is not corrosive to metals. Iron, mild steel, copper or aluminium are suitable for plant and containers. Small quantities for laboratory use may be kept in glass containers.[2.54]

Care must be taken with storage, process and drying plant to ensure that hazard-free conditions exist regarding fire and explosion risks and exposure of personnel to toluene. [As with other flammable solvents the organization of storage facilities has to take into account two separate requirements. These are that the stored material be protected from a fire elsewhere on the premises and that the premises be protected from a fire involving the stored materials. On an industrial scale, sufficient physical separation and water spray curtains are often used effectively. On a laboratory scale, flammable solvents need to be stored in fire resistant, thermally insulated cupboards.

It is important to appreciate that a non-combustible barrier is not synonymous with a fire barrier. A fire barrier has to protect against both conducted and radiant heat. For example, a metal cupboard is useless on both these counts.]

Storage tanks in the open should be bunded and there should be a ramped sill at the doorways of storerooms in order to prevent egress of spilled material.[55] Such storerooms should be well ventilated, flammable liquid stores, free from source of ignition. Toluene should be stored away from chemicals with which it is incompatible (see Section 14). The construction and siting of process buildings and plant should be such as to prevent the spread of escaping liquid throughout the building or site.[55] Large items of equipment should be earthed and tools or equipment used should be of the non-sparking/flameproof type.[56]

OSHA[57] recommends that engineering controls such as process enclosure, general dilution ventilation, local exhaust ventilation and the use of personal protective equipment should be used, as appropriate, to keep personnel exposures at or below permissible limits. Care should be taken to prevent recirculation of toluene exhaust due to the proximity of air conditioning intakes or windows. Repeated or prolonged skin contact with liquid toluene should be prevented by personnel being required to use, as appropriate, impervious clothing, gloves, splash proof safety goggles, face shields (eight inch minimum) and such other protective clothing as may be necessary. Adequate emergency eye-wash facilities should be available as should facilities for washing toluene from the skin with soap or mild detergent and water.

[Personnel should not be allowed to enter areas containing potentially hazardous concentrations of toluene unless all relevant legislation and official guidelines have been strictly adhered to. Suitable respiratory protective equipment should be used.] Such exposed personnel should be able to communicate with each other and with those (*i.e.*, more than one person) monitoring the work from outside the hazardous area. Monitors should be suitably equipped to effect rescue (possibly with lifelines to exposed personnel) and should themselves ensure, so far as is possible, that one of their number remains outside the hazard area in the event of an emergency.

Toluene vapour can cause narcosis. A concentration of 2,000 ppm is immediately dangerous to life and health.[58]

Commercial toluene may contain benzene as a toxic impurity and it should be borne in mind that the higher vapour pressure of benzene will ensure that the vapour mixture in equilibrium with the liquid will contain a higher proportion of benzene than exists in the liquid itself.

N.B. the caveat in Section 13 (final paragraph).

13. FIRE HAZARDS[58-60]

Toluene is a highly flammable liquid of boiling point 110.6 °C, specific gravity about 0.87 and water solubility about 0.05%. The flash point (closed cup) is 4 °C, the auto-ignition temperature is about 480 °C and the lower and upper explosive limits in air are 1.2 and 7.1% v/v (see Section 17). The vapour density of toluene is about three times that of air. The vapour, if in sufficient quantity, may therefore flow a considerable distance to a source of ignition and then 'flash back'.

Containers of toluene can explode when heated and, if feasible, should be cooled with flooding quantities of water applied from as far as possible until well after the fire is extinguished.

Small fires may be fought with foam, dry powder, carbon dioxide or vaporizing liquid. For larger fires water spray, fog or foam may be used. Water-spray should be applied carefully over the higher boiling liquid to avoid excessive frothing, particularly as the water–toluene azeotrope boils at only about 85 °C (see Section 17). Jets of water will result in spreading the immiscible and less dense burning liquid. Immediate withdrawal is imperative in the case of rising sound from a venting safety device or any discolouration of a storage tank due to the fire. Personnel should keep upwind of the fire to avoid breathing any toxic vapours. Combustion products may include carbon monoxide as well as acrid smoke and irritating fumes.

Ref. 60 notes that where a fire is fed by an uncontrolled flow of combustible liquid the decision on how or if to fight it will depend on the size and type of fire anticipated and must be carefully considered. This may call for special engineering judgement, particularly in large-scale applications.

[Notwithstanding the foregoing review of various authorities noted in Sections 12 and 13, it is essential that managers and others responsible for the planning, implementation and overseeing of personnel and plant safety should be familiar with the legal constraints and official guidelines applicable to them and that they liaise with their local emergency services in the planning of plant and storage facilities and in the preparation of contingency plans for dealing with fires and other emergencies. Managers should regularly monitor staff knowledge of, and ability to implement, emergency procedures and should ensure that equipment provided for use in emergencies is regularly inspected and maintained.]

14. HAZARDOUS REACTIONS[58,61]

Bretherick[61] includes the following:

At −80 °C, solid bromine trifluoride reacts violently with toluene.

A mixture of dinitrogen tetraoxide and toluene exploded; possibly initiated by unsaturated impurities.

Lack of proper control in nitration of toluene with mixed acids may lead to runaway or explosive reaction. A contributory factor is the oxidative formation and subsequent nitration and decomposition of nitrocresols.

When tetranitromethane is mixed with hydrocarbons in approximately stoicheiometric proportions, a highly sensitive explosive mixture requiring careful handling is formed. The explosion of only 10 g of a mixture with toluene caused 10 deaths and 20 severe injuries, though here an excess of toluene was present.

Uranium hexafluoride reacts very vigorously with toluene with the separation of carbon.

OSHA[58] notes the following:

Violent or explosive reaction may occur with 1,3-dichloro-5,5-dimethyl-2,4-imidazoline-2,4-dione.

An explosive hydrocarbon–metal perchlorate complex may be formed with silver perchlorate.

There is the possibility of explosion with allyl chloride in the presence of dichloroethylaluminium or ethylaluminium sesquichloride.

There is explosive reaction with nitronium perchlorate.

15. EMERGENCY MEASURES IN THE CASE OF ACCIDENTAL SPILLAGE[58,62]

Only necessary personnel should be allowed within the hazard area. Suitable apparel giving protection against the inhalation of high vapour concentrations, against excessive skin contact and against eye splashes (see Section 12) must be used. All ignition sources should be eliminated and any leak stopped if this can be done without risk. Water spray may be used to reduce vapours.[58]

For spillages up to approximately 25 l, the liquid may be absorbed onto sand or vermiculite, followed by transfer in suitable containers [*e.g.*, covered buckets] to a safe, open area where atmospheric evaporation can take place. The site of the spillage should be washed thoroughly with water and biodegradable detergent.[62] Washings should be discharged to a sewer and not to a 'soakaway' nor a land drain.

Warren[62] notes that "*ideally* all hydrocarbons and related flammable organic chemicals should be burned in an incinerator with an afterburner."

OSHA[58] recommends that larger or massive ground spills should be bunded far ahead of the spill pending later disposal. Bunding may be achieved using soil, sandbags or barriers such as polyurethane or concrete. The spill may be immobilized with universal gelling agent and the vapour and fire hazard can be reduced with fluorocarbon water foam. Where an environmental hazard is caused, the relevant authorities, including the water authority, must be informed.

In the case of spillage onto water, natural barriers or oil spill control booms can be used to limit spill motion and its dispersion. The spilled material can be thickened by applying detergents, soaps, alcohols or other surface active agents. The application of a universal gelling agent to immobilize the trapped spill will increase the efficiency of removal.[58]

16. FIRST AID

General

The casualty should be removed from the danger of further exposure into the fresh air. Rescuers should ensure their own safety from inhalation and skin contamination where necessary. Conscious casualties should be asked for information about what has happened. Bear in mind that the casualty may lose consciousness at any time. Where necessary, continued observation and care should be ensured. First aiders should take care not to become contaminated. Note that toluene is flammable.

Inhalation

The conscious casualty should be kept at rest. The unconscious casualty should be placed in the recovery position and an open airway maintained. If breathing or heart beat stops, resuscitation should be commenced immediately. Medical aid should be obtained or the casualty removed to hospital immediately.

Skin contamination

Remove contaminated clothing. Wash with copious amounts of water, or soap and water, for at least 15 minutes. If necessary, seek medical aid. In the case of persistent skin irritation refer for medical advice.

Eye contamination

Irrigate with copious amounts of water for at least ten minutes but avoid further damage to the eye from the use of excessive force. Seek immediate medical aid. Remember the possible presence of contact lenses, which may be affected by some solvents and may impede decontamination of the eye.

Ingestion

If the lips or mouth are contaminated rinse thoroughly with water. Do not induce vomiting. Give further supportive treatment as for inhalation. In the case of ingestion of significant amounts wash out the mouth with water. Obtain immediate medical attention. Contact a hospital or poisons centre at once for advice. Remember that symptoms may develop many hours after exposure so continued care and observation may be necessary.

In all cases note information on the nature of exposure, and give this to ambulance or medical personnel. Information for doctors is provided in some technical literature issued by manufacturers.

17. PHYSICO-CHEMICAL PROPERTIES

17.1 General[2,53,63]

Toluene is a colourless, flammable, refractive liquid, with a benzene-like odour.[2,53]

At 25 °C, toluene is infinitely miscible in acetone, carbon tetrachloride, benzene, ether, n-heptane, and ethanol.[63] The solubility of toluene in water is 0.047 g/100 g water.[2]

17.2 Melting Point[53]

Merck[53] gives the freezing point of toluene as

$-95\,°C$.

17.3 Boiling Point[64,65]

Weast[64] gives the boiling point of toluene as

110.6 °C.

Toluene forms azeotropes with a number of compounds. Some binary ones formed with some organic compounds and water are as follows[65]:

Wt% toluene	Second component	Wt% second component	Boiling pt azeotrope (°C)
50	Formic acid	50	85.8
55 – 56	Isobutanol	44 – 45	101.2
79.8 – 86.5	Water	13.5 – 20.2	84.1 – 85

17.4 Density/Specific Gravity[53,63]

Merck[53] quotes the specific gravity of toluene at 20 °C (referred to water at 4 °C) as 0.866.

The critical density of toluene is 0.288 g/ml, and its critical volume is 3.473 ml/g.[63]

17.5 Vapour Pressure[63,64]

Vapour pressures for toluene below one atmosphere are tabulated by Weast[64] as follows:

Temperature in °C	Vapour pressure in mmHg
-26.7	1
6.4	10
31.8	40
51.9	100
89.5	400

Pressures at and above atmospheric pressure are also reported by Weast[64]:

Temperature in °C	Pressure in atm
110.6	1
136.5	2
178.0	5
215.8	10
262.5	20

The critical temperature and critical pressure of toluene are given by Dreisbach[63] as 320.8 °C and 30,400 mmHg [40 atm], respectively.

17.6 Vapour Density[60]

3.1 (relative to air = 1).

17.7 Flash Point[60]

4 °C, closed cup test.

17.8 Explosive Limits[60]

The limits of flammability of toluene in air at normal atmospheric temperature and pressure are given by the National Fire Protection Association[60] as:

Lower limit: 1.2% v/v
Upper limit: 7.1% v/v

The NFPA[60] give the ignition temperature of toluene as 480 °C. However, they point out that different test conditions (and also different definitions of "ignition temperature") can result in widely varying values being quoted. They therefore recommend that any value should be treated as an approximation.

17.9 Viscosity[63.64]

The viscosity of liquid toluene is reported by Weast[64] as follows:

Temperature in °C	Viscosity in cp
0	0.772
17	0.61
20	0.590
30	0.526
40	0.471
70	0.354

Dreisbach[63] quotes the kinematic viscosity (= absolute viscosity/density) of toluene as:

Temperature in °C	Kinematic viscosity in cSt
20	0.67778
40	0.56457
60	0.45825
80	0.39119

17.10 Concentration Conversion Factors

At 25 °C and 760 Torr (1 atm),

1 ppm = 3.77 mg/m^3, and
1 mg/m^3 = 0.265 ppm

18. TOXICITY

18.1 General[3,55,66-68]

Sandmeyer in ref. 3 observes that toluene closely resembles benzene in its toxicological properties but is devoid of the latter's chronic haematopoietic effects. Toluene's acute toxicity to humans is somewhat more intense than benzene; a vapour concentration in air of about 1,000 ppm gives rise, after two to three hours, to vertigo, difficulty in maintaining equilibrium and intense frontal headache. Higher concentrations may result in a narcotic coma.[55]

ILO[55] observes that the symptoms of chronic toxicity are those habitually encountered with exposure to the commonly used solvents. These include mucous membrane irritation, euphoria, headaches, vertigo, nausea, loss of appetite and alcohol intolerance. These symptoms generally appear at the end of the day, are more severe at the end of the week and lessen or disappear during weekend or holiday periods. Toluene has never been proved to be hepatotoxic.[55] Occasional reports of renal damage in glue sniffers have appeared, characterized by a form of distal tubular acidosis.[66]

Toluene exerts a stronger irritant action on the skin and mucous membranes than does benzene, and severe dermatitis may result from its drying and defatting action following prolonged or repeated contact. It is readily absorbed by inhalation, ingestion, and somewhat through skin contact. If ingested, toluene may present a lung aspiration hazard.[3]

It should be noted that, prior to the 1950s, benzene was a common contaminant of commercial toluene and the purity of the toluene used must be considered in evaluating the effects of exposures.[66]

Human exposure to levels above 100 ppm may lead to impairment of coordination, of mental alertness and of reaction times thus leading to accident proneness. This effect is potentiated in those taking drugs, such as diazepam, which interfere with microsomal enzyme activity.

The odour of toluene is described as sweet, pungent, benzene-like[3] and aromatic.[67] Ruth[68] describes the odour of toluene from petroleum as "rubbery, mothballs", having an odour threshold ranging from 8 mg/m³ [approx. 2 ppm] to 150 mg/m³ [approx. 40 ppm]; and describes toluene from coke as having a floral, pungent odour with the odour threshold ranging from about 17.5 mg/m³ [approx. 4.5 ppm] to 260 mg/m³ [approx. 70 ppm].

18.2 Toxicokinetics[5,66,69]

Toluene is readily absorbed through the respiratory tracts of humans and experimental animals. The blood/air partition coefficient is approximately 15. The amount of toluene absorbed is proportional to the concentration in inspired air, length of exposure and pulmonary ventilation (and thus on the subject's level of physical activity). In humans this uptake amounted to about 50% of the amount of the toluene inspired.[5]

Absorption of liquid toluene through the skin occurs less readily although the amount may nevertheless be significant. In an experiment with humans, an absorption rate of 14 to 23 mg/cm²/hour was found. It was found though that a maximum toluene concentration of 170 µg/l in the blood of subjects who immersed one hand in toluene for 30 minutes was only 26% of the concentration (650 µg/l) in the blood of subjects who inhaled 100 ppm of toluene vapour for 30 minutes. It has been suggested that some of the toluene that penetrates the stratum corneum may subsequently be released to the air rather than entering the systemic circulation. Toluene appears to pass slowly into the bloodstream through the skin. It has been observed that the maximum levels of toluene in venous blood were maintained for about fifteen minutes following cessation of exposure, and elimination of toluene in alveolar air sometimes increased during the first twenty minutes following the termination of exposure of both hands to liquid toluene.[5] Absorption of toluene *vapour* through the skin in humans however probably amounts to less than 5% of the total uptake *via* the respiratory tract under the same conditions of exposure.[5]

Absorption of toluene from the gastrointestinal tract appears to occur more slowly than through the respiratory tract. Nevertheless, in animal experiments the absorption appeared fairly complete based on the amounts of toluene and metabolites subsequently excreted in urine or exhaled air.[5]

Following absorption, toluene is rapidly distributed, with the highest levels observed in adipose tissue followed by bone marrow, adrenals, kidneys, liver, brain and blood.[66] The tissue/blood partition coefficients for fatty tissues in the rabbit are very high being 113 for adipose tissue and 35 for bone marrow. For other tissues the coefficients ranged from about 1 to 3.[5] Equilibrium with the tissues may take at least two to three hours to achieve however, as indicated by the slow approach of peripheral venous concentrations to steady state as compared with arterial concentrations. Concentrations in peripheral venous blood though do not reflect the discharge of toluene to the tissues as fully as would concentrations in central venous blood.[5]

The tissues of a teenage boy who died from sniffing glue had toluene levels of: heart blood 11 mg/kg; liver 47 mg/kg; brain 44 mg/kg; kidney 39 mg/kg. In mice exposed to a relatively high concentration of toluene vapour of 3,950 ppm (15 mg/l) for three hours in a dynamic exposure chamber, the concentrations of toluene in liver, brain and blood rose continuously throughout the exposure period, reaching 625 mg/kg in the liver, 420 mg/kg in the brain and 200 mg/kg in the blood at the end of exposure.[5]

In an attempt to simulate solvent abuse, such as glue sniffing, by humans, mice were intermittently exposed to 10,600 ppm of toluene vapour in cycles of five minutes on, ten minutes off; or ten minutes on, twenty minutes off for a total of three hours. Tissue and blood levels were produced approximately three times higher than by a single ten minute exposure to 10,600 ppm and were similar to those produced by three hour exposure at this level.[5]

Ref. 5 notes studies showing that alveolar concentrations appear to reflect arterial concentrations during exposure to 100 – 200 ppm of toluene, both at rest and at various intensities of exercise, the ratio of the two concentrations remaining about the same. Ref. 66 observes that by measuring toluene concentration in alveolar air during exposure, it is possible to estimate the arterial blood concentration. For example, after exposure at rest to 300 mg/m³ [approx. 80 ppm] for thirty minutes, the relative uptake averaged 52%, the alveolar concentration was 28% of the inspired air concentration and the arterial concentration amounted to 0.7 mg/litre of blood.

Exercise increases the total uptake of toluene although the relative uptake (compared with the increased amount available due to increased pulmonary ventilation) decreases.[5,69] The higher alveolar concentration (concomitant with higher blood concentration) correlates ($r^2 = 0.72$) with this decreased percentage uptake by an inverse function, *viz.*

$$\text{Percentage uptake} = 72.9 - \frac{0.63 \text{ alveolar conc'n (mg/l)} \times 100}{\text{inspired concentration (mg/l)}}$$

The alveolar concentration is as determined after thirty minutes exposure to 100 ppm during rest or various levels of exercise (50, 100 or 150 W on a bicycle ergometer).

The major portion of inhaled or ingested toluene is metabolized by side chain oxidation to benzoic acid which in turn conjugates with glycine to form hippuric acid which is then excreted in the urine. Colman in ref. 5 citing various studies, observes that, irrespective of the route of administration, dose or species, 60 – 75% of toluene absorbed by inhalation or orally administered could be accounted for as hippuric acid in the urine. Much of the remaining toluene (9 – 18%) is exhaled unchanged and 2% or less appears in the urine as conjugated cresols and benzylmercapturic acid.

o-Cresol and possibly *p*-cresol are minor metabolites formed by toluene undergoing ring hydroxylation by mixed function oxidases, probably *via* arene oxide intermediates. Their excretion in the urine is as sulphate or glucuronide conjugates.

The suggestion that arene oxides are intermediates in the metabolism of toluene to cresols derives from studies involving labelled toluene. These putative intermediates are of concern because they are highly reactive and may bind to cellular macromolecules. Colman points out however that very little toluene is metabolized by this pathway and that studies conducted indicate that the binding of toluene metabolites to proteins and nucleic acids does not occur to any significant extent. No evidence of covalent binding to tissue components was detected by autoradiography of mice that inhaled ^{14}C toluene.[5]

Metabolism of toluene probably occurs primarily in the liver and is rapid, the excretion of hippuric acid in the urine being elevated within thirty minutes of the initiation of inhalational exposure. The initial step in the metabolism to benzoic acid appears to be side chain hydroxylation to benzyl alcohol by the microsomal mixed-function oxidase system. Although benzaldehyde has not been found in the urine or expired air of animals given toluene orally, it has been assumed that benzyl alcohol is metabolized to benzaldehyde and thence to benzoic acid by the action of alcohol dehydrogenase in both stages. Reaction with glycine then produces hippuric acid, or reaction to a lesser extent with glucuronic acid produces benzoylglucuronide.[5]

The maximum rate of hippuric acid formation from benzoic acid appears limited by the availability of glycine and it has been estimated, assuming 60% retention of the inhaled concentration of toluene, that the conjugation capacity may be saturated at a toluene concentration of 780 ppm during light work (pulmonary ventilation of 10 l/min) or 270 ppm during heavy work (pulmonary ventilation of 30 l/min).

Most inhalationally or orally absorbed toluene is excreted from humans or animals within twelve hours of cessation of exposure. In experimental animals, elimination of toluene and its metabolites from most tissues, including the brain, was rapid while elimination from fat and bone marrow was slower.

In humans, the cessation of inhalational exposure is followed by a desaturation process that appears to consist of three exponential phases (half-lives 1.95, 35.2 and 204 minutes) for toluene concentrations in peripheral venous blood, and similarly for toluene concentrations in alveolar air (half-lives 1.59, 26.5 and 221 minutes). Great variability among individuals makes toluene concentrations in expired air or peripheral venous blood unreliable indicators of toluene uptake or of exposure levels. Similarly, although the excretion of hippuric acid in the urine is roughly proportional to toluene exposure, individual variation makes measurement of hippuric acid concentration or excretion rate unreliable for individuals as such[5] [though it is a useful measure of group exposure].

18.3 Human Acute Toxicity[3,5,54,58,66,67]

Ref. 66 observes that toxicity studies on human beings have primarily involved individuals exposed to toluene *via* inhalation either in experimental or occupational settings or during episodes of intentional abuse of solvent mixtures containing toluene. Toluene is the hydrocarbon solvent most frequently implicated as the cause of adverse effects associated with such deliberate inhalation ('glue sniffing').[5] In such abuse, excessive levels of toluene are inhaled over a short time, repeated inhalation being associated with the development of tolerance and psychological dependence. The concentration of toluene inhaled under some of the conditions of abuse (often in a very confined space) can approach 30,000 ppm, *i.e.*, saturation concentration at 20 °C.[5]

The primary effect of toluene is on the central nervous system. Euphoria in the induction phase is followed by disorientation, tremulousness, changes of mood, tinnitus, diplopia, hallucinations, dysarthria, ataxia, convulsions and coma.[66]

Bosch in ref. 5 observes that Von Oettingen *et al.* in 1942 provided what is generally acknowledged to be the most complete description of the effects of pure toluene (benzene $\leqslant 0.01\%$) on the CNS. In single eight hour exposures, three humans subjects were subjected in an exposure chamber to concentrations of toluene ranging from 50 – 800 ppm. The exposures were no more than two per week to allow sufficient time in between for recovery. The effects as summarized in ref. 5 were as follows (the number of subjects affected being noted in parentheses):

Concentration	*No. of Exposures*	*Effects*
0 ppm (control)	7	No complaints or objective symptoms, except occasional moderate tiredness toward the end of each exposure, which was attributed to lack of physical exercise, unfavourable illumination, and monotonous noise from fans
50 ppm	2	Drowsiness with a very mild headache in 1 subject. No after-effects
100 ppm	4	Moderate fatigue and sleepiness (3), and a slight headache on one occasion (1)
200 ppm	3	Fatigue (3), muscular weakness (2), confusion (2), impaired coordination (2), paraesthesia of the skin (2), repeated headache (1), and nausea (1) at the end of the exposure. In several instances, the pupils were dilated, pupillary light reflex was impaired, and the fundus of the eye was engorged. After-effects included fatigue, general confusion, moderate insomnia, and restless sleep in all 3 subjects.
300 ppm	2	Severe fatigue (3), headache (2), muscular weakness and incoordination (1), and slight pallor of the eyeground (2). After-effects included fatigue (3) and insomnia (1)
400 ppm	2	Fatigue and mental confusion (3), headache, paraesthesia of the skin, muscular weakness, dilated pupils, and pale eyeground (2).

Concentration	No. of Exposures	Effects
		After-effects were fatigue (3), skin paraesthesia (1), headache (1), and insomnia (2)
600 ppm	1	Extreme fatigue, mental confusion, exhilaration, nausea, headache and dizziness (3), and severe headache (2) after 3 hours of exposure. After 8 hours' exposure, the effects included considerable incoordination and staggering gait (3), and several instances of dilated pupils, impaired pupillary light reflex and pale optic discs; after effects included fatigue and weakness, nausea, nervousness and some confusion (3), severe headache (2), and insomnia (2). Fatigue and nervousness persisted on the following day
800 ppm	1	Rapid onset of severe fatigue and, after 3 hours, pronounced nausea, confusion, lack of self-control, and considerable incoordination and staggering gait in all 3 subjects. Also, pupillary light reflex was strongly impaired (1) and optic discs were pale (2). All 3 subjects showed considerable after effects, lasting at least several days, which included severe nervousness, muscular fatigue, and insomnia

ACGIH[67] notes the opinion of one author that the foregoing study did not justify a 200 ppm exposure limit although Von Oettingen *et al.* had concluded that at that level there were unlikely to be any discernible untoward effects on health. In another study noted by ACGIH,[67] *inter alia*, workers found that experimental human subjects exposed at 200 ppm for seven hours showed increases in reaction time, decrease in pulse rate and in systolic blood pressure. These workers also considered 200 ppm to be a too high MAC. Ref. 54 notes that the accident proneness engendered by the impairment of coordination, mental alertness and reaction times at toluene levels above 100 ppm is potentiated in those taking drugs which interfere with microsomal enzyme activity, *e.g.*, diazepam.

Ref. 66 observes that acute exposure to high levels of toluene, such as 37,500 mg/m³ [approx. 10,000 ppm] or higher for a few minutes, during industrial accidents, has been characterized by initial CNS excitative effects, followed by impairment of consciousness, eventually resulting in seizures and coma. Sandmeyer in ref. 3 observes that tests on an employee who was found unconscious after an exposure to high vapour concentrations for 18 hours indicated hepatic and renal involvement with myoglobinuria, with all effects reversible within six months.

Toluene abuse by 'sniffing' also is characterized by the progressive development of CNS symptoms though not all the symptoms described have been exhibited in any single sniffer nor in any single episode of sniffing. The initial excitatory stage is typically characterized by drunkenness, dizziness, euphoria, delusions, nausea and vomiting, and, less commonly, visual and auditory hallucinations. With continued exposure symptoms indicative of CNS depression become evident, *viz*, confusion and disorientation, headache, blurred vision and reduced speech, drowsiness, muscular incoordination, ataxia, depressed reflexes and nystagmus. In extreme cases, loss of consciousness, possibly with convulsions, occurs. Depending on the intensity of the exposure, the severity and duration of these effects varies greatly. The duration may vary from fifteen minutes to a few hours. Of sudden deaths from solvent sniffing, and attributed to a direct effect of the solvent itself rather than to suffocation, toluene was implicated in 10 out of 122 cases. Severe cardiac arrhythmia resulting from light plane anaesthesia was offered as the most likely explanation for the cause of sudden sniffing deaths.

The autopsy on an adolescent who had died through sniffing model aeroplane glue containing toluene revealed the following: The cut surfaces of the lungs were found to be extremely frothy and congested, with diminished amounts of crepitation throughout the lung tissue. Other gross observations included some petechial haemorrhages in the larynx and upper trachea, firmness and congestion in the spleen and a dark, red-brown colour and congestion in the liver. No haemorrhages, obstructions or ulcerations were seen anywhere in the gastrointestinal tract and all other organs were unremarkable.[5] Congestion in various organs, swelling of the brain, subseromucous petechiae and pulmonary oedema were associated with nineteen other cases of death from acute intoxication by a thinner of which the major component was indicated as being toluene.[5]

OSHA[58] observes that acute prolonged dermal exposure to liquid toluene may result in irritation, scaling, cracking and dermatitis. Paraesthesias of the skin may result from vapour exposure. Toluene exerts a stronger irritant action on skin and mucous membranes than does benzene.[3] Toluene can penetrate the skin to some extent (see Section 18.2).

Liquid toluene splashed in the eyes caused temporary corneal damage and conjunctival irritation, with complete recovery within 48 hours. Toluene vapour may cause a noticeable sensation of eye irritation at 300 – 400 ppm, mild irritation and lacrymation at 400 ppm, but even at 800 ppm the irritation is slight. Extremely high concentrations may cause blurring of vision and lacrymation but a level which is high enough to produce narcosis can exist without associated eye irritation.[58]

18.4 Animal Acute Toxicity[3,5,66,70 – 73]

Sandmeyer in ref. 3 notes that toluene appears to be the least acutely orally toxic of the alkylbenzenes but is somewhat more toxic than benzene.

Smyth *et al.*[70,71] using a commercial grade (unspecified) of toluene, report the oral LD_{50} for the rat as 6.55 g/kg. Kimura *et al.*[72] give the acute oral LD_{50}s for three groups of rats, ACS Analytical Grade toluene being used. The results are reported along with the 95% confidence limits, thus:

14 day old (both sexes)	2.6	(2.2 – 3.2) g/kg
Young adults (male)	5.6	(5.0 – 6.4) g/kg
Adults (male)	6.5	(5.4 – 7.7) g/kg

The results for 14-day old rats were significantly different from those for adult rats ($p < 0.05$).

Ref. 66 notes another study giving the oral LD_{50} for the rat as 5.90 g/kg.

The single skin penetration LD_{50} for rabbits is given by Smyth *et al.*[70,71] as 12.3 g/kg. Sandmeyer in ref. 3 quotes a study giving the dermal LD_{50} value as 14 g/kg for the rabbit and observes that this indicates that toluene is practically not absorbed through the intact skin.

Sandmeyer in ref. 3 notes a report by Von Oettingen that toluene has an irritant action on the skin and that in the eye it causes rapid and intense turbidity of the cornea and toxic effects on the conjunctiva. Ryan and Bosch in ref. 5 note another report that application to the rabbit cornea has caused slight to moderate irritation.

Ryan and Bosch in ref. 5 note the intraperitoneal LD_{50}s for mice as being 1.15 g/kg for males and 1.64 g/kg for females in separate studies. In rats, an intraperitoneal injection of 0.65 g/kg produced apathy while 1.5 – 1.7 g/kg caused death from respiratory failure, a cause of death noted in various of the intraperitoneal studies. Other studies have shown 1.7 g/kg to be a lethal intraperitoneal dose in rats, mice and guinea pigs.[5]

Ryan and Bosch in ref. 5 observe that, in all species studied, the progressive symptoms typically found after increasingly higher inhalational doses of toluene are irritation of the mucous membranes, incoordination, mydriasis, narcosis, tremors, prostration, anaesthesia and death.

These reviewers also note that an inhalational exposure to 24,400 ppm of toluene for 1.5 hours produced 60% mortality in rats and 10% mortality in mice. Exposure of these species to 12,200 ppm for 6.5 hours produced 50% mortality in rats and 100% mortality in mice. It is stated that the two species are probably equally sensitive.

In another study noted by Ryan and Bosch in ref. 5 an increase in respiratory rate and a decrease in respiratory volume was observed in six dogs exposed to 850 ppm toluene (0.01% benzene content) for one hour.

Bruckner and Peterson[73] found that CNS depression was manifested more rapidly in four week old mice than in eight and twelve week old animals exposed to 2,600 ppm, 5,200 ppm and 12,000 ppm. Four week old rats were reported as being slightly more sensitive to toluene narcosis than were eight or twelve week old animals (data not shown in the paper).

Ryan and Bosch in ref. 5 also note the results of various studies including the following:

LC_{17}, rats	–	4,000 ppm for 4 h (technical grade toluene)
LC_{50}, mice	–	6,942 ppm for 6 h (99.5% purity)
LC_{50}, Swiss mice	–	5,320 ppm for 7 h (< 0.01% benzene present)

Sandmeyer in ref. 3 notes that inhalation of 1,500 – 2,500 ppm of toluene by the rat in 15 to 35 minutes resulted in recoveries of 0.27 mg in the blood, 0.64 mg in the liver and 0.87 mg in the brain.

Ryan and Bosch in ref. 5 note that more sensitive detection methods have revealed an effect on simple behavioural parameters and the CNS at lower levels. EEG changes were seen in rats exposed to 1,000 ppm during sleep. After a single exposure of male rats to 800 ppm for four hours, unconditioned reflexes and simple behavioural parameters began to fail.

18.5 Human Sub-acute and Chronic Toxicity Including Reproductive Effects[5,66,74-79]

General

In a study by Andersen *et al.*,[74] sixteen young, healthy, male students were exposed over four days for six hours per day to a regimen of various concentrations of toluene in air under controlled conditions. Each of the subjects had experienced each of four exposure levels (0, 10, 40 and 100 ppm) by the end of the study during which various functions were observed. The authors' summary states that

> ... the toluene exposures did not affect nasal mucus flow or lung function. At 100 ppm irritation was experienced in the eyes and nose. There was a significant deterioration in the perceived air quality and a significant increased odour level during all exposures to toluene. The test battery investigated visual perception, vigilance, psychomotor functions, and higher cortical functions and comprised five-choice, rotary pursuit, screw-plate, Landolt's rings, Bourdon Wiersma, multiplication, sentence comprehension, and word memory tests. In these eight tests measuring 20 parameters, no statistically significant effects of the toluene exposure occurred. For three tests (multiplication errors, Landolt's rings, and the screw plate test) there was a borderline significance (0.05% < p < 0.10%). The subjects felt that the tests were more difficult and strenuous during the 100 ppm exposure, for which headache, dizziness, and feeling of intoxication were significantly more often reported. The exposures to 10 and 40 ppm did not result in any adverse effects.

Tähti *et al.*[75] studied 46 workers exposed to various concentrations of toluene in air ranging from 20 – 200 ppm, in many cases for 10 – 20 years. An additional 32 workers from the same (tarpaulin preparing) factory had not been exposed. The authors state that the urinary hippuric acid at the end of work shifts showed good correlations to toluene concentrations in air. Most of the biological parameters measured showed no correlation with toluene exposure. The blood leucocyte count showed slight positive correlations with toluene exposure but nevertheless remained inside the range of normal values. The occurrence of chronic diseases, drug-using habits, and drinking and smoking habits did not show any correlations with toluene exposure.

Bosch in ref. 5 observes that though earlier reports of occupational exposures to toluene (generally pre-1950) ascribed myelotoxic effects to the solvent, most of recent evidence indicates that toluene is not toxic to the blood or bone marrow. [The earlier results may have been due to a higher benzene level in the toluene.]

Liver enlargement was reported (in 1942) in a study of painters exposed to 100 – 1,000 ppm of toluene for two weeks to more than five years. This effect was not associated with clinical or laboratory evidence of disease nor corroborated in later studies of workers. Chronic occupational exposure to toluene has generally not been associated with abnormal liver function although reductions in serum bilirubin and alkaline phosphatase, and increases in gamma glutamyl transpeptidase have been noted.[5]

Bosch in ref. 5 further notes that neither [the above noted] exposures of aeroplane painters to mean concentrations of 100 – 1,000 ppm of toluene for two weeks to five years, nor, in another study, exposure of female shoemakers to 60 – 100 ppm for over three years, resulted in any abnormal urinalysis findings.

Spencer and Schaumburg[76] observe that a consistent pattern of neurological damage has been reported in individuals repeatedly inhaling toluene for its euphoric properties, and progressive and irreversible changes in brain structure and function have been found in others deliberately inhaling toluene-based paints. The brain damage from toluene abuse seems to appear insidiously after 1 – 20 years of repeated exposure to estimated concentrations of several thousand parts per million. Such chronic abuse of pure toluene produces irreversible cerebellar, brainstem, and pyramidal-tract dysfunction. However comparable deficits have not been reported in workers occupationally exposed to toluene for 3 – 32 years at levels (50 – 150 ppm) close to an hygienic standard of 100 ppm; at these levels, only evidence of acute reversible changes in brain function was seen. Spencer and Schaumburg observe though that there have been no attempts to monitor auditory function.

Cavanagh in ref. 77 observes that on regular inhalational abuse of toluene at high levels over long periods, epileptic seizures have been noted with abnormalities of the EEG, but that this complication is in fact rather uncommon. In one case, optic neuropathy was reported in which there was a decrease in the amplitude of the visual evoked potentials and from which recovery was made. Cavanagh also notes more serious and frequent reports of dysarthria, a cerebellar type of ataxia, and movement disorders associated with varying degrees of dementia which may still be found present more than a year later. Various investigations all attest to the strong likelihood of an atrophic process in the brain but Cavanagh adds that in the apparent absence of post-mortem material the underlying process can only be guessed at. However, the number of cases published, the regularity of the signs and symptoms and the lack of recovery in most cases, all point to structural damage having occurred. Also noted is that, of three women who inhaled high toluene levels regularly throughout their pregnancies, one of them had an infant with a cerebellar defect.

Grasso et al.,[78] in their literature review, concluded that the evidence seen failed to confirm that long-term industrial exposure to toluene leads to chronic organic or functional damage of the nervous system in man. Grasso et al. point out however that the data seen did not entirely refute that view, and that controversy is likely to remain until well designed studies have been performed.

Ref. 66 notes that no adequate epidemiological studies on populations exposed to toluene are available.

Reproductive Effects in Man

Ref. 66 observes that data on human beings are not adequate for the evaluation of the teratogenicity of toluene. Subjective complaints of dysmenorrhoea and disturbances in menstruation have been reported in female workers exposed concurrently to toluene, benzene, xylene and other unspecified solvents. Ref. 66 concludes though that the limited data available do not specifically associate occupational exposure to toluene with reproductive effects in male and female workers. A similar conclusion is drawn by Bosch in ref. 5. Barlow and Sullivan[79] give a detailed analysis of these studies and draw attention to the need for further studies of occupationally exposed women. As noted above, of three women who engaged in inhalational abuse of toluene vapour during their pregnancies, one gave birth to an infant with a cerebellar defect.[77]

Carcinogenicity to Man

No reports on toluene showing carcinogenicity to man were found.

18.6 Animal Sub-acute and Chronic Toxicity Including Reproductive Effects[5,66,79]

General

Ryan and Bosch in ref. 5 observe that toluene vapour levels of 1,000 ppm have little effect on gross manifestations of CNS depression. They note that in the pilot study for a longer (12 month) CIIT study (*v.i.*), the inhalation of 1,000 ppm toluene vapour for six hours/day, five days/week for thirteen weeks did not produce observable behavioural effects in rats.[5]

Ryan and Bosch in ref. 5 further note the following studies.

Continuous inhalational exposure to 107 ppm toluene for ninety days or repeated exposure to 1,085 ppm toluene for eight hours/day, five days/week for six weeks did not adversely affect the liver, kidney, lungs, spleen or heart in thirty rats, thirty guinea pigs, four dogs or six monkeys. No significant change was observed in haemoglobin, haematocrit or leucocyte count. All animals (except two of thirty treated rats) survived exposure. All, except the monkeys, gained in body weight.[5]

Male rats exposed by inhalation to 1,000 ppm of toluene vapour six hours/day, five days/week for six months exhibited no treatment-related effects. Body weight gain, haematological parameters (RBC and WBC counts, haemoglobin, mean corpuscular volume, haematocrit, sedimentation rate), and tissue histology (lungs, liver, spleen, kidney, genitals and other unspecified 'principal' organs) were assessed.[5]

Ryan and Bosch[5] note that, prior to the early 1940s, toluene was believed to be myelotoxic. However benzene was a contaminant of toluene in these early studies and could well have accounted for the findings. More recent studies, presumably using toluene of greater purity, have generally indicated a lack of myelotoxicity due to this solvent, although a few have indicated a positive effect, usually weak and/or reversible. Ryan and Bosch observe that there is no unanimity on the point but they believe that the suggestion made by NRC in 1980 that the few positive findings may indicate subtle, unrecognized haematopoietic responses is sound. For example, the effect of toluene in reducing haematocrit levels and increasing mean corpuscular haemoglobin concentrations in female Fischer rats and not in male rats in the CIIT study [*v.i.*] is of interest in view of a study on the female rabbit with benzene. In this benzene study, the decrease in erythrocytes, haemoglobin content, white blood cells and mean corpuscular haemoglobin concentration, and increase in mean corpuscular volume observed in the female was simulated in the oestradiol propionate-treated orchidectomized male.

Other studies noted by Ryan and Bosch in ref. 5 include:

The inhalation by guinea pigs of 4,000 ppm toluene, purified by distillation, for four hours/day, was lethal within a few days [*sic*] to two of three animals. The third animal was severely prostrated. Under the same regimen, guinea pigs exposed to 1,250 ppm for six days/week survived three weeks of exposure although they were severely afflicted. Marked pulmonary inflammation was observed under these conditions. At 1,000 ppm, the effects after 35 exposures were of slight toxic degenerative changes in the liver and kidney.

In a study in which two dogs were exposed by inhalation for eight hours/day, six days/week to 2,000 ppm of pure toluene for four months and subsequently to 2,660 ppm for two months, slight nasal and ocular irritation occurred at 2,000 ppm. In the terminal phase, paralysis of the extremities occurred preceded by motor incoordination. Death occurred on days 179 and 180. There was no effect on gain in body weight, on the bone marrow nor on the adrenal, thyroid nor pituitary glands. Congestion in the lungs, haemorrhage in the liver, a decrease of lymphoid follicles and haemosiderosis of the spleen were observed. Glomeruli of the kidney were hyperaemic, and albumin was found in the urine.

A study in which rats were exposed by inhalation to 1,000 ppm toluene eight hours/day for one week, increased the SGOT and SGPT activity and induced metabolic acidosis.[5]

Ref. 66 notes a study reporting increasing numbers of casts in the collecting tubules of rat kidneys during inhalation of toluene (99.9% purity) vapour at concentrations of 200, 600, 2,500 and 5,000 ppm, seven hours/day, five days/week for five or fifteen weeks. A few casts were seen after the third week at the 600 ppm level and earlier at the higher dose levels. In the blood, there was a temporary decrease in WBC count at the 5,000 ppm level.

The long-term inhalational toxicity of toluene was studied by CIIT using Fischer-344 rats. Four groups of 120 male and 120 female rats were exposed to 0, 30, 100 and 300 ppm for six hours/day, five times/week for 24 months. No effect on haematology, clinical chemistry, body weight or histopathology was noted except for two haematological parameters in females; reduced haematocrit levels were found in females exposed to the 100 and 300 ppm levels and increased mean corpuscular haemoglobin concentration was found in females exposed to the 300 ppm level.[5,66]

Ryan and Bosch in ref. 5 observe that oral administration to rats of 590 mg/kg/day for six months resulted in no adverse effects.

Ref. 66 states that repeated (10 – 20) applications of undiluted toluene to the rabbit ear or shaven abdomen over two to four weeks produced slight to moderate irritation and increased capillary permeability locally.

Carcinogenicity to Animals

Bosch in ref. 5 observes that toluene has been used extensively as a solvent for assessing the carcinogenic potential of lipophilic chemicals applied topically to the shaved skin of animals and that the results of control experiments with pure toluene have been uniformly negative.

Ref. 66 notes studies in which the promoting effect of toluene in Swiss mice following initiation with 7,12-dimethylbenz(*a*)anthracene (DMBA) was examined. An indication that toluene had some weak promoter activity was not confirmed.

The above mentioned CIIT study (see 18.6 General) showed no carcinogenic potential. However, Bosch in ref. 5 observes that the study has been considered inadequate for carcinogenicity evaluation because a maximum tolerated dose (MTD) was not achieved in either this two-year study at 300 ppm or in a 90-day pilot study at 1,000 ppm, and the low mortality of rats in the CIIT study (14.6%) differed from that (up to 25%) normally associated with monitoring these animals under barrier conditions. In addition a number of factors militate against the strain of rats used (Fischer-334) being appropriate for the study of a chemical that may be myelotoxic.

Reproductive Effects in Animals

Ref. 66 notes a study in which degeneration of germinal cells in the testes was reported in four out of twelve rats exposed by inhalation to 750 mg/m³ [approx. 200 ppm] toluene for eight hours/day, six days/week for one year. Absolute testicular weight at one year was lower in rats exposed to 375 mg/m³ [approx. 100 ppm] and 750 mg/m³ [200 ppm] in comparison with controls, and there was a trend toward a decrease in the relative testes weight.

Rats exposed to 1,000 mg/m³ [approx. 265 ppm] toluene for eight hours/day on days 1 – 21 of pregnancy, or to 1,500 mg/m³ [approx. 400 ppm] for eight hours/day on days 1 – 8 or days 9 – 14 of pregnancy showed no significant effects on implants/dam, live foetuses/dam, dead or resorbed foetuses/dam or malformations. There were no signs of maternal toxicity at 1,000 mg/m³ [265 ppm]. Foetal body weight was significantly reduced by 13% when dams were exposed to 1,000 mg/m³ [265 ppm] throughout pregnancy but not in the 1,500 mg/m³ [400 ppm] groups exposed in early or mid-pregnancy. A significant increase in retarded ossification was seen in the 1,000 mg/m³ [265 ppm] groups exposed throughout pregnancy and in the 1,500 mg/m³ [400 ppm] group exposed on days 1 – 8 of pregnancy. In the foetuses of dams exposed to 1,500 mg/m³ [400 ppm] on days 9 – 14 of pregnancy (a maternally toxic dose) there were significant increases in fused and extra ribs.[66]

When rats were exposed to 6,000 mg toluene/m³ [approx. 1,600 ppm] for 24 hours/day during days 1 – 8, 7 – 14, 9 – 14 or 9 – 21 of pregnancy no teratogenic effects were found, but a definite embryotoxic effect, related to the exposure duration, was noted. In the groups exposed on days 7 – 14, 17% of implants died or were resorbed, a significant difference from controls. In the group exposed from days 9 – 21, foetal and placental weights were decreased and ossification was retarded, with minor skeletal anomalies.[66,79] There was maternal mortality and toxicity in all treated groups, so that the findings do not indicate a specific adverse reproductive effect.

It was concluded, from a study in which mice were exposed by inhalation to 500 mg/m³ [approx. 130 ppm] for 24 hours/day from days 6 – 13 of pregnancy, that toluene was not teratogenic under these conditions.[66]

In a study in which mice were exposed to toluene at 375 mg/m³ [approx. 100 ppm] or 3,750 mg/m³ [approx. 1,000 ppm], six hours/day for days 1 – 17 of gestation, it was found that observations on mouse embryos, foetuses and postnatal growth did not differ significantly from controls. At the higher exposure there was a slight increase (32.6% compared with 19.2% in the control litter) in the incidence of resorbed foetuses and rudimentary 14th ribs, and an increase in the incidence of extra 14th ribs.[66]

When rabbits were exposed to 500 mg/m³ [approx. 130 ppm] or 1,000 mg/m³ [approx. 265 ppm] toluene for 24 hours/day on days 6 – 20 of pregnancy, spontaneous abortions occurred at 1,000 mg/m³ [265 ppm], but no teratogenic effects were noted at either concentration.[66]

When toluene in cottonseed oil was administered to mice by gavage at 260, 430 or 870 mg/kg/day on days 6 – 15 of gestation, a significant increase in embryonic lethality but no maternal toxicity occurred at all dose levels, and a significantly increased incidence of cleft palate occurred in the 870 mg/kg/day group (this effect not appearing to be due merely to a general retardation of growth rate). In another group exposed to 870 mg/kg/day on days 12 – 15 however, decreased maternal weight gain was the only effect noted.[66]

The effects of toluene in drinking water were studied on mice exposed continuously, prenatally and postnatally where the test animals and dams were given water containing 16, 80 or 400 mg toluene/litre during pregnancy and lactation. After weaning, the test mice were exposed to the same regimens as the dams. Maternal fluid consumption, offspring mortality rate, development of eye or ear opening, or surface-righting response were not affected. Offspring exposed to 400 mg/l showed decreased habituation of open-field activity at 35 days of age. Rotorod performance measured at 45 – 55 days of age was depressed in all exposed groups. Postnatal exposure alone did not produce similar results.[66]

Ref. 66 observes that, although it is generally accepted that toluene readily crosses the placenta, it does not appear to be teratogenic in mice, rats or rabbits. It is foetotoxic though, causing a reduction in foetal weight in mice and rats and retarded ossification, with some increase in minor skeletal abnormalities, at doses that are below those toxic for the dam as well as at toxic doses.[66]

18.7 Mutagenicity[5,66]

Ref. 66 notes studies evaluating the ability of toluene to induce DNA damage by comparing its differential toxicity for wild-type and DNA repair-deficient *E. coli* and *S. typhimurium*. The tests produced negative results.

Toluene was non-mutagenic in the Ames *Salmonella* assay when tested with strains TA1535, TA1537, TA1538, TA98 and TA100; and in the *E. coli* WP2 reversion to trp$^+$ prototrophy assay.[66]

Toluene, at 0.05 – 0.30 µl/ml, with and without mouse liver activation, failed to induce specific locus forward mutations in the L5178y Thymidine Kinase (TK) mouse lymphoma cell assay.[66]

Toluene failed to elicit positive mutagenic responses when tested, with or without metabolic activation, for its ability to induce reversions to isoleucine independence in *S. cerevisiae* strain D7, mitotic gene conversion to tryptophan independence in strains D4 and D7, and mitotic crossing over at the ade2 locus in strain D7.[66]

Pure liquid toluene did not induce recessive lethal mutations when fed in doses of 500 and 1,000 mg/kg body weight to *Drosophila melanogaster*.[66]

Bosch in ref. 5 observes that toluene was negative in the micronucleus test in mice and in the mouse dominant lethal assay. Sister-chromatid exchange (SCE) frequencies were not altered in Chinese hamster ovary cells or in human lymphocytes cultured with toluene *in vitro*.

Ref. 66 observes that there are discrepancies in findings related to chromosome damage in peripheral lymphocytes among workers exposed to toluene. It further notes that an unequivocal evaluation of the genetic effects of occupational toluene exposure, based on available studies, cannot be made. Reasons given are, the relatively small number of subjects analysed, variation in the extent of exposure between these studies and insufficient information on possible exposure to other chromosome-damaging agents such as benzene, tobacco smoke, *etc*. [No useful conclusions result from the combination of these human studies.]

18.8 Summary

[Toluene is of fairly low acute oral and dermal toxicity, but high inhaled vapour concentrations may cause headache, dizziness and narcosis. Effects of long-term over-exposure are generally similar, accompanied at high concentrations by kidney damage and possibly neurological changes. Haematological changes may also occur if there is significant contamination with benzene. Systemic effects of acute or repeated inhalational exposure to toluene are generally slight at 50 ppm, significant at 100 ppm and severe at 200 ppm. The liquid is defatting to skin, particularly following prolonged or repeated contact, leading to dermatitis, and is strongly irritating to eyes, while vapour concentrations above 100 – 400 ppm may also cause mild eye irritation. Toluene appears not to be carcinogenic or mutagenic. No substantiated evidence has been found of adverse effects on reproduction except for some foetotoxicity at doses generally toxic to the mother, and no reports of teratogenicity have been traced. Absorbed toluene is partly exhaled unchanged and partly excreted rapidly in the urine after metabolism mainly to hippuric acid.]

19. MEDICAL / HEALTH SURVEILLANCE

A decision on the need for, and content of, medical surveillance should be based on an assessment of the possibility and extent of exposure in the work operation. In addition, medical examination may be directed to identifying any pre-existing or newly arising condition in the individual workers which might either be aggravated by subsequent exposure, or might confuse any subsequent medical assessment in the event of excessive exposure or an illness not related to exposure. A particular aspect to be considered is the identification of sensitive subjects not adequately protected by the control limit in operation. A professional medical judgement is required on continuance of employment in the specified process. The following information is relevant in the case of this solvent.

The need for and content of medical surveillance should be related to an assessment of the exposure and of the possible presence of impurities, notably benzene, in the toluene being handled in the workplace. Subject to these considerations, pre-employment and periodic medical examinations may consist of the following:

Pre-Employment Medical Examination

A complete history and general medical examination should be undertaken to ensure that any pre-existing condition which might be complicated by exposure to toluene is evaluated. Particular attention should be paid to disease of skin and mucosae, evidence of alcoholism, blood, liver and central nervous system. Information may be sought on concurrent drug treatment. Haematological tests may be indicated.

Periodic Medical Examination

Medical examinations similar to the pre-employment medical examinations are recommended at yearly intervals or if excessive exposure has occurred. Several biological tests for assessing exposure to toluene have been considered: direct measurements of toluene in blood and in exhaled air; and indirect measurements by determination of metabolites in urine. Good correlation has been found between daily time-weighted average exposure (TWA) for toluene, using passive dosimeters, and corresponding urinary metabolites. Hippuric acid and o-cresol in shift-end urine samples are measured using hplc. Values must be corrected per gram of creatinine. Of these, the hippuric acid measurement is that usually employed. Although hippuric acid originating from food sources is normally found in urine of subjects not exposed to toluene, this rarely exceeds 1.0 mol hippuric acid per mol creatinine but, in group studies, a correlation has been shown between total shift exposure and hippuric acid in urine at shift-end. Only a weak correlation was found in individuals, due to differences in absorption and biotransformation of toluene and the variability in intake of benzoic acid in food. Urine samples at shift-end are therefore mainly of value for assessment of exposure of groups, because changes in exposure are rapidly reflected in changes in hippuric acid excretion. Measurement of o-cresol is not yet validated as a measure of exposure for routine use.

The possibility of exposure outside work should not be forgotten. There are no known delayed effects after a single exposure.

20. OCCUPATIONAL EXPOSURE LIMITS

[The Committee felt that a time-weighted average limit of 100 ppm was appropriate for toluene. The Committee could find no evidence on which to base a STEL but in the light of the acute effects of several hours exposure to toluene a STEL of 150 ppm may be prudent.

It should be noted that other toxic volatile substances may be present in this solvent and should be monitored separately.]

REFERENCES

1. National Institute for Occupational Safety & Health. 'Registry of Toxic Effects of Chemical Substances (RTECS)'. DHHS (NIOSH) Publication No.84-101-6. US Department of Health & Human Services, April, 1986.
2. Marsden, C. and S. Mann, Eds. 'Solvents Guide'. 2nd ed. rev. Cleaver-Hume Press, 1963.
3. Clayton, G.D. and F.E. Clayton, Eds. 'Patty's Industrial Hygiene and Toxicology'. 3rd ed. rev. Wiley Interscience, 1982.
4. Graedel, T.E. 'Chemical Compounds in the Atmosphere'. Academic Press, 1978.
5. US Environmental Protection Agency. 'Health assessment document for toluene'. Report No.EPA-600/8-82-008F. Environmental Criteria & Assessment Office, EPA, Washington D.C., 1983.
6. Committee on Alkyl Benzene Derivatives of the Board on Toxicology & Environmental Health Hazards, Assembly of Life Sciences, National Research Council. 'The Alkyl Benzenes'. National Academy Press, Washington D.C., USA, 1981.
7. Grayson, M., Exec. Ed. 'Kirk-Othmer Encyclopedia of Chemical Technology'. 3rd ed. John Wiley & Sons, 1979.
8. Grasselli, J.G. and W.M. Ritchey, Eds. 'Atlas of Spectral Data and Physical Constants for Organic Compounds'. 2nd ed. CRC Press, 1975.
9. National Institute for Occupational Safety & Health. 'NIOSH Manual of Analytical Methods'. 3rd ed. DHHS (NIOSH) Publication No.84-100. US Department of Health & Human Services, 1984.
10. Health & Safety Executive (UK), Occupational Medicine & Hygiene Laboratory. 'Methods of Determination of Hazardous Substances'. MDHS 36, April 1984.
11. Health & Safety Executive (UK), Occupational Medicine & Hygiene Laboratory. 'Methods of Determination of Hazardous Substances'. MDHS 40, June 1984.
12. Grob, K. and G. Grob. *Journal of Chromatography*, **62**, No.1(1971):1.
13. Wathne, B.M. *Atmospheric Environment*, **17**, No.9(1983):1713.
14. Tyras, H. *et al. Chemia Analityczna (Warsaw)*, **29**, No.3(1984):281.
15. Goering, H.W. *Plaste und Kautschuk*, **28**, No.4(1981):226.
16. Kettrup, A. *et al. Erdoel und Kohle, Erdgas, Petrochemie*, **35**, No.10(1982):475.
17. V'Yunov, K.A. *et al. Lakokrasochnye Materialy i Ikh Primenenie*, 6(1984):47.
18. Mukhtarova, M. *Khigiena i Zdraveopazvane*, **23**, No.2(1980):173.
19. Ciupe, R. *Revista de Chimie (Bucharest)*, **32**, No.6(1981):584.
20. Ciupe, R. *et al. Revue Roumaine de Chimie*, **30**, No.11-12(1985):1053.
21. Imamura, K. and T. Fujii. *Bunseki Kagaku*, **28**, No.9(1979):549.
22. Tsibul'skii, V.V. *et al. Zhurnal Analiticheskoi Khimii*, **34**, No.7(1979):1364.
23. Treister, T.E. *Gigiena i Sanitariya*, **49**, No.11(1984):52.
24. Bianchi, A. and G. Muccioli. *ICP (Industria Chimica e Petrolifera)*, **9**, No.4(1981):77.
25. Diaz-Rueda, J. *et al. Applied Spectroscopy*, **31**, No.4(1977):298.
26. Drugov, Y.S. and G.V. Murav'eva. *Zhurnal Analiticheskoi Khimii*, **37**, No.7(1982):1302.
27. Snell, J. *Pollution Engineering*, **14**, No.6(1982):36.
28. Dmitriev, M.T. and G.M. Kolesnikov. *Gigiena i Sanitariya*, **47**, No.2(1982):59.
29. Firsova, O.V. *et al. Khimicheskie Volokna*, 3(1980):56.

30. Ioffe, B.V. *et al. Journal of Chromatography*, 186(1979):851.
31. Vitenberg, A.G. and I.A. Tsibul'skaya. *Gigiena i Sanitariya*, **47**, No.2(1982):60.
32. Yasouka, T. *et al. Proceedings of the Faculty of Science of Tokai University*, 15(1979):81.
33. Brown, V.R. and D.J. Kroes. *Anal. Instrum.*, 15(1977):83.
34. Poltronieri, G. *Inquinamento*, **22**, No.3(1980):97.
35. Mierzwinski, A. and Z. Witkiewicz. *Chemia Analityczna (Warsaw)*, **30**, No.3(1985):429.
36. Ho, M.H. *et al. Analytical Chemistry*, **52**, No.9(1980):1489.
37. Ho, M.H. *et al. Analytical Chemistry*, **55**, No.11(1983):1830.
38. Ghimenti, G. *et al. Annali dell'Istituto Superiore di Sanita*, **14**, No.3(1978):583.
39. Cocheo, V. *et al. American Industrial Hygiene Association Journal*, **43**, No.12(1982):938.
40. Miyaura, S. *et al. Eisei Kagaku*, **29**, No.2(1983):83.
41. Anthony, R.M. *et al. Journal of Analytical Toxicology*, **2**, No.6(1978):262.
42. Bellanca, J.A. *et al. Journal of Analytical Toxicology*, **6**, (1982):238. 'Detection and quantitation of multiple volatile compounds in tissues by GC and GC/MS'.
43. Peterson, R.G. and J.V. Bruckner. *Journal of Chromatography*, **152**, No.1(1978):69.
44. Dworzanski, J.P. and M.T. Debowski. *Chemia Analityczna (Warsaw)*, **26**, No.2(1981):319.
45. Takahashi, S. and Y. Fukui. *Nippon Hoigaku Zasshi*, **33**, No.4(1979):352.
46. Van Roosmalen, P.B. and I. Drummond. *British Journal of Industrial Medicine*, 35(1978):56.
47. Niinuma, Y. *et al. Sangyo Igaku*, **24**, No.3(1982):322.
48. Poggi, G. *et al. International Archives of Occupational & Environmental Health*, **50**, No.1(1982):25.
49. Bieniek, G. and T. Wilczok. *British Journal of Industrial Medicine*, **38**, No.3(1981):304.
50. Sollenberg, J. and A. Baldesten. *Journal of Chromatography*, **132**, No.3(1977):469.
51. Lowenheim, F.A. and M.K. Moran. 'Faith, Keyes and Clark's Industrial Chemicals'. 4th ed. John Wiley & Sons, 1975.
52. Commission of the European Communities. 'Environmental Chemicals Data and Information Network (ECDIN)' Databank. Commission of the European Communities Joint Research Centre, ECDIN Group, I-21020 Ispra (Varese), Italy, 1986.
53. Windholz, M. *et al.*, Eds. 'The Merck Index'. 10th ed. Merck & Co., 1983.
54. The Royal Society of Chemistry. Laboratory Hazards Data Sheet No.15: 'Toluene', 1983.
55. International Labour Office. 'Encyclopaedia of Occupational Health and Safety'. 3rd rev. ed. ILO, 1983.
56. Sax, N.I., Ed. 'Hazardous Chemicals Information Annual', No.1. Van Nostrand Reinhold Information Services, New York, 1986.
57. National Institute for Occupational Safety & Health/Occupational Safety & Health Administration. 'NIOSH/OSHA Occupational health guideline for toluene'. DHHS (NIOSH) Publication. US Department of Health & Human Services/US Department of Labor, 1978.
58. Occupational Safety & Health Administration. Material safety data sheets from the Occupational Health Services database, OSHA, Washington D.C., 1986.
59. Bretherick, L., Ed. 'Hazards in the Chemical Laboratory'. 4th ed. Royal Society of Chemistry, UK, 1986.
60. National Fire Protection Association. 'Fire Protection Guide on Hazardous Materials'. 9th ed. NFPA, Massachusetts, USA, 1986.
61. Bretherick, L. 'Handbook of Reactive Chemical Hazards'. 3rd ed. Butterworths, 1985.
62. Warren, P.J., Ed. 'Dangerous Chemicals Emergency Spillage Guide'. 1st ed. Jensons (Scientific) Ltd., Leighton Buzzard, UK, 1985.
63. Dreisbach, R.R. 'Physical Properties of Chemical Compounds'. Advances in Chemistry Series No.15, American Chemical Society, 1955.
64. Weast, R.C., Ed.-in-Chief. 'CRC Handbook of Chemistry and Physics' ('The Rubber Handbook'). 67th ed. CRC Press, 1986-1987.
65. Horsley, L.H. 'Azeotropic Data III'. Advances in Chemistry Series No.116, American Chemical Society, 1973.
66. World Health Organization. 'Environmental Health Criteria 52: Toluene'. UNEP/ILO/WHO, Geneva, 1985.

67. American Conference of Governmental Industrial Hygienists. 'Documentation of the threshold limit values for chemical substances in the workroom environment'. ACGIH, 1986.

68. Ruth, J.H. *American Industrial Hygiene Association Journal,* 47(1986):A-142. 'Odor thresholds and irritation levels of several chemical substances: a review'.

69. Carlsson, A. *Arbete och Halsa,* 1982:3. 'Uptake, distribution and elimination of methylene chloride and toluene'.

70. Smyth, H.F. *et al. American Industrial Hygiene Association Journal,* 30(1969):470. 'Range-finding toxicity data: list VII'.

71. Smyth, H.F. *et al. American Industrial Hygiene Association Journal,* 23(1962):95. 'Range-finding toxicity data: list VI'.

72. Kimura, E.T. *et al. Toxicology & Applied Pharmacology,* 19(1971):699. 'Acute toxicity and limits of solvent residue for sixteen organic solvents'.

73. Bruckner, J.V. and R.G. Peterson. *Toxicology & Applied Pharmacology,* 61(1981):27. 'Evaluation of toluene and acetone inhalant abuse'.

74. Andersen, I. *et al. Scandinavian Journal of Work, Environment & Health,* **9,** No.5(1983):405. 'Human response to controlled levels of toluene in six-hour exposures'.

75. Tähti, H. *et al. International Archives of Occupational & Environmental Health,* **48,** No.1(1981):61. 'Chronic occupational exposure to toluene'.

76. Spencer, P.S. and H.H. Schaumburg. *Scandinavian Journal of Work, Environment & Health,* **11** Suppl., 1(1985):53. 'Organic solvent neurotoxicity'.

77. World Health Organization/Nordic Council of Ministers. Environmental Health Series 5(1985). 'Organic solvents and the central nervous system'. Report on a joint WHO/Nordic Council of Ministers Working Group.

78. Grasso, P. *et al. Food & Chemical Toxicology,* **22,** No.10(1984):819. 'Neurophysiological and psychological disorders and occupational exposure to organic solvents'.

79. Barlow, S.M. and F.M. Sullivan. 'Reproductive Hazards of Industrial Chemicals'. 2nd printing. Academic Press, 1984.

CONTENTS

1. CHEMICAL ABSTRACTS NAME

benzene, dimethyl-
(10th Collective Index)

2. SYNONYMS AND TRADE NAMES[1-6]

agrisolve 50
apco 467
benzene, dimethyl
dimethylbenzene
dimethylbenzene, isomer
dimethylbenzene (9CI)
ksylen (Polish)
methyltoluene
NCI-C55232
OHS25150
socal aquatic solvent 3501
U239
UN 1307
violet 3
xilene (Italian)
xiloli (Italian)
xilolo (Italian)
xyleen (Dutch)
xylele (German)
xylene
xylene isomer mixture
xylenes, mixed-
xylène (French)
xylenen (Dutch)
xylol
xylole (German)

3. CHEMICAL ABSTRACTS SERVICES REGISTRY NUMBER

1330-20-7	xylene
[95-47-6	*o*-xylene]
[106-42-3	*p*-xylene]
[108-38-3	*m*-xylene]

4. NIOSH NUMBER

ZE2100000	xylene
ZE2190000	xylene (mixed) [No equivalent CAS number]
[ZE2450000	*o*-xylene]
[ZE2625000	*p*-xylene]
[ZE2275000	*m*-xylene]

5. CHEMICAL FORMULA

C_8H_{10}
(Molecular weight 106.18)

6. STRUCTURAL FORMULA

7. OCCURRENCE[7-9]

Sandmeyer in ref. 7 states that xylene occurs in many petroleum products, in coal naphthas and as an impurity in petrochemicals such as benzene, toluene and similar materials. All three isomers have been identified among the volatile products in tobacco smoke and in the atmosphere, particularly if the latter is urban.[8]

ECETOC[8] states that total US emissions to the environment in 1978 were estimated as 480,000 tonnes. Gasoline appeared as the largest source of emission, with the other major source deriving from the use of xylene as a solvent. A calculation of distribution gave, air 99.1%, water 0.7%, soil 0.1%, sediment 0.1% and fish ≪ 0.01%. Measurements taken at different times following a crude oil spillage in the Gulf of Mexico demonstrated the rapid disappearance of xylenes from water under natural conditions. ECETOC[8] notes studies on the biodegradation of xylenes in the hydrosphere. Where *m*- and *p*-xylene were completely eliminated in seven days, *o*-xylene took 11 – 12 days. In each case the corresponding methylbenzyl alcohol was detected as an intermediate.

No effect on activated sludge was seen at concentrations of up to 300 mg/kg sludge/day. However, a concentration of 500 mg/l of any isomer was toxic to unacclimatized activated sludge during the first 24 hours of aeration.[8]

No specific studies on the breakdown of xylenes in the soil are available but certain micro-organisms have been shown to be capable of oxidizing xylenes.[8]

Xylene levels in air are variable, ranging from being barely detectable in Leningrad to a 266 ppb measurement in Hamburg. As the largest source of xylene is gasoline emissions, urban air tends to have higher concentrations.[8]

Xylene levels in water, varying from 6 μg/l to 1 mg/l, have been found in petroleum refinery effluents in various countries. Generally, surface waters contain very low levels except near fuel processing activities (*e.g.*, up to 830 ppb in one US fuel processing location). Concentrations of 2 and 9 ng/l of *m*- and *p*-xylene have been found in Los Angeles rain water.[8]

Various other sources of xylenes in the atmosphere given by Graedel[9] are fish oil manufacture, forest fires, various chemical manufactures, lacquer manufacture, turbines and vulcanization.

ECETOC[8] observes that studies have shown that xylenes are likely to disappear rapidly in the air by phototransformation with the hydroxyl radical but that they will react only slowly with ozone. The photoreaction products with OH. are formic and acetic acids which, after absorption in the hydrosphere, are further degraded to carbon dioxide and water.

8. COMMERCIAL AND INDUSTRIAL XYLENE[6,10-13]

Commercial xylene is an isomeric mixture also containing ethylbenzene except where the xylene has been obtained by toluene disproportionation.[6]

The composition is variable but the components are generally within the following ranges[10]:

o-xylene	18.0 – 24.4%
m-xylene	45.5 – 52.2%
p-xylene	17.1 – 20.3%
ethylbenzene	8.6 – 13.2%

These individual components exhibit many similar physical properties, *e.g.*, the density and boiling points of each are[6]:

	Density at 25°C	Boiling pt. (°C)
o-xylene	0.8802	144.41
m-xylene	0.8642	139.12
p-xylene	0.8610	138.37
ethylbenzene	0.8671	136.19

(See also Section 17.)

Unpurified xylene may contain thiophene and pseudocumene.[11] It may also contain benzene, but is not in general use in the UK if the benzene content is greater than 1%.

Ref. 12 notes standards for several grades of xylene, *viz*:

3° solvent xylene

British Standard Specification, BS 458/2:1963 for 3° solvent xylene requires: specific gravity 0.860 – 0.875; boiling range (5% – 95%) not to exceed 3.0 °C and to lie between 137.5 and 144.5 °C; neutral; free from mercaptans, H_2S, and undissolved water; flash point not below 22.8 °C; max. residue 10 mg per 100 ml; tests for free sulphur and colour. (National Benzole and Allied Products Association 10B:1960)

5° solvent xylene

British Standards Specification, BS 458/4:1963 for 5° solvent xylene requires: specific gravity 0.860 – 0.875; boiling range (5 – 95%) not to exceed 5.0 °C and to lie between 137.0 and 145.5 °C; neutral; free from mercaptans, H_2S and undissolved water; flash point not below 22.8 °C; max. residue 10 mg per 100 ml; tests for free sulphur and colour. (National Benzole and Allied Products Association 11A:1960).

ASTM specifications D 364-65 for industrial grade xylene requires: specific gravity 0.860 – 0.871; min. initial boiling point 123 °C, max. 5% at 130 °C, min. 90% at 145 °C, max. dry point 155 °C; acidity not more than 0.005% by weight; free of H_2S, SO_2, and undissolved water; tests for acid wash and corrosion.

Ref. 12 also notes that a purified xylene of high flash point [about 30 °C] is also available. Ref. 13 notes a nitration grade (boiling range 137.2 – 140.5 °C) and states that in some cases one or other of the industrial isomers are partially removed for use in chemical production. [All three isomers and ethylbenzene are used individually as industrial raw materials.]

9. SPECTROSCOPIC DATA[14]

Infrared, Raman, ultraviolet, ^{1}H NMR and mass spectral data have been tabulated for *o*-xylene and *p*-xylene.[14] Infrared, Raman, ultraviolet, both ^{1}H and ^{13}C NMR, and mass spectral data have been tabulated for *m*-xylene and ethylbenzene.[14]

10. MEASUREMENT TECHNIQUES[15-43]

The method recommended for the determination of xylene in workshop atmospheres is the NIOSH[15] activated charcoal trapping method for collection and concentration, followed by solvent extraction of the charcoal and a gas chromatographic (gc) analysis of the extract.

The collection tube is 7 cm long with a 4 mm internal diameter. It contains a total of 150 mg of activated charcoal (20 – 40 mesh) divided into a front section of 100 mg and a rear section of 50 mg separated by a small plug of urethane foam. Air is sampled by means of a small pump at a flow-rate of not more than 200 ml per minute to obtain an optimum sample of 12 l, with a maximum of 23 l. The entire apparatus is portable so that it may be carried in a pocket with the sampling tube in the breathing zone (normally a coat lapel). The apparatus can also be static.

Carbon disulphide is used to extract the xylene, which is separated on a 3 m × 2 mm glass column packed with 10% OV-275 on Chromosorb W-AW (100 – 120 mesh) using flame ionization detection.

The method is designed to separate the following aromatic hydrocarbons: xylene, toluene, benzene, cumene, styrene, α-methylstyrene, vinyltoluene, *p*-t-butyltoluene, ethylbenzene and naphthalene. For full details of the method see ref. 15.

The sampling method uses a small portable apparatus with no liquids and is therefore easy to handle and maintain. The analytical method is relatively specific with the additional advantage of flexibility of operating conditions.

Air samples are normally taken by means of suction using an electric pump capable of operating at low flow-rates (*i.e.*, 30 – 300 ml per minute). If higher flow-rates are used, breakthrough can occur (*i.e.*, the xylene is not completely adsorbed and some of it passes through the adsorbent). It is also convenient either to establish the trapping device in a static position, or to have it attached to the user in such a way that the breathing zone is sampled (this usually means attachment to the worker's lapel).

It should be noted that suitable personal samplers may be commercially available.

Most gc techniques more often than not separate the *o*-, *m*- and *p*-xylenes. This is often of academic interest only and the concentrations of each are added together to give the total xylene content. After metabolism in the body the corresponding isomeric methylhippuric acids are excreted in urine and there appears to be more interest in the separation of these acids.

The most popular trapping medium for xylene is activated charcoal. Wood-charcoal cigarette-filters (two 25 mg units in series) were used to trap xylene, toluene and benzene at a flow-rate of 2.5 ml per minute for eight days.[16] This trapping technique was later used, followed by carbon disulphide extraction and gc separation of the extract.[17] The 50 m capillary column was coated with UCON LB 55OX-33 and temperature-programmed from 20 to 120 °C. Flame ionization detection was used and xylene was measured in the range 1 – 210 µg per cubic metre of air.

Adsorption on to activated charcoal was found to be more efficient than that on to alumina, the adsorptive capacity of which was more prone to be affected by moisture.[18]

In an estimation of the efficiency of various solvents for extraction of trapped *p*-xylene on activated charcoal, carbon disulphide was found to be the best.[19]

In an examination of the workshop atmosphere during the use of solvents from chlorinated rubber and epoxy paint production, 20 l of air, containing xylene and other solvents, were trapped in a tube containing 1.5 g activated charcoal[20] The xylene was extracted with carbon disulphide and gc analysis was performed on a 3 m × 4 mm column packed with 10% OV-101 on Chromosorb W AW-DMCS (80 – 100 mesh). Temperature-programming was carried out between 80 and 130 °C and detection was by flame ionization, the detection limit being 25 mg per cubic metre of air.

The vented gases from a fume cupboard outlet in a laboratory were examined for xylene and other volatile organic compounds by trapping on activated charcoal, followed by carbon disulphide extraction, prior to analysis.[21]

Heptane has been used to extract xylene trapped on activated charcoal.[22] The subsequent gc separation was carried out on a column packed with poly(ethanediol adipate) on Porolith at 100 °C and using flame ionization detection.

Xylene and toluene have been contained on activated charcoal after the passage of air from a paint and varnish industrial workshop.[23] Pentane was used to extract the compounds and gc was used to separate them on a 50 m capillary column coated with PFMS-4. The column was temperature-programmed from 50 to 90 °C at 1 °C per minute and detection was by flame ionization.

Air volumes of 1 – 3 l were passed at a flow-rate of up to 500 ml per minute through a 7 cm × 4 mm tube containing 50 – 100 mg activated charcoal.[24] Dimethylformamide was used to extract the trapped xylene plus other organic compounds and the extract was subjected to gc on a 3 m × 4 mm column packed with 10% Carbowax 20M on Chromosorb W (80 – 100 mesh). The column temperature was 120 °C and detection was by flame ionization.

Workplace air was sampled for organic compounds, including xylene, by trapping on activated charcoal contained in a 20 cm × 4 mm tube.[25] After extraction, the extract was analysed by gc at 86 °C on a column packed with 10% dioctyl phthalate on Chromosorb W (80 – 100 mesh).

A 7.5 cm × 3 mm tube containing activated charcoal was used to trap xylene and other organic compounds from air, applying a flow-rate of 100 ml per minute.[26] The tube was then placed in a desorption chamber and heated to 190 °C and the headspace was swept on to a gc for analysis. A 5 m × 4 mm column packed with 10% SP-1000 on Chromosorb W AW-DMCS (100 – 120 mesh) was used for the analysis, the column temperature being programmed between 100 and 160 °C. Detection was made by flame ionization and the detection limit was 2 ppm.

Having trapped xylene on 50 – 200 mg activated charcoal from an air sample, it was extracted with toluene–chloroform.[27] After evaporation of the extract to give an approximate concentration of 0.1 mg per litre it was burnished on to a piece of acid-etched silver foil and subjected to secondary-ion mass spectrometry (ms). The detection limit for this novel method was approximately 2 ng xylene.

Activated charcoal is almost invariably used as the trapping medium for xylene from air. However, Tenax-GC has been used to trap xylene from 20 – 100 l of Parisian air.[28] The xylene was desorbed by heating to 250 °C and determined by gc-ms using a Porapak Q column.

Poly(phenylmethylsiloxane) on diatomite has been used as a trapping medium for xylene and toluene in air.[29] After heating, the headspace was analysed by gc on a 2 m × 3 mm column packed with 15% 1,2,3-tris-(2-cyanoethoxy)-propane on Chromaton N AW-DMCS. Flame ionization detection was used and the limit of detection for xylene was 0.01 mg per cubic metre.

Air samples were passed at flow-rates of 350 – 450 ml per minute through 80% acetic acid solution.[30] Potassium hydroxide solution (40%) was added for neutralization and the mixture heated to 25 °C and the headspace taken for gc analysis of xylene and toluene. The column was 1.25 m × 3 mm, packed with 10% nitrilotripropionitrile on Chromosorb P (60 – 80 mesh) and it was operated at 80 °C with flame ionization detection.

After a similar trapping method, the headspace was separated on a 20% Apiezon L on Chromosorb W column, giving a detection limit for xylene of 8 µg per cubic metre of air.[31]

Xylene and toluene in workshop air from polythene production was sampled in a 2 l gas burette which was subsequently heated to 40 °C and the headspace analysed by gc.[32] The 6 m × 4 mm column was packed with 10% Carbowax 20M – terephthalic acid on Chromosorb P (60 – 80 mesh) and was operated at 120 °C and using flame ionization detection. The detection limit for xylene was of the order of 1 ppm.

In another method, air containing xylene and toluene was sampled for 10 – 15 minutes at a flow-rate of 100 – 200 ml per minute through water contained in a glass ampoule.[33] The ampoule was sealed and ultimately broken open in the laboratory and the contents were analysed by gc on a 3 m × 2 mm column packed with 20% poly(ethanediol adipate) on Celite C-22. The column was operated at 100 °C and detection was by flame ionization, the optimum range for xylene being 1 – 50 mg per cubic metre of air.

If trapping and concentration is to be avoided, the analytical technique must be ultra-sensitive. In this connection, air contaminated with xylene and toluene was analysed directly by gc-ms using $m/e = 91$.[34] The limits of detection for *o*-, *m*-, and *p*-xylenes were 4, 3.5 and 3 ppb, respectively.

Xylene may also be determined on a 2 m × 4 mm column packed with 1.18% β-cyclodextrin supported on Celite.[35] The column was operated at 60 °C and detection was by flame ionization.

Solid state electrolytic sensors have been used to determine the concentration of pollution gases, including xylene, in air.[36] The sensors each consist of two electrodes to which a potential of 1.3 – 2.5 V is applied. The sensitivity of the method is of the order of a few ppm.

There has been considerable interest in the fate of xylene in the bodies of personnel exposed to its vapour, particularly in its metabolism to methylhippuric acid which is excreted in the urine. The metabolites were extracted from acidified urine with ethyl acetate, benzoylproline as internal standard was added to the extract and methylation carried out using diazomethane.[37] The methyl esters were separated by gc on a 2 m × 2 mm column packed with 2% OV-225, operated at 210 °C and using N-P thermionic detection.

In a similar method, the urine of workers exposed to xylene and toluene vapour was acidified, sodium chloride added and the metabolites methylated after the addition of heptadecanoic acid as internal standard.[38] The methyl esters of the methylhippuric acids were separated from the other methyl esters by gc on a 1.5 m × 2 mm column packed with 3% OV-101 on diatomite Q AW-DMCS (100 – 120 mesh). The column was operated at 170 °C and detection was by flame ionization.

Heptadecanoic acid was also added to urine as an internal standard in the determination of 3- and 4-methylhippuric acids.[39] Ammonium sulphate plus 5% sulphuric acid were also added prior to extraction of the urine with ethyl acetate and the extract was trimethylsilylated and the derivatives were separated by gc on a 1.2 m × 2 mm column packed with 3% OV-1 on Gas Chrom Q (80 – 100 mesh). The column was temperature-programmed up to 215 °C and detection was by flame ionization. The limit of detection was 4 mg per litre for each compound.

In the determination of the methylhippuric acids in the urine of workers exposed to xylene, this was carried out either directly or after extraction with ethyl acetate.[40]

The analysis was carried out by high-performance liquid chromatography (hplc) at 40 °C on a 15 cm × 4 mm column packed with Li Chrosorb RP-18 (5 μm) using water–methanol–acetic acid (40:10:0.1) as mobile phase and with detection at 272.4 nm.

Butyl chloride–isopropanol was used to extract metabolites from urine of workers exposed to xylene and toluene.[41] The extract containing the metabolites, including hippuric acid, *m*-methylhippuric acid, phenylglyoxylic acid and mandelic acid, was evaporated to dryness and the residue dissolved in water–acetonitrile. After the addition of 4-hydroxybenzoic acid as internal standard, this solution was subjected to hplc on a 25 cm × 2.6 mm column packed with ODS SIL X, using detection at 225 nm.

3-Methylhippuric acid in urine from workers exposed to *m*-xylene was extracted with chloroform prior to thin layer chromatographic analysis.[42] This was carried out on silica gel using chloroform–water–acetic acid (4:1:1) as mobile phase and 4-dimethylaminobenzaldehyde solution in acetic anhydride as locating agent. The detection limit was 6 μg per ml.

In an examination of general body tissue for xylene and toluene residues, brain, liver, kidney, lung tissue and blood samples were sealed in phials and heated to 65 °C for 40 minutes before releasing the headspace for gc analysis.[43] This was carried out on a 50 m capillary column coated with OV-101 with temperature-programming from 40 to 250 °C at 12 °C per minute. Detection was by flame ionization or mass spectrometry, 0.1 – 15.8 ppm being the range of residues found.

11. CONDITIONS UNDER WHICH XYLENE IS PUT ON THE MARKET

Production[44]

Ransley in ref. 44 observes that since World War II, the main source of xylenes has been from the reforming of petroleum fractions. Coal had been an earlier source and could become so again where a bountiful supply may be developed for fuel and petrochemical processes.

As with toluene, xylene is obtained from certain petroleum fractions by catalytic reforming. The xylene-containing fraction separated from the reformate may be used as such or it may be further concentrated by distillation before being used either as the isomeric mixture, or to undergo further treatment to separate the isomers in whole or in part.

Ransley in ref. 44 gives a general scheme for a plant producing xylene (and, incidentally, benzene by distillation and dealkylation of toluene formed during the process). The 65 – 175 °C cut from a straight-run petroleum fraction or from an isocracker is used as feed to a reformer. Heart cutting and extraction units are followed by a toluene recovery unit, after which, if required, *o*-xylene (and higher benzene homologues) are separated. *p*-Xylene is next separated, following which the raffinate or mother liquor is passed to an isomerization unit where an acid catalyst (such as the Friedel–Crafts type or silica–alumina, both amorphous and zeolitic) enables a fresh isomeric equilibrium mixture to be achieved which is then recycled to the *o*- and *p*-isomer separation stages.

This makes for economic viability whereas a once-through procedure for recovery of xylenes is generally not economic due to the considerable pre-treatment required before the xylene isomer separation stages are reached. Eventually though, impurities or diluents accumulate. Ethylbenzene, if present, will accumulate unless rejected by distillation (but in most processes this is not done because of high energy costs) and so pass eventually to the isomerization process. In this the ethylbenzene can transform to some extent, and its ability to isomerize to xylenes and also to disproportionate and undergo hydrogenolysis can be used to ensure that its concentration in the feed back to the xylene isomers separation unit is kept to a minimum. In fact, disproportionation is difficult to suppress and occurs in all isomerization processes and such conversion to xylenes can help offset any loss of xylenes due to side reactions.[44]

Ransley in ref. 44 states that an efficient commercial method to separate *m*-xylene, and which also greatly simplifies and reduces the cost of separating the other C_8 isomers is the Mitsubishi Gas-Chemical Company (MGCC) process. In this the C_8-aromatic mixture is treated with HF–BF_3 in which *m*-xylene selectively dissolves to form a 1:1 molecular complex, *m*-xylene–HBF_4. From this complex *m*-xylene can be recovered or, if isomerized by heating to below 100 °C, can yield an equilibrium mixture favouring *p*-xylene over the *o*-isomer.

Xylene is transported in rail and road tankers, drums and glass bottles.

Uses[6.8.10 – 12.44.45]

Ref. 10 observes that the largest outlet for xylene at present is gasoline octane improvement. This situation will probably continue because the xylene-rich fraction remaining after benzene and toluene are removed from reformates is returned to the gasoline blending pool.

Considerable quantities of mixed xylenes also find use as solvents. Ransley in ref. 44 observes that the use of mixed xylenes as solvents in the paint and coatings industry is likely to decline, particularly in the US, as efforts are increased to reduce hydrocarbon emissions into the atmosphere. The paint industry is moving more and more towards water-based latex paints for environmental reasons and ease both in use and in clearing up.

Ransley in ref. 6 notes that, in the US, the uses to which mixed-xylenes production is put are *p*-xylene production (50 – 60%), gasoline blending (10 – 25%), *o*-xylene production (10 – 15%), solvents (10%), ethylbenzene (3%), *m*-xylene production (1%).

Digests of xylene production figures for various countries are not always specific as to whether total xylenes (including separated isomers) or mixed xylenes alone are being referred to. Ref. 44 gives US production of mixed xylenes as 3,877 kilotons p.a. in 1979 and 3,713 kilotons in 1980. Ref. 45 states that US mixed xylene production in the first half of 1985 came to 342.9 million gallons [approx. 1,130 kilotonnes], this being down 25% from the previous year. The approximate world production of *o*-, *p*- and mixed xylenes in 1984 was 15,431 kilotonnes.[8]

Xylene is a solvent for ethyl and benzyl cellulose, ester gum, copal ester, benzyl abietate, rubber, gutta percha resin, methyl methacrylate, polyisobutylene, rubber chloride dammar, elemi, coumarone, colophony, fluid silicones, melamine–formaldehyde and urea–formaldehyde resins, castor, linseed and other oils. It attacks polyethylene and PVC but not nylon. As xylene is a good solvent for paraffin, Canada balsam and polystyrene, it is used in histology. Of the individual isomers, *o*-xylene is used in the manufacture of phthalic anhydride, a precursor for phthalate esters; *p*-xylene is used to produce terephthalic acid and dimethyl terephthalate, intermediates in the production of polyester fibre; *m*-xylene is used to produce isophthalic acid, an intermediate in the production of polyester resins; ethylbenzene is used to produce styrene [8,11,12,44]

12. STORAGE, HANDLING AND USE PRECAUTIONS[3,46–48]

Mixed xylenes form a flammable, non water-miscible liquid boiling over a range of about 137 – 145 °C and having a vapour density about 3.7 times that of air.

Care must be taken with storage, process and drying plant to ensure that hazard-free conditions exist regarding fire and explosion risks and exposure of personnel to xylene. The material is not corrosive to metals. Iron, mild steel, copper or aluminium are suitable for plant and containers. Small quantities for laboratory use may be kept in glass containers.[46,47] Care should be exercised with valves and gaskets as xylene has a deleterious action on many forms of rubber.[46]

[As with other flammable solvents the organization of storage facilities has to take into account two separate requirements. These are that the stored material be protected from a fire elsewhere on the premises and that the premises be protected from a fire involving the stored materials. On an industrial scale, sufficient physical separation and water spray curtains are often used effectively. On a laboratory scale, flammable solvents need to be stored in fire resistant, thermally insulated cupboards.

It is important to appreciate that a non-combustible barrier is not synonymous with a fire barrier. A fire barrier has to protect against both conducted and radiant heat. For example, a metal cupboard is useless on both these counts.]

Storage tanks in the open should be bunded. Drums should be stored in a well-ventilated area away from sources of ignition, heat and sun. There is no limitation on storage period but the maximum storage temperature should not exceed 40 °C. In storerooms there should be a ramped sill at the doorways to prevent egress of spilled material. Such storerooms should be well-ventilated, flammable liquids stores, free from sources of ignition. Xylene should be stored away from chemicals with which it is incompatible (see Section 14). The construction and siting of process buildings and plant should be such as to prevent the spread of escaping liquid throughout the building or site. Large items of equipment should be earthed and tools or equipment used should be of the non-sparking/flameproof type.

OSHA[48] recommends that engineering controls such as process enclosure, general dilution ventilation, local exhaust ventilation and the use of personal protective equipment should be used, as appropriate, to keep personnel exposures at or below permissible limits. Repeated or prolonged skin contact with liquid xylene should be prevented by requiring personnel to use, as appropriate, impervious clothing, gloves, splash-proof safety goggles, face shields (eight inch minimum) and such other protective clothing as may be necessary. Adequate emergency eye-wash facilities should be available as should facilities for washing xylene from the skin with soap or mild detergent and water.

[Personnel should not be allowed to enter areas containing potentially hazardous concentrations of xylene unless all relevant legislation and official guidelines have been strictly adhered to. Suitable respiratory protective equipment should be used.] Such exposed personnel should be able to communicate with each other and with those (*i.e.*, more than one person) monitoring the work from outside the hazardous area. Monitors should be suitably equipped to effect rescue (possibly with lifelines to exposed personnel) and should themselves ensure, so far as is possible, that one of their number remains outside the hazard area in the event of an emergency.

Xylene vapour causes narcosis and is irritant to the respiratory tract. A concentration of 10,000 ppm is immediately dangerous to life and health.[3] The liquid is a skin irritant if not removed promptly. (See also Section 18.1 Toxicity. *General*.)

Note that a relatively small content of benzene impurity in the liquid may result in a hazardous concentration in the vapour due to benzene's higher vapour pressure (see Section 18.1).

N.B. the caveat in Section 13 (final paragraph).

13. FIRE HAZARDS[3,49,50]

Xylene is a flammable liquid boiling over a range of about 137 – 145 °C. It is immiscible with water and has a specific gravity of about 0.87. The flash point (closed cup) of the isomers range from 27 – 32 °C and the auto-ignition temperatures from about 463 – 528 °C. The lower and upper explosive limits in air are 1% and 7% v/v.[50] The vapour density is about 3.7 times that of air (see Section 17). The vapour, if in sufficient quantity, may therefore flow a considerable distance to a source of ignition and then 'flash back'.

Containers of xylene can explode when heated and, if feasible, should be cooled with flooding quantities of water applied from as far as possible until well after the fire is extinguished.

Small fires may be fought with foam, dry powder, carbon dioxide or vaporizing liquid. For larger fires, foam, carbon dioxide or dry chemical can be used. Water spray may be used to reduce the rate of burning but should be applied carefully over the higher boiling liquid in order to avoid excessive frothing. Jets of water will result in spreading the immiscible and less dense burning liquid. Immediate withdrawal is imperative in the case of rising sound from a venting safety device or any discolouration of a storage tank due to the fire. Personnel should keep upwind of the fire to avoid breathing any toxic vapours. Combustion products may include carbon monoxide as well as acrid smoke and irritating fumes.

Ref. 50 notes that where a fire is fed by an uncontrolled flow of combustible liquid the decision on how or if to fight it will depend on the size and type of fire anticipated and must be carefully considered. This may call for special engineering judgement, particularly in large-scale applications.

[Notwithstanding the foregoing review of various authorities noted in Sections 12 and 13, it is essential that managers and others responsible for the planning, implementation and overseeing of personnel and plant safety should be familiar with the legal constraints and official guidelines applicable to them and that they liaise with their local emergency services in the planning of plant and storage facilities and in the preparation of contingency plans for dealing with fires and other emergencies. Managers should regularly monitor staff knowledge of, and ability to implement, emergency procedures and should ensure that equipment provided for use in emergencies is regularly inspected and maintained.]

14. HAZARDOUS REACTIONS[3,51]

OSHA[3] notes that an intense reaction occurs with nitric acid in the presence of sulphuric acid and that with oxidizing agents generally there is the possibility of hazardous reaction.

Bretherick[51] notes that an attempt to chlorinate xylene with 'dichlorohydantoin' (1,3-dichloro-5,5-dimethyl-2,4-imidazolidindione) caused a violent explosion. The haloimide undergoes immediate self-accelerating decomposition in the presence of solvents, but safe conditions (including lower temperatures and progressive addition of reagent to match its consumption) can be developed for its use.

Bretherick[51] also notes with regard to *p*-xylene that, in its liquid-phase aerobic oxidation in acetic acid to produce terephthalic acid, it is important to eliminate the inherent hazards of this fuel–air mixture. Bretherick further notes that the effects of temperature, pressure and presence of steam on the explosive limits of the mixture have been studied.

15. EMERGENCY MEASURES IN THE CASE OF ACCIDENTAL SPILLAGE[3,52]

Only necessary personnel should be allowed within the hazard area. Suitable apparel giving protection against the inhalation of high vapour concentrations, against excessive skin contact and against eye splashes must be used (see Section 12). All ignition sources should be eliminated and any leak stopped if this can be done without risk. Water spray may be used to reduce vapours.[3]

For spillages up to approximately 25 l, the liquid may be absorbed onto sand or vermiculite, followed by transfer in suitable containers [*e.g.*, covered buckets] to a safe open area where atmospheric evaporation can take place. The site of the spillage should be washed thoroughly with water and biodegradable detergent.[52] Washings should be discharged to a sewer and not to a 'soakaway' nor a land drain.

Warren[52] notes that "*ideally* all hydrocarbons and related flammable organic chemicals should be burned in an incinerator with an afterburner."

OSHA[3] recommends that larger or massive ground spills should be bunded far ahead of the spill pending later disposal. Bunding may be achieved using soil, sandbags or barriers such as polyurethane or concrete. The spill may be immobilized with universal gelling agent and the vapour and fire hazard can be reduced with fluorocarbon water foam. Where an environmental hazard is caused, the relevant authorities, including the water authority, must be informed.

In the case of spillage onto water, natural barriers or oil spill control booms can be used to limit spill motion and its dispersion. The spilled material can be thickened by applying detergents, soaps, alcohols or other surface active agents. The application of a universal gelling agent to immobilize the trapped spill will increase the efficiency of removal.[3]

16. FIRST AID

General

The casualty should be removed from the danger of further exposure into the fresh air. Rescuers should ensure their own safety from inhalation and skin contamination where necessary. Conscious casualties should be asked for information about what has happened. Bear in mind that the casualty may lose consciousness at any time. Where necessary continued observation and care should be ensured. First aiders should take care not to become contaminated. Note that xylene is flammable.

Inhalation

The conscious casualty should be kept at rest. The unconscious casualty should be placed in the recovery position and an open airway maintained. If breathing or heart beat stops, resuscitation should be commenced immediately. Medical aid should be obtained or the casualty removed to hospital immediately.

Skin contamination

Remove contaminated clothing. Wash with copious amounts of water, or soap and water, for at least 15 minutes. If necessary, seek medical aid. In the case of persistent skin irritation refer for medical advice.

Eye contamination

Irrigate with copious amounts of water for at least ten minutes but avoid further damage to the eye from the use of excessive force. Seek medical aid. Remember that contact lenses may be worn and that these may be affected by some solvents and may impede decontamination of the eye.

Ingestion

If the lips or mouth are contaminated rinse thoroughly with water. Do not induce vomiting. Give further supportive treatment as for inhalation. In the case of ingestion of significant amounts wash out the mouth with water. Obtain immediate medical attention. Contact a hospital or poisons centre at once for advice. Remember that symptoms may develop many hours after exposure so continued care and observation may be necessary. Aspiration of xylene after vomiting may cause severe respiratory irritation.

In all cases note information on the nature of exposure and give this to ambulance or medical personnel. Information for doctors is provided in some technical literature issued by manufacturers.

17. PHYSICO-CHEMICAL PROPERTIES

[Information on the major impurity, ethylbenzene, is included where relevant.]

17.1 General[46,53-55]

Commercially available xylene is a mixture of three isomers: *ortho-*, *meta-* and *para*-xylene, with *meta*-xylene predominating.[53] Ethylbenzene is nearly always present as an impurity. The xylene isomers and ethylbenzene are colourless, flammable liquids with a similar characteristic odour, and generally similar physical properties.[46,53] Solid *p*-xylene (melting point 13.3 °C) exists as colourless plates or monoclinic prisms.[53,54]

All three xylene isomers and ethylbenzene are infinitely soluble in acetone, benzene, carbon tetrachloride, ethanol, ether and n-heptane,[55] and practically insoluble in water.[53]

17.2 Melting Point[54]

Melting points of the xylene isomers and ethylbenzene are as follows[54]:

o-Xylene	−25.2 °C
m-Xylene	−47.9 °C
p-Xylene	13.3 °C
Ethylbenzene	−95 °C

17.3 Boiling Point[54,56]

The boiling points of *o*-, *m*- and *p*-xylene and ethylbenzene are given by Weast[54] as follows:

o-Xylene	144.4 °C
m-Xylene	139.1 °C
p-Xylene	138.3 °C
Ethylbenzene	136.2 °C

Xylenes and ethylbenzene form azeotropes with a number of compounds. Details of binary ones formed with formic acid and iso-amyl alcohol are set out in the table below[56]:

First component	Wt% first component	Second component	Wt% second component	Boiling pt azeotrope ($°C$)
o-Xylene	26	Formic acid	74	95.5
m-Xylene	28.2	Formic acid	71.8	92.8
p-Xylene	30.0	Formic acid	70.0	~95
Ethylbenzene	32	Formic acid	68	~94.0
o-Xylene	<48	iso-Amyl alcohol	>52	127
m-Xylene	48	iso-Amyl alcohol	52	125 – 126
p-Xylene	48	iso-Amyl alcohol	52	125 – 126
Ethylbenzene	51	iso-Amyl alcohol	49	125.9

17.4 Density/Specific Gravity[53–55]

The specific gravities of the xylene isomers and ethylbenzene at 20 °C (referred to water at 4 °C) are[53,54]:

o-Xylene	0.8801
m-Xylene	0.8642
p-Xylene	0.86104
Ethylbenzene	0.8670

(See also Section 8).
Critical densities and critical volumes are as follows[55]:

	Critical density in g/ml	Critical volume in ml/g
o-Xylene	0.28	3.58
m-Xylene	0.27	3.67
p-Xylene	0.29	3.48
Ethylbenzene	0.29	3.448

17.5 Vapour Pressure[54]

The temperatures corresponding with selected vapour pressures of xylene and ethylbenzene below atmospheric pressure are as follows[54]:

Pressure:	1 mm	10 mm	40 mm	100 mm	400 mm
o-Xylene	−3.8 °C	32.1 °C	59.5 °C	81.3 °C	121.7 °C
m-Xylene	−6.9 °C	28.3 °C	55.3 °C	76.8 °C	116.7 °C

Pressure:	1 mm	10 mm	40 mm	100 mm	400 mm
p-Xylene	−8.1 °C	27.3 °C	54.4 °C	75.9 °C	115.9 °C
Ethylbenzene	−9.8 °C	25.9 °C	52.8 °C	74.1 °C	113.8 °C

The following table shows critical temperatures and pressures[54]:

	Critical temperature in °C	Critical pressure in atm
o-Xylene	359	35.7
m-Xylene	346	34.7
p-Xylene	345	33.9
Ethylbenzene	343.9	36.9

17.6 Vapour Density[50]

The vapour density for all three xylene isomers and ethylbenzene is reported as:

3.7 (relative to air = 1)[50]

17.7 Flash Point[46,50]

The flash points of the three xylene isomers and of ethylbenzene are reported as follows[50]:

o-Xylene	32 °C, closed cup
m-Xylene	27 °C, closed cup
p-Xylene	27 °C, closed cup
Ethylbenzene	21 °C, closed cup

Marsden and Mann[46] note closed cup flash points of xylene mixtures as ranging from 24 – 29.5 °C.

17.8 Explosive Limits[50]

The limits of flammability of the three xylene isomers and ethylbenzene in air at normal atmospheric temperature and pressure are given by the National Fire Protection Association as follows[50]:

	Lower limit	Upper limit
o-Xylene	1.0% v/v	7.0% v/v
m-Xylene	1.1% v/v	7.0% v/v
p-Xylene	1.1% v/v	7.0% v/v
Ethylbenzene	1.0% v/v	6.7% v/v

The NFPA[50] also give the ignition temperatures as follows:

o-Xylene	463 °C
m-Xylene	527 °C
p-Xylene	528 °C
Ethylbenzene	432 °C

However, they point out that different test conditions (and also different definitions of "ignition temperature") can result in widely varying values being quoted. They therefore recommend that any value should be treated as an approximation.

17.9 Viscosity[46,54,55]

The viscosity of liquid *o*-, *m*- and *p*-xylene and ethylbenzene at selected temperatures is given by Weast[54]:

	Temperature in °C	*Viscosity in cp*
o-Xylene	0	1.105
	16	0.876
	20	0.810
	40	0.627
m-Xylene	0	0.806
	15	0.650
	20	0.620
	40	0.497
p-Xylene	16	0.696
	20	0.648
	40	0.513
Ethylbenzene	17	0.691

The viscosity of commercial mixed isomers is quoted as 0.65 cp at 20 °C.[46]

The kinematic viscosity of the three xylene isomers and ethylbenzene is given by Dreisbach[55] as follows:

	Temperature in °C	*Viscosity in cSt*
o-Xylene	20	0.919
	40	0.724
	60	0.592
	80	0.497

	Temperature in °C	Viscosity in cSt
m-Xylene	20	0.714
	40	0.581
	60	0.488
	80	0.419
p-Xylene	20	0.748
	40	0.602
	60	0.502
	80	0.428
Ethylbenzene	20	0.7823
	40	0.6305
	60	0.525
	80	0.447

17.10 Concentration Conversion Factors

At 25 °C and 760 Torr (1 atm),

1 ppm = 4.34 mg/m³, and
1 mg/m³ = 0.23 ppm.

(These conversion factors apply to both xylene and ethylbenzene.)

18. TOXICITY

18.1 General[7,48,57]

Sandmeyer in ref. 7 observes that whereas most experimental work indicates that xylene is more toxic than toluene, some work contradicts this. Sandmeyer also observes that the possible presence of benzene as an impurity may render earlier studies unreliable.

In assessing the toxicity of xylene to humans note should be taken of a possible benzene content particularly when dealing with the vapour. Schulz in ref. 57 observes that a quite small benzene content can be a hazard. At 26 °C the vapour pressure of benzene is approximately 100 mmHg while at that temperature the vapour pressures of the xylene isomers are 10 mmHg or less. Thus the vapour that is in equilibrium with a liquid mixture of benzene and xylenes at this temperature has a benzene content at least ten times that existing in the liquid mixture itself. Where the TWA limit for benzene is 10 ppm, xylene containing 1% benzene could be regulated by the benzene standard as well as by whatever the xylene standard might dictate. [In the UK, xylene containing more than 1% benzene is not in general use.]

Xylene can affect the body by inhalation, ingestion or by contact with the eyes or skin. It may enter the body through the skin.[48]

Xylene can have various effects on the liver, kidneys and gastrointestinal tract. Another major problem with xylene toxicity is its narcotic effect, causing symptoms such as muscular weakness, lack of coordination and mental confusion. A death from pulmonary pneumonia resulted from a worker being overcome by an estimated 10,000 ppm vapour in air and remaining undiscovered for 18.5 hours. Xylene has an irritant effect on the skin and mucous membranes. Brief exposure of humans to 200 ppm in air has caused irritation to eyes, nose and throat. If not removed promptly, liquid xylene on the skin can cause erythema, dryness, defatting and, on very prolonged contact, may cause the formation of vesicles. The liquid has caused conjunctivitis and corneal burns by direct eye contact.

Several studies are noted in the review literature as having been obscured by lack of experimental detail and/or controls.

18.2 Toxicokinetics[5,7,8,58 – 63]

Jori *et al.*[5] observe that the kinetics and metabolism of the three xylene isomers are sufficiently established in man and in animals, particularly the rat and the rabbit, and that there are only minor differences in their patterns.

ECETOC[8] notes that, as with many other lipophilic solvents, xylenes are absorbed into the blood on inhalation of the vapour, ingestion of the liquid, or contact with the intact skin.

m-Xylene vapour absorption through the skin was estimated at 0.006 μmole/cm^2/hour when exposure was to 600 ppm, while more than three times normal absorption was experienced by one subject with 'atopic dermatitis'.[8]

ECETOC[58] notes that vapour of the impurity, ethylbenzene, is not absorbed through the skin but the liquid is very easily absorbed. The vapour is well absorbed when inhaled.

Percutaneous *m*-xylene absorption following immersion of both hands in the liquid has been variously reported as between 2 and 2.45 μg/cm^2/min, with the total absorption after 15 minutes being roughly the same as the total pulmonary absorption from inhalation of 100 ppm for 15 minutes.[8]

Sandmeyer in ref. 7 observes that the human system readily absorbs ingested xylene as is evident from accidental ingestions.

In experiments on the pulmonary uptake of technical xylenes (40.4% ethylbenzene, 49.4% *m*-xylene, 8.8% *o*-xylene and 1.4% *p*- xylene), male subjects were exposed to either 200 or 100 ppm in air, while at rest for 30 minutes and during light work for 90 minutes. In both experiments it was estimated that uptake by the lungs was about 60% of the amount inhaled and that the amount of unchanged xylene expired was 4 – 5% of the retained dose. Similar results have been reported for *m*-xylene alone.[8] The blood-gas partition ratio for xylene has been variously calculated as 29 to 1 and 42 to 1.

Like the other alkylbenzenes, xylene is distributed rapidly into all tissues, especially into the adrenal glands, bone marrow, brain, spleen and adipose tissue.[59] Jori *et al.*[5] note that most tissues reach equilibrium with blood within a few hours of exposure but that as *m*-xylene is actively transformed by the parenchymal organs (which thus remove it from the blood), the pulmonary retention remains constant at around 60% throughout the exposure time. *m*-Xylene in the blood is mainly protein-bound; little enters the blood cells.

Jori *et al.*[5] observe that uptake by, like elimination from, tissues takes only a few minutes for parenchymal organs, a few hours for muscle and several days for adipose tissue. Elimination from fat takes about 58 hours.

In spite of the distribution coefficient of *m*-xylene between fat and air being about 3,600, it has been calculated that only 4% of total pulmonary uptake is distributed to adipose tissue. This low percentage is a consequence of active metabolism removing *m*-xylene from the blood.

Animal studies on the absorption and distribution of xylenes include one by Fabre *et al.*[60] in which the following xylene levels were reported in the organs of rabbits exposed to 3 mg/l in air, eight hours/day, six days/week for 130 days; adrenal 148, bone marrow 130, spleen 115, brain 100, blood 91, kidneys 86, liver 75, heart 60, lungs 50, thyroid gland 19 and hypophysis 10.5 (µg/g). ECETOC[8] notes a study in which investigation by low-temperature whole-body autoradiography following ten minute inhalational exposure of mice to ^{14}C *m*-xylene (10 µl containing 10 µCi) found that, immediately following the exposure, high levels of 'volatile' radioactivity were found in the body fat, bone marrow, brain (white matter), spinal cord and nerves, liver and kidney. It persisted in nerve tissue for up to one hour and in body fat for up to eight hours following inhalation. Xylene metabolites ('non-volatile' radioactivity) appeared rapidly in the blood, lung, liver and kidney. Biliary excretion was inferred from the increasing amount of non-volatile radioactivity in the intestinal contents two to eight hours after exposure.

In a human study, xylenes were identified along with other low molecular weight volatile organic materials before or following childbirth in both the maternal and cord blood of 11 female patients whose only known exposure was to traces in the environment or catheter tubing.[61] The actual xylene levels were not given in the study but it is evident from this that xylene can cross the placenta. Barlow and Sullivan[62] state that xylene crosses the placenta in mice and may be metabolized in the foetus.

In man, over 95% of xylene retained by inhalation is excreted as methylhippuric acid and 1 – 2% as xylenols.[8] ECETOC[8] note that the excretion of methylhippuric acids in urine has been found to be proportional to the concentration of xylenes in the air and the duration of exposure provided all-shift or 24-hour urine samples were used. Samples taken over a period shorter than the actual exposure can be less reliable because of the rapid excretion of methylhippuric acids. It appears that, in man, none or at most very little material appears as metabolite toluic acid glucuronide.[5]

Two metabolic pathways are proposed for the xylenes. One is oxidation of one methyl group to carboxy *via* the alcohol and aldehyde and involving alcohol dehydrogenase. The toluic acid thus formed then conjugates with glycine to form methylhippuric acid (toluric acid) or, in animals, may conjugate with glucuronic acid. The other pathway involves the formation of xylenols and involves the action of mixed oxidase function.[5]

ECETOC[8] notes a suggestion that, in man, ethylbenzene impurity is preferentially oxidized at the aromatic nucleus thereby competitively inhibiting the oxidation of the aromatic ring in xylene. Another study, however, reported that the simultaneous exposure of man to ethylbenzene and *m*-xylene resulted in a reduction in the 24 hour urinary excretion of the metabolites of both materials compared with the metabolite excretion following separate exposures. In both combined and separate exposures the profile of urinary metabolites of both materials was the same. Also, ECETOC[58] note studies wherein the main metabolites in man exposed to ethylbenzene by inhalation were found to be mandelic acid and phenylglyoxylic acid, the metabolic conversion proceeding mainly through side chain oxidation whereas ring oxidation was of minor quantitative importance. One study found that the urinary metabolites of ethylbenzene in exposed volunteers were mandelic acid (64% of retained ethylbenzene), phenylglyoxylic acid (25%), and 1-phenylethanol (5%).

With xylene metabolism, glycine conjugation of toluic acid may be the rate-limiting step. Riihimäki *et al.*[63] point out that by analogy with the body's limited ability to conjugate benzoic acid with glycine (limited by the mobilization of endogenous glycine), it may be presumed that the body's capacity for conjugating toluic acid with glycine is similar. If it is assumed that the maximum rate of glycine mobilization of about 190 µmol/min is the limiting step in xylene metabolism, it can be calculated that the capacity is fully occupied in a sedentary man whose pulmonary ventilation is 10 l/min with 60% retention and inhaling xylene at 780 ppm. This suggests that man's metabolic capacity for xylene is not likely to be challenged under normal work conditions. Saturation of metabolism might occur though under conditions of heavy work at 270 ppm.[8,63]

In animals, metabolism of xylenes also is principally by side chain oxidation to toluic acids. Small amounts of xylenols have been found in the urine of rats administered *o*-, *m*- and *p*-xylenes, 2-methylbenzyl alcohol has been found in small amounts in the urine of rats administered *o*-xylene. 3-Methylbenzyl alcohol was tentatively identified when the treatment was with *m*-xylene but no 4-methylbenzyl alcohol was found when the treatment was with *p*-xylene.[8]

A decreased hepatic glutathione concentration was reported in rats administered xylenes by intraperitoneal injection. *o*-Xylene was the most effective, causing an approximately 75% reduction.[8] Urinary thioether excretion was also enhanced, being most pronounced following *o*-xylene administration where *o*-methylbenzylmercapturic acid (10 – 20% of the administered dose) was identified. This suggests a reactive intermediate resulting from side chain oxidation as being responsible for the hepatic glutathione depletion rather than this being caused by oxidation of the aromatic nucleus as had earlier been suggested.[8]

Various workers have reported the induction of hepatic, and to a lesser extent renal, microsomal cytochrome P450 and related enzyme activities. The induction appeared to be of the 'phenobarbitone type' with *p*-xylene being the least potent of the three isomers. Simultaneous exposure to ethanol has been found to potentiate the induction of microsomal enzymes by xylenes.

As noted above, absorbed ethylbenzene is mainly excreted in the urine as mandelic and phenylglyoxylic acids, these accounting for 90% of the ethylbenzene absorbed from the lungs. Only 4 – 5% of the retained ethylbenzene is estimated to be exhaled unchanged.[58]

In animals, ethylbenzene is oxidized mainly by side chain oxidation, as with humans. However with most animal species the transformation continues up to benzoic acid leading to the excretion of hippuric acid following conjugation with glycine. This conjugate is normally one of the main urinary metabolites, together with mandelic acid, in the rat, dog and rabbit.[58]

18.3 Human Acute Toxicity[3,7,8,64–67]

Sandmeyer in ref. 7 observes that ingestion of xylene causes severe gastrointestinal distress and that aspiration of regurgitated liquid into the lungs causes chemical pneumonitis, pulmonary oedema and haemorrhage. A case is noted where the ingestion of small quantities of xylene produced urinary dextrose and urobilinogen excretion with toxic hepatitis. This was reversed in 20 days.

ECETOC[8] notes a case where the ingestion of food, probably contaminated with technical xylene, induced deep coma, acute pulmonary oedema, hepatic impairment and haematemesis. The three xylene isomers and ethylbenzene were detected in the blood and in exhaled vapours. Charcoal haemoperfusion treatment was started after 26 hours of coma and the man recovered within two hours.

ECETOC[8] also notes that eye irritation and photophobia was experienced by six furniture polishers exposed to xylene [conditions unknown]. The symptoms disappeared after a few hours. Minute corneal vacuoles were found. These healed completely in a few days leaving no scars.

Sandmeyer in ref. 7 observes that conjunctivitis and corneal burns have been reported following direct eye contact with liquid xylene. Nelson *et al.*[64] state that xylene vapour at 200 ppm in air caused eye, nose and throat irritation in the majority of about ten persons exposed for 3 – 5 minutes and that, for day-long working conditions, most exposed subjects selected 100 ppm as the highest tolerable concentration.

When in direct contact with the intact skin, xylene is an irritant and causes defatting. This may lead to dryness, cracking, blistering or dermatitis.[7]

ECETOC[8] notes a study in which thirteen volunteers immersed both hands in *m*-xylene for twenty minutes. A burning sensation of the exposed skin resulted after ten minutes and this subsided within ten minutes post-exposure. The exposed skin became very erythematous but returned to normal appearance within a few hours.

Carpenter *et al.*[65] exposed volunteers to mixed xylenes ranging from 110 to 690 ppm in air over fifteen minutes. One of seven subjects exposed at 230 ppm and one of six exposed at 460 ppm experienced slight lightheadedness but without loss of equilibrium or coordination. A number of sensory thresholds were determined and it was considered that 110 ppm should not be objectionable to most people.

Sandmeyer in ref. 7 states that when xylene vapour is inhaled at high concentrations, human acute signs may include a flushing and reddening of the face and a feeling of increased body heat owing to the dilation of the superficial blood vessels. Other symptoms that may be apparent are disturbed vision, dizziness, tremors, salivation, cardiac stress, CNS depression, confusion and coma as well as respiratory difficulties.

Sandmeyer also notes that xylene, as an air pollutant, has been classified as a cilia toxin and mucus coagulating agent.

Crooke[66] includes xylene among solvents abused by glue sniffers, possibly with sudden, lethal results usually following physical exertion. Sandmeyer in ref. 7 quotes observations that such death may be due to sensitization of the myocardium to epinephrine so that even endogenous hormones precipitate sudden and fatal ventricular fibrillation, or respiratory arrest and consequent asphyxia.

In a case noted by Morley *et al.*,[67] one of three workers who were overcome by an estimated 10,000 ppm of xylene vapour in a tank and were not found until 18.5 hours after starting work, died shortly after from pulmonary pneumonia. The remaining two men recovered but suffered temporary hepatic impairment and one of them had evidence of temporary renal impairment. This last man had spent his lunch break at home and had then returned to the tank after which he remembered nothing more until recovering consciousness in hospital.

Necropsy on the deceased man was carried out 12 hours after death and showed a well-nourished man with pronounced cyanosis of the head and extremities. The heart showed slight left ventricular hypertrophy. Both lungs were heavy and plum-coloured. The cut surfaces were deeply congested and copious amounts of fluid exuded from them. The liver was congested but otherwise appeared macroscopically normal.

Histological examination of the lungs showed severe congestion with focal intra-alveolar haemorrhage and acute pulmonary oedema with pink-staining amorphous material within the alveolar lumina. The liver showed congestion, with swelling and vacuolation of many cells, confined mainly to centrilobular areas. The brain showed microscopical petechial haemorrhages in both grey and white matter and haemorrhages in Virchow–Robin spaces. There was evidence of anoxic neuronal damage with swelling and loss of Nissl substance.

ECETOC[8] notes studies in which impairment of balance had been reported in volunteers exposed for a short time (hours) to 100 – 400 ppm *m*-xylene. These concentrations also influenced reaction time and manual coordination. No dose-response relationship was found between blood xylene levels and these effects.

ECETOC[8] notes that some studies have been performed in order to assess the effects of combined exposure to xylene and other solvents as such situations often exist in the workplace. *m*-Xylene did not influence changes induced by 1,1,1-trichloroethane in body sway, reaction time and visually evoked potentials of male volunteers exposed by inhalation [other details not cited].

It has been found that alcohol may potentiate CNS effects induced by *m*-xylene in male volunteers, and the ingestion of 0.8 g alcohol/kgbw before or after the inhalation of 250 – 300 ppm of *m*-xylene for four hours has been reported to cause dizziness, nausea or red flush in susceptible individuals.[8]

OSHA[3] states that a concentration of 10,000 ppm of xylenes in air is immediately dangerous to life and health.

18.4 Animal Acute Toxicity[7.8.65.68]

Carpenter *et al.*[65] report that groups of ten young male rats were exposed to mixed xylenes at concentrations of 580, 1,300, 2,800, 6,000 and 9,900 ppm for four hour periods. No deaths occurred at or below 2,800 ppm and ten out of ten died at 9,900 ppm in 2.25 hours. At 6,000 ppm rats were prostrate in 30 minutes and four died within three to five hours. All survivors were prostrate as the exposure ended but recovered promptly and appeared normal throughout the 14 day observation period. At 2,800 ppm, after the usual signs of respiratory irritation, all rats became prostrate between 2 and 3.5 hours but recovered within one hour of exposure ceasing although coordination remained poor. At 1,300 ppm rats had poor coordination after two hours but this did not persist post-exposure. No signs of distress were noted at 580 ppm. The LC_{50} calculated from the mortalities during a 14 day observation period is 6,700 (5,100 – 8,500) ppm. In the opinion of the pathologist, the only findings that were significant to the deaths due to single inhalational exposure to mixed xylenes at high concentration were two cases each of atelectasis, haemorrhage and interlobular oedema of the lungs. All other lesions were considered as common sporadic occurrences.

In the same study, four male cats exposed to a mean concentration of 9,500 ppm died within two hours. Careful observation had been kept for signs of effects on the CNS. The response was a time-related pattern of salivation, ataxia, tonic and clonic spasms, and anaesthesia followed by death, these toxic signs being suggestive of a CNS effect. Histological evaluation of the tissues revealed scattered sporadic lesions which were considered to be unrelated to the treatment.[65]

ECETOC,[8] referring to respiratory irritation on inhalation, note that the exposure of the mice to 1,450 ppm of *o*- and *p*-xylene halved the respiratory rate of mice in a manner characteristic of sensory irritants.

Sandmeyer in ref. 7 notes a study in which the single-dose oral LD_{50} in rats was 4.3 g/kg. In another study using the three isomers mixed with ethylbenzene, an oral LD_{50} of about 10 ml/kg [approx. 8.6 g/kg] was determined.

ECETOC[8] states that the administration of a single oral dose of 3.5 – 10 g/kg of 'commercial xylene' to female rats resulted in mild congestion of the cells of the liver, the kidneys and the spleen. Female rats, dosed by gavage with a mixture of xylene isomers at 5.95 g/kg or higher, showed immediate sluggishness, followed by dullness, anaesthesia, narcosis, coma and death.

Aschan *et al.*[68] note that human visual disturbances can be compared with exaggerated rotary and positional nystagmus observed in the rabbit, both being the result of vestibular disturbance.

ECETOC[8] notes a study in which two drops of mixed xylenes in the rabbit eye induced slight conjunctival irritation and very slight and transient injury. However, Sandmeyer in ref. 7 notes that 13.8 mg [16 µl] in the rabbit eye produced marked turbidity and severe irritation of the conjunctiva, with lacrymation and oedema, and undiluted liquid caused corneal damage to the cat.

18.5 Human Sub-acute and Chronic Toxicity Including Reproductive Effects[5,7,8,62,69,70]

General

Sandmeyer in ref. 7 observes that the signs and symptoms resulting from chronic exposure resemble those from acute mishaps but are in part systemically more severe.

A number of chronic poisonings noted by reviewers are also noted by them as being obscured by there having been concurrent exposure to other solvents or by there being no indication of duration or exposure levels. WHO[69] notes that there is a surprising lack of systematic research in the field of long-term occupational exposure. The reported effects of chronic xylene exposure are summarized by WHO, *viz*: headache, irritability, fatigue, lassitude and sleepiness during the day, and sleep disorders at night. Although anaemia, leukopenia or leucocytosis have been mentioned as features of chronic xylene poisoning, and single cases of blood dyscrasias attributed to contact with xylene have been reported, there is no evidence of xylene being haematotoxic [unless excessive benzene is present]. Dyspeptic disorders are among the common symptoms of xylene over-exposure. Urobilinogen in urine, and changes in the activity of serum transaminases in workers exposed to xylene suggest possible hepatic involvement.[69] Xylene-induced hepatomegaly has been reported in man and its cause has been ascribed to the *ortho*-isomer.[5]

ECETOC[8] notes various studies relating to sub-chronic exposure. In one study, six men were exposed to 100 ppm *m*-xylene with peak concentrations of 200 – 400 ppm. The exposure was for six hours/day for five days followed by a two day interruption and then a further three days exposure. Disturbance in reaction times was reported but balance was apparently affected only at 400 ppm. These effects appeared to correlate with blood xylene levels. Tolerance against toxic effects seemed to develop over five subsequent exposure days. Two studies involved eight men being exposed to *m*-xylene for five days plus a further day following the weekend break. In one of those studies the exposure was to 90 ppm with peaks of 200 ppm. In the other it was of 64 ppm with peaks of 400 ppm. The volunteers exercised four times per day on a bicycle ergonometer. At 90 ppm there were acute deleterious effects on reaction time and manual coordination. Impairment of balance was seen only on the first days of exposure and re-exposure, possibly due to the development of tolerance. At 64 ppm the findings were similar except that no impairment of balance was noted. Various measurements on four of the subjects indicated a slight lowering of the vigilance level and that slight physical exercise seemed to antagonize the effects of xylene.

ECETOC, *inter alios*, notes a study of 45 workers who were employed between six months and five years in xylene production from petroleum. Exposure to xylene varied between 15 and 40 ppm in air and exposure to other hydrocarbons also occurred. One third of the workers complained of headaches, irritability, insomnia, dyspepsia and tachycardia. Twenty percent suffered from a 'neurasthenic- or asthesio-autonomic syndrome' and 15% from 'autonomic vascular dysfunction'. A decreased glycogen and peroxidase content of neutrophils was also reported.[8]

ECETOC[8] also observes that in recent years a number of epidemiological studies that have been carried out on painters and other groups exposed to a wide range of solvents, including xylenes, have led some investigators to conclude that prolonged exposure to such materials may cause permanent effects on the CNS. ECETOC notes the conclusions of Grasso *et al*.[70] however, who report that the evidence reviewed by them failed to confirm that long-term industrial exposure to solvents (with the exception of carbon disulphide) leads to chronic organic or functional damage to the nervous system in man. Grasso *et al*. note though that controversy is likely to remain until well designed studies have been performed. ECETOC[8] notes that the report of an international workshop on neurobehavioural effects of solvents held in October 1985 came to similar conclusions.

ECETOC[8] further notes a recent (1985) investigation into the long-term effects of solvents in 50 workers from a paint factory where xylene was one of the most frequently used of eight solvents. The authors of this study concluded that there was evidence of exposure-related 'brain dysfunction and neurasthenic/ emotional problems' based on a 4% decrease in cerebral blood flow, 'increased power in delta and beta bands of EEGs', poor performance in neuropsychological tests (14% of exposed workers) and psychiatric interviews.

Reproductive Effects in Man

Various reports reviewed by Barlow and Sullivan[62] and ECETOC[8] are noted by these same reviewers as being obscured, either by the presence of other solvents, or possible predisposition to the observed condition, or by insufficiency of data. Effects noted in these cases are of menstrual disorders and CNS and skeletal defects in offspring. No reports of adverse reproductive effects attributable to xylene over-exposure have been found.

Carcinogenicity to Man

No reports on xylene showing carcinogenicity to man were found.

18.6 Animal Sub-acute and Chronic Toxicity Including Reproductive Effects[5,8,60,62,65,71-77]

General

Xylene-induced hepatomegaly has been reported in animals and its cause has been ascribed to the *ortho*-isomer.[5]

Fabre *et al.*[60] report a study in which rats and rabbits were exposed to a 690 ppm mixture of xylene isomers in air for eight hours/day, six days/week. After 130 days, no significant deviations from normal were found in the peripheral blood. Similar exposures of rabbits to 1,150 ppm for 55 days produced a decrease in red and white cell counts and an increase in the platelet count in the blood. Both concentrations caused narcosis, delayed dyspnoea and ataxia; the higher concentration also caused conjunctival irritation, anorexia and weight loss from the end of the first week.

In a thirteen week inhalation study using rats and beagle dogs, no statistically significant ill effects were noted at the highest concentration used (810 ppm). Exposure had been for six hours/day, five days/week. After 65 days of exposure some rats from this study were taken and subjected to a four hour challenge period in which they inhaled a concentration of 6,700 ppm of mixed xylenes. No statistically significant differences from similarly challenged, previously unexposed rats were noted in the median 'time till death' and no protective acclimatization was found resulting from the exposure to 810 ppm for 65 days at six hours/day.[65]

ACGIH[71] notes a 1960 publication which stated that animal studies suggest that xylene is not a myelotoxicant. NIOSH[72] notes a 1968 study whose authors suggested that earlier reports of aplastic anaemia caused by toluene and xylene could be attributed to benzene contamination and that their own results presented a substantial argument for the lack of myelotoxicity of toluene and xylene. NIOSH[72] cites other upholders of this view (including the ACGIH source).

ECETOC[8] notes a study on rabbits in which 10 – 20 applications of undiluted xylenes of 95% purity (19% *o*-, 52% *m*-, 24% *p*-) either to the ears or shaved abdomen (occluded) for two or four weeks resulted in moderate to marked erythema, oedema and superficial necrosis at both sites.

ECETOC[8] also notes that the instillation of mixed xylenes into the rabbit eye, once/day for two days and subsequently three times/day for three days, produced swelling of the eyelids but no corneal vacuolisation.

Browning[73] notes a 1956 report that lesions in the form of fine vacuoles have been observed in the cornea of cats exposed to the vapour of commercial xylene [concentration and duration not stated].

Reproductive Effects in Animals

As noted in Section 18.2, Barlow and Sullivan[62] state that xylene crosses the placenta in mice and may be metabolized in the foetus.

API[74] reports a study in which 80 male and 160 female rats were inhalationally exposed to mixed xylenes. Exposure was to 0, 60, 250 and 500 ppm for six hours per day for 131 pre-mating days, 20 mating days, during gestation and, for those females allowed to litter, through 21 days of lactation. Twenty control females and twelve females given 500 ppm xylene were killed on day 20 of pregnancy for evaluation of teratogenic potential. The only effects observed in parent animals were lower mating indices in both sexes at 250 ppm and in one of two groups of females at 500 ppm. [In view of the inconsistent findings at 500 ppm and a high mating index in controls, the finding at 250 ppm appears to be incidental.] Also at 500 ppm, kidney weights were reportedly increased at day 21 of gestation. No narcosis nor similar systemic effects were reported. Amongst those foetuses evaluated for teratogenic potential, mean foetal weight was slightly reduced at 500 ppm and was associated with a marginal reduction in ossification. The birth weight of offspring at this dose level was unaffected, but from day 14 post-partum mean pup weight was slightly reduced at 500 ppm. [Both of these effects are considered to reflect embryo-foetal toxicity rather than teratogenicity.]

ECETOC[8] notes that recent teratology and multigeneration studies do not indicate that xylene has any teratogenic potential. In a recent review of data on developmental effects on rodents, Hood and Ottley[75] agree with ECETOC[8] in concluding that the likelihood that xylenes present a special hazard to the human conceptus, (*i.e.*, major effects at doses well below the maternal toxic dose) does not follow from the data at hand. These reviewers note though that, at sufficiently high doses to cause maternal toxicity, xylenes appear to be capable of causing adverse effects, mainly foetotoxicity, in developing mammals following maternal exposure by any of several routes.

Carcinogenicity to Animals

Reports by Maltoni *et al.*[76,77] note gavage studies in which a mixture of xylene isomers at 500 mg/kg in olive oil was administered to groups of 40 male and 40 female Sprague–Dawley rats, 4 – 5 days/week for 104 weeks, following which the animals were observed for their remaining lifespan.

After 134 weeks, survival of the xylene treated animals was 6/40 males and 4/40 females. (Control survival was 2/50 males and 5/50 females.) Four carcinomas of the oral cavity were reported in the xylene-treated rats. There was none in the controls. (Historical control data showed no carcinomas of the oral cavity among a total of 3,326 animals subjected to various control treatments.) One carcinoma of the stomach and four "acanthomas and dysplasias" were also reported in xylene-treated rats but the controls had none. There were insignificantly small increases in the incidence of malignant mammary tumours (xylene treated = 7, controls = 4) and in leukaemias (xylene treated = 6, controls = 3). The interim total number of malignant tumours in xylene-treated rats was 34% after 134 weeks compared with 15% in the controls. The majority of these, except for carcinomas of the mouth and stomach are noted as being relatively common in untreated rats.

After 141 weeks (the end of the study), of the 78 xylene-treated rats that had been alive when the first malignant tumour was observed in the experiment (at week 33), 36 had a total of 44 malignant tumours. Of the 94 controls that had been alive at that same time, 21 animals had a total of 23 tumours. The majority of these increases occurred in female animals [and only the female result was statistically significant]. "Haemolymphoreticular neoplasias" were also reported (eight in the xylene-treated group and four in the controls). [This last difference was not statistically significant.] [The implications of the overall findings cannot be evaluated from the available data, where terminal incidences of quite different tumour types are summated.]

ECETOC[8] also notes a recently audited NTP gavage study (reported in 1986) in which groups of 50 F344 rats of each sex were dosed with a mixture of xylene isomers and ethylbenzene in corn oil at 0, 250 and 500 mg/kg/day, five days/week for 103 weeks. Using the same schedule, groups of 50 B6C3F1 mice of each sex were administered 0, 500 or 1,000 mg/kg/day of xylenes. ECETOC observes that an abstract of the study reports that there was no evidence of systemic toxicity. At no site was the incidence of non-neoplastic or neoplastic effects in dosed rats and mice considered to be related to xylene treatment. This study included the same dose level (for rats) as the study referred to above and higher dose levels were experienced by the mice. It did not confirm the findings of the Maltoni study and supported the view that xylenes do not possess carcinogenic potential in rats and mice.

In another study noted by ECETOC[8] in which the effect of irritant materials on benzo(a)pyrene-induced tumour production was investigated, mixed xylenes and 0.05% benzo(a)pyrene or 1% benzo(a)pyrene in xylene were applied to mouse skin weekly for up to 28 weeks. The xylenes did not influence the tumour yield either when applied together, or alternately, with benzo(a)pyrene, and did not produce tumours when applied alone.

18.7 Mutagenicity[78,79]

Gerner-Smidt and Friedrich[78] found that no increases in chromosomal aberrations or of sister-chromatid exchanges were observed in human leucocytes exposed to xylenes, *in vitro*.

Haley in ref. 79 notes that no mutagenic activity has been seen in various activated or non-activated *S. typhimurium* strains exposed to xylenes, to ethylbenzene or to various xylene metabolites (*o-*, *m-*, *p*-xylenols and methylbenzyl alcohols), and that xylenes were non-mutagenic when tested *in vitro* with *E. coli* and strains H17 and M45 of *B. subtilis*.

Haley also notes the following: No gene reversion of *E. coli* strain SD-4-73 was caused by exposure to 2,4-dimethylphenol; that commercial xylene was mutagenic in the *Drosophila* sex-linked recessive lethal test whereas *o-* and *m*-xylene and ethylbenzene were not; that when rats were inhalationally exposed to 300 ppm, six hours/day for 9–18 weeks, no increases in the frequency of chromosomal aberrations caused by the xylenes [*sic*] were produced; that although genetic testing of mixed xylenes is complicated by the presence of 17% ethylbenzene, this chemical was not mutagenic when tested with various activated or non-activated strains of *Saccharomyces cerevisiae* or *S. typhimurium*; ethylbenzene, when tested *in vitro* with cultured human lymphocytes, caused both toxicity and a significant increase in sister-chromatid exchanges at a dose of 10 mM.[79]

18.8 Summary

[Xylene, as separate or mixed isomers, is of fairly low acute oral and dermal toxicity, though it can be absorbed percutaneously, but high inhaled vapour concentrations cause primarily narcosis and irritation of the upper respiratory tract, perhaps with vasodilation. Toxicological differences between the isomers are slight and of little practical consequence. Prolonged or repeated over-exposure may additionally cause liver, kidney or gastrointestinal disturbance, possibly with neurological changes. Heavy benzene contamination may cause long-term haematological changes.

The liquid is a strong eye irritant, and causes defatting of the skin which may progress to dermatitis if contact is prolonged or repeated, while high vapour concentrations also cause eye irritation.

No reports of adverse reproductive effects attributable to xylene over-exposure, other than some foetotoxicity only at significant maternally toxic doses, have been found. There is a report of increased tumour incidence in females only following long-term high-dose oral administration to rats, of doubtful interpretation and not confirmed by a second test. Most mutagenicity tests were negative.

Absorbed xylene is rapidly distributed around the body particularly into fatty tissues, and generally rapidly eliminated after metabolism mainly to methylhippuric acid, though some may persist for several days in fatty tissues.

From the point of view of acute eye and upper respiratory irritation 100 ppm in air is reported to be the highest tolerable concentration, while slight symptoms of acute narcosis are reported at levels down to 100 ppm in air. Cumulative exposure to concentrations in air down to 64 ppm caused incoordination and slowed reaction time. The toxic effects may be potentiated by ethanol ingestion.]

19. MEDICAL / HEALTH SURVEILLANCE

A decision on the need for, and content of, medical surveillance should be based on an estimate of the possibility and extent of exposure in the work operation. In addition, medical examination may be directed to identifying any pre-existing or newly arising condition in the individual workers which might either be aggravated by subsequent exposure, or might confuse any subsequent medical assessment in the event of excessive exposure or an illness not related to exposure. A particular aspect to be considered is the identification of sensitive subjects not adequately protected by the control limit in operation. A professional medical judgement may be required on continuance of employment in the specified process. The following information is relevant in the case of this solvent.

Pre-employment Medical Examination

A complete history and a physical examination may be undertaken to identify any pre-existing condition which might be aggravated by subsequent exposure. Examination of skin, respiratory tract, liver, kidneys, blood and central and peripheral nervous systems have been recommended, but the need for these should be related to the assessment of the possible exposure. Solvents classed as xylene may contain impurities including possibly benzene; evaluation of this possibility should be taken into account in deciding on medical tests necessary.
The possibility of exposure outside work should not be forgotten.

Periodic Medical Examination

A medical review should be carried out annually or following excessive exposure. A correlation has been shown between concentration of exposure to xylene and excretion of methylhippuric acid in urine. The test may be used for group comparison with environmental estimations.

20. OCCUPATIONAL EXPOSURE LIMITS

[The Committee felt that a time-weighted average limit of 100 ppm was appropriate for xylene. A STEL of 150 ppm is recommended on the basis of short term human toxicity.

It should be noted that other toxic volatile substances may be present in this solvent and should be monitored separately.]

REFERENCES

1. Sax, N.I., Ed.-in-Chief. Dangerous Properties of Industrial Materials Report 6 No.5(1986):93. 'Xylene'. Van Nostrand Reinhold.
2. National Institute for Occupational Safety & Health. 'Registry of Toxic Effects of Chemical Substances (RTECS)'. DHHS (NIOSH) Publication No.84-101-6. US Department of Health & Human Services, April, 1986.

3. Occupational Safety & Health Administration. Material safety data sheets from the Occupational Health Services database, OSHA, Washington D.C., 1986.
4. Council of Europe. 'Dangerous Chemical Substances and Proposals Concerning their Labelling'. ('The Yellow Book'.) 4th ed. Maisonneuvre, 1978.
5. Jori, A. *et al. Ecotoxicology & Environmental Safety*, 11(1986):44. 'Ecotoxicological profile on xylenes'.
6. Grayson, M., Exec. Ed. 'Kirk-Othmer Concise Encyclopedia of Chemical Technology'. 3rd ed. John Wiley & Sons, 1985.
7. Clayton, G.D. and F.E. Clayton, Eds. 'Patty's Industrial Hygiene and Toxicology'. 3rd ed. rev. Wiley Interscience, 1982.
8. European Chemical Industry Ecology & Toxicology Centre. 'Joint assessment of commodity chemicals' No.6: Xylenes. ECETOC, 1986.
9. Graedel, T.E. 'Chemical Compounds in the Atmosphere'. Academic Press, 1978.
10. Lowenheim, F.A. and M.K. Moran. 'Faith, Keyes and Clark's Industrial Chemicals'. 4th ed. John Wiley & Sons, 1975.
11. International Labour Office. 'Encyclopaedia of Occupational Health and Safety'. 3rd rev. ed. ILO, 1983.
12. Durrans, T.H. 'Solvents'. 8th ed. rev. by E.H. Davies. Chapman and Hall, 1971.
13. 'Hawley's Condensed Chemical Dictionary'. 11th ed. rev. by N.I. Sax and R.J. Lewis. Van Nostrand Reinhold, 1987.
14. Grasselli, J.G. and W.M. Ritchey, Eds. 'Atlas of Spectral Data and Physical Constants for Organic Compounds'. 2nd ed. CRC Press, 1975.
15. National Institute for Occupational Safety & Health. 'NIOSH Manual of Analytical Methods'. 3rd ed. DHHS (NIOSH) Publication No.84-100. US Department of Health & Human Services, 1984.
16. Grob, K. and G. Grob. *Journal of Chromatography*, **62**, No.1(1971):1.
17. Wathne, B.M. *Atmospheric Environment*, **17**, No.9(1983):1713.
18. Kettrup, A. *et al. Erdoel und Kohle, Erdgas, Petrochemie*, **35**, No.10(1982):475.
19. Mariotti, M. *Bollettino dei Chimici dei Laboratori Provinciali*, **28**, No.7(1977)(III):193.
20. Nowicka, K. *Chemia Analityczna (Warsaw)*, **27**, No.3-4(1982):277.
21. Snell, J. *Pollution Engineering*, **14**, No.6(1982):36.
22. Goering, H.W. *Plaste und Kautschuk*, **28**, No.4(1981):226.
23. V'Yunov, K.A. *et al. Lakokrasochnye Materialy i Ikh Primenenie*, 6(1984):47.
24. Tyras, H. and J. Stufka-Olczyk. *Chemia Analityczna (Warsaw)*, **29**, No.3(1984):281.
25. Mukhtarova, M. *Khigiena i Zdraveopazvane* **23**, No.2(1980):173.
26. Ciupe, R. *Revista de Chimie (Bucharest)*, **32**, No.6(1981):584.
27. Ross, M.M. and R.J. Colton. *Analytical Chemistry*, **55**, No.1(1983):150.
28. Hagemann, R. *et al. Analusis*, **6**, No.9(1978):401.
29. Dmitriev, M.T. and G.M. Kolesnikov. *Gigiena i Sanitariya*, **47**, No.2(1982):59.
30. Ioffe, B.V. *et al. Journal of Chromatography*, 186(1979):851.
31. Vitenberg, A.G. and I.A. Tsibul'skaya. *Gigiena i Sanitariya*, **47**, No.2(1982):60.
32. Bianchi, A. and G. Muccioli. *ICP (Industria Chimica e Petrolifera)*, **9**, No.4(1981):77.
33. Tsibul'skii, V.V.*et al. Zhurnal Analiticheskoi Khimii*, **34**, No.7(1979):1364.
34. Imamura, K. and T. Fujii. *Bunseki Kagaku*, **28**, No.9(1979):549.
35. Sybilska, D. and T. Koscielski. *Journal of Chromatography*, **261**, No.3(1983):357.
36. Poltronieri, G. *Inquinamento*, **22**, No.3(1980):97.
37. Morin, M. *et al. Journal of Chromatography*, **210**, No.2(1981):346.
38. Dworzanski, J.P. and M.T. Debowski. *Chemia Analityczna (Warsaw)*, **26**, No.2(1981):319.
39. Van Roosmalen, P.B. and I. Drummond. *British Journal of Industrial Medicine*, 35(1978):56.
40. Niinuma, Y. *et al. Sangyo Igaku*, **24**, No.3(1982):322.
41. Poggi, G. *et al. International Archives of Occupational & Environmental Health*, **50**, No.1(1982):25.
42. Bieniek, G. and T. Wilczok. *British Journal of Industrial Medicine*, **38**, No.3(1981):304.
43. Bellanca, J.A. *et al. Journal of Analytical Toxicology*, 6(1982):238. 'Detection and quantitation of multiple volatile compounds in tissues by GC and GC/MS'.

44. Grayson, M., Exec. Ed. 'Kirk-Othmer Encyclopedia of Chemical Technology'. 3rd ed. John Wiley & Sons, 1979.
45. Standen, A., Exec. Ed. 'Kirk-Othmer Encyclopedia of Chemical Technology'. 2nd ed. John Wiley & Sons, 1967.
46. Marsden, C. and S. Mann, Eds. 'Solvents Guide'. 2nd ed. rev. Cleaver-Hume Press, 1963.
47. The Royal Society of Chemistry. Laboratory Hazard Data Sheet No.16: 'Xylene', 1983.
48. National Institute for Occupational Safety & Health/Occupational Safety & Health Administration. 'NIOSH/OSHA Occupational health guideline for xylene'. DHSS (NIOSH) Publication. US Department of Health & Human Services/US Department of Labor, 1978.
49. Bretherick, L., Ed. 'Hazards in the Chemical Laboratory'. 4th ed. Royal Society of Chemistry, UK, 1986.
50. National Fire Protection Association. 'Fire Protection Guide on Hazardous Materials'. 9th ed. NFPA, Massachusetts, USA, 1986.
51. Bretherick, L. 'Handbook of Reactive Chemical Hazards'. 3rd ed. Butterworths, 1985.
52. Warren, P.J., Ed. 'Dangerous Chemicals Emergency Spillage Guide'. 1st ed. Jensons (Scientific) Ltd., Leighton Buzzard, UK, 1985.
53. Windholz, M. *et al.*, Eds. 'The Merck Index'. 10th ed. Merck & Co., 1983.
54. Weast, R.C., Ed.-in-Chief. 'CRC Handbook of Chemistry and Physics' ('The Rubber Handbook'). 67th ed. CRC Press, 1986-1987.
55. Dreisbach, R.R. 'Physical Properties of Chemical Compounds'. Advances in Chemistry Series No.15, American Chemical Society, 1955.
56. Horsley, L.H. 'Azeotropic Data III'. Advances in Chemistry Series No.116, American Chemical Society, 1973.
57. Utidjian, H.M.D. *Journal of Occupational Medicine*, **18,** No.8(1976):567. 'Criteria documents: recommendations for a xylene standard'.
58. European Chemical Industry Ecology & Toxicology Centre. 'Joint assessment of commodity chemicals' No.7: Ethylbenzene. ECETOC, 1986.
59. Fishbein, L. The Science of the Total Environment 43 No.1-2(1985):165. 'An overview of environmental and toxicological aspects of aromatic hydrocarbons III: Xylene'.
60. Fabre, R. *et al. Archives des Maladies Professionelles de Medecine du Travail et de Securite Sociale*, **21,** No.6(1960):301. 'Recherches toxicologiques sur les solvants de remplacement du benzene – IV. Etude des xylenes'.
61. Dowty, B.J. *et al. Pediatric Research*, 10(1976):696. 'The transplacental migration and accumulation in blood of volatile organic constituents'.
62. Barlow, S.M. and F.M. Sullivan. 'Reproductive Hazards of Industrial Chemicals'. 2nd printing. Academic Press, 1984.
63. Riihimäki, V. *et al. Scandinavian Journal of Work, Environment & Health*, 5(1979):217. 'Kinetics of *m*-xylene in man. General features of absorption, distribution, biotransformation and excretion in repetitive inhalation exposure'.
64. Nelson, K.W. *et al. Journal of Industrial Hygiene & Toxicology*, 25(1943):282. 'Sensory response to certain industrial solvent vapors'.
65. Carpenter, C.P. *et al. Toxicology & Applied Pharmacology*, 33(1975):543.'Petroleum hydrocarbon toxicity studies – V. Animal and human response to vapors of mixed xylenes'.
66. Crooke, S.T. *Texas Medicine*, 68(1972):67. 'Solvent inhalation'.
67. Morley, R. *et al. British Medical Journal*, 3(1970):442. 'Xylene poisoning: A report on one fatal case and two cases of recovery after prolonged unconsciousness'.
68. Aschan, G. *et al. Acta Oto-laryngologica*, **84,** No.5-6(1977):370. 'Xylene exposure: electronystagmographic and gas chromatographic studies in rabbits'.
69. World Health Organization. Technical Report Series No. 664: 'Recommended health-based limits in occupational exposure to selected organic solvents'. WHO, Geneva, 1981.
70. Grasso, P. *et al. Food & Chemical Toxicology*, **22,** No.10(1984):819. 'Neurophysiological and psychological disorders and occupational exposure to organic solvents'.

71. American Conference of Governmental Industrial Hygienists. 'Documentation of the threshold limit values for chemical substances in the workroom environment'. ACGIH, 1986.
72. National Institute for Occupational Safety & Health. 'Criteria for a recommended standard: occupational exposure to xylene'. Report No. NIOSH-75-168. US Department of Health, Education & Welfare, 1975.
73. Browning, E. 'Toxicity and Metabolism of Industrial Solvents'. Elsevier, 1965.
74. American Petroleum Institute. 'Parental and fetal reproduction inhalation toxicity study in rats with mixed xylenes'. API, Washington D.C., 1983. (Summary.)
75. Hood, R.D. and M.S. Ottley. *Drug & Chemical Toxicology*, 8(1985):281. 'Developmental effects associated with exposure to xylene: A review'.
76. Maltoni, C. *et al. American Journal of Industrial Medicine*, 7(1985):415. 'Experimental studies on benzene carcinogenicity at the Bologna Institute of Oncology: current results and ongoing research'.
77. Maltoni, C. *et al. Acta Oncologica*, **4**, No.3(1983):141. 'Benzene as an experimental carcinogen: up-to-date evidence'.
78. Gerner-Smidt, P. and U. Friedrich. *Mutation Research*, 58(1978):313. 'The mutagenic effect of benzene, toluene and xylene studied by the SCE technique'.
79. Haley, T.J. Dangerous Properties of Industrial Materials Report 6, No.6(1986):2. 'Xylene'. Van Nostrand Reinhold.